Economic Dynamics of Global Energy Geopolitics

Ahmet Salih Ikiz
Muğla Sıtkı Koçman University, Turkey

A volume in the Advances in Finance, Accounting, and Economics (AFAE) Book Series

IGI Global
DISSEMINATOR OF KNOWLEDGE

Published in the United States of America by
IGI Global
Engineering Science Reference (an imprint of IGI Global)
701 E. Chocolate Avenue
Hershey PA, USA 17033
Tel: 717-533-8845
Fax: 717-533-8661
E-mail: cust@igi-global.com
Web site: http://www.igi-global.com

Library of Congress Cataloging-in-Publication Data

Names: Ikiz, Ahmet Salih, 1968- editor.
Title: Economic dynamics of global energy geopolitics / Ahmet Salih Ikiz, editor.
Description: Hershey, PA : Engineering Science Reference, [2018]
Identifiers: LCCN 2017030322| ISBN 9781522542032 (hardcover) | ISBN 9781522542049 (ebook)
Subjects: LCSH: Energy policy--Economic aspects. | Energy consumption--Political aspects. | Energy industries--Political aspects. | Energy development--Political aspects. | Geopolitcs.
Classification: LCC HD9502.A2 E2185 2018 | DDC 333.79--dc23
LC record available at https://lccn.loc.gov/2017030322

This book is published in the IGI Global book series Advances in Finance, Accounting, and Economics (AFAE) (ISSN: 2327-5677; eISSN: 2327-5685)

British Cataloguing in Publication Data
A Cataloguing in Publication record for this book is available from the British Library.

Advances in Finance, Accounting, and Economics (AFAE) Book Series

ISSN:2327-5677
EISSN:2327-5685

Editor-in-Chief: Ahmed Driouchi, Al Akhawayn University, Morocco

MISSION

In our changing economic and business environment, it is important to consider the financial changes occurring internationally as well as within individual organizations and business environments. Understanding these changes as well as the factors that influence them is crucial in preparing for our financial future and ensuring economic sustainability and growth.

The **Advances in Finance, Accounting, and Economics (AFAE)** book series aims to publish comprehensive and informative titles in all areas of economics and economic theory, finance, and accounting to assist in advancing the available knowledge and providing for further research development in these dynamic fields.

COVERAGE

- Economic Downturn
- Economics of Agriculture and Biotechnology
- Evidence-Based Studies
- E-finance
- Statistical Analysis
- Development Economics
- Accounting information systems
- Economic Indices and Quantitative Economic Methods
- Economics of Migration and Spatial Mobility
- Bankruptcy

IGI Global is currently accepting manuscripts for publication within this series. To submit a proposal for a volume in this series, please contact our Acquisition Editors at Acquisitions@igi-global.com or visit: http://www.igi-global.com/publish/.

The Advances in Finance, Accounting, and Economics (AFAE) Book Series (ISSN 2327-5677) is published by IGI Global, 701 E. Chocolate Avenue, Hershey, PA 17033-1240, USA, www.igi-global.com. This series is composed of titles available for purchase individually; each title is edited to be contextually exclusive from any other title within the series. For pricing and ordering information please visit http://www.igi-global.com/book-series/advances-finance-accounting-economics/73685. Postmaster: Send all address changes to above address.

Titles in this Series

For a list of additional titles in this series, please visit:
http://www.igi-global.com/book-series/advances-finance-accounting-economics/73685

Perspectives, Trends, and Applications in Corporate Finance and Accounting
Constantin Zopounidis (Technical University of Crete, Greece) Apostolos G. Christopoulos (National and Kapodistrian University of Athens, Greece) and Petros Kalantonis (Technological Education Institute (TEI) of Piraeus, Greece)
Business Science Reference • ©2018 • 354pp • H/C (ISBN: 9781522561149) • US $215.00

Microfinance and Its Impact on Entrepreneurial Development, Sustainability, and Inclusive Growth
Ramesh Chandra Das (Katwa College, India)
Business Science Reference • ©2018 • 385pp • H/C (ISBN: 9781522552130) • US $245.00

Maintaining Sustainable Accounting Systems in Small Business
Luísa Cagica Carvalho (Open University, Portugal & University of Évora, Portugal) and Elisa Truant (University of Turin, Italy)
Business Science Reference • ©2018 • 387pp • H/C (ISBN: 9781522552673) • US $215.00

Agricultural Finance and Opportunities for Investment and Expansion
Augustine Odinakachukwu Ejiogu (Imo State University, Nigeria)
Business Science Reference • ©2018 • 236pp • H/C (ISBN: 9781522530596) • US $195.00

Regaining Global Stability After the Financial Crisis
Bruno Sergi (Harvard University, USA) Filip Fidanoski (University of New South Wales, Australia) Magdalena Ziolo (University of Szczecin, Poland) and Vladimir Naumovski (University American College Skopje, Macedonia)
Business Science Reference • ©2018 • 383pp • H/C (ISBN: 9781522540267) • US $215.00

Employment Protection Legislation in Emerging Economies
Samir Amine (Universite du Quebec en Outaouais, Canada)
Business Science Reference • ©2018 • 301pp • H/C (ISBN: 9781522541349) • US $195.00

For an entire list of titles in this series, please visit:
http://www.igi-global.com/book-series/advances-finance-accounting-economics/73685

IGI Global
DISSEMINATOR OF KNOWLEDGE

701 East Chocolate Avenue, Hershey, PA 17033, USA
Tel: 717-533-8845 x100 • Fax: 717-533-8661
E-Mail: cust@igi-global.com • www.igi-global.com

To my Spiritual Father
Professor Ethem Ruhi Fığlalı
Inaugural Rector of Muğla University

Table of Contents

Detailed Table of Contents

This chapter addresses geopolitics and sustainability and their interrelations with peace and stable economic conditions. Historical events in energy and geopolitics often suggest economic and social patterns which aim for sustainable growth but result in impoverished living conditions. Through analyzing the downward spirals in supply and demand, behaviors, and lifestyles, economies have the potential to offer diverse strategies for management of change and development. Exploring cultural and economic development, striking a balanced and normative approach which acknowledges the existence of societal dogmas, and finding alignment in educational, political, economic, and spiritual values are keys to sustainability.

On the road ahead for Egypt, the formidable obstacle is a two-headed monster: character and leadership. Now, after two revolutions, it is time to analyze this monster. Egypt must deal with this as success in development is simply not possible without first slaying this monster. As Egypt embarks on its immediate future, targeting its immediate political, economic, and social priorities, it must not overlook the crucial objective of determining the character of the nation, and the responsibility of its leadership. This is a prerequisite that if not met will delay further progress and development.

Foreign direct investment (FDI) is the engine of growth of all countries, contributing to the inflow of financial capital, technology, skills, employment, to the establishment of production of modern goods and services, which enables a national economy to become more competitive in the global market. However, the developing or transition countries often lack the capital to finance their own development. Analysis of 21 developed and developing oil-producing countries from 2008 to 2014 show that the most important factors for attracting investment in the oil and gas sector are the discovery of attractive fields, the creation of a developed and modern infrastructure, increases proven reserves of hydrocarbons, and of corruption. Less important but still factors are a stable currency, an open trade regime, favorable business conditions, as well as lower taxes on oil-producing business.

Energy politics are one of the effective ways to interact with other states' political systems. When external relations are important for internal energy security, then energy politics are covered with other political strategies. As the case for European Union, energy strategies structures several sectors and one of them is energy supply security and promotion of sustainable use of energy worldwide. Thus, security of supply is an important part in EU energy politics based on competitiveness and sustainability. In the case of regional cooperation with Central Asia, EU is trying to be self-sufficient for its internal market to build its future safe in energy. Thus, security on energy supply becomes on the fore. So, EU, as a prominent actor in global political system, has growing interest on Central Asian states because of their high energy potential. But, regional cooperation and assistance in the region for energy politics are not enough, and EU needs to promote and take further steps toward democratization and development cooperation embedded in energy politics.

The energy sector holds a crucial strategic importance for development and its sustainability. However, the energy reserves in Turkey are insufficient for the increasing energy demand. For this reason, the need for alternative energy sources has emerged. The fact that fossil fuels used in energy production will be exhausted and their damaging effects on the environment has made it inevitable for the world to use renewable energy. In Turkey, the following are the main energy sources used in electricity production: hydraulic sources, brown coal, natural gas, coal and fuel oil. The use of renewable energy sources in electricity production has shown a considerable development in the world over the past few decades. However, these energy resources have yet to be exploited to any large extent in Turkey. In this chapter, the range of potential renewable energy resources available for electricity production in Turkey will be analyzed in order to determine the necessary technical substructure to make these renewable energy resources more attractive.

Energy is an issue of strategic importance to the European Union and Turkey. Both are dependent to the outside. EU is the world's largest energy importer and second largest energy consumer after the United States. Turkey's alignment with the EU's energy policy is extremely important for EU in terms of increasing the diversity and quality of its energy resources. Turkey's strategic location makes Turkey a land of passage for transporting oil and gas to Europe. This geopolitical importance is an important opportunity for both sides. In this chapter, EU's position on energy in the world has been explained first. Then the energy situation and energy policy in EU has been examined. Secondly, Turkey's energy policy and compatibility to EU together with complementary role to EU on Energy has been presented.

This chapter discusses the possibility of developing an energy market in Turkey. Turkey currently serves as an energy transit corridor, with the Baku-Tbilisi-Ceyhan (BTC) and Kirkuk-Ceyhan pipelines and with its seaborne oil trade, where large volumes are carried by tankers through its straits. Turkey also has the potential to become an energy market with new projects connecting producers in Central Asia, the Middle East, and the Caucasus, and major consumers of oil and natural gas in Europe and other regions of the world. Two recent megaprojects, The Trans Anatolian Natural Gas Pipeline Project (TANAP) and Turkish Stream will move Turkey closer to fulfilling this dream. Turkey, however, needs to meet some requirements to be considered a mature energy market. These are related, among others, to factors such as its infrastructure, storage capacity, market reforms, and easy market access for private firms to actively participate in the energy market.

The object of the study is the asymmetry of interests in geopolitics between developed and developing countries. In the context of the global crisis, the issue of de-dollarization is relevant from the political and economic points of view. What will be the behavior of small oil countries in this situation is a big problem. Also, for them, the question remains how to get off the oil needle in a painless way? The ways of solving the above-mentioned problem by other states are examined in detail, and comparative analysis is conducted in the case of applying these methods to Azerbaijan.

Central Asia has gained extraordinary importance in recent years in the framework of global energy security. China is the most significant example of a power that looks to its periphery as a viable option for energy supply. In Central Asia, Chinese companies are dynamic players having even broken the long Soviet and Russian monopoly over regional pipelines. This chapter examines the importance of the region within China's energy security, while not overlooking the potential contribution of the China-Pakistan economic corridor in the energy transit. In addition, Central Asia is likely to help China reduce the energy deficit in Xinjiang, through the import of hydroelectricity generated in Tajikistan and Kyrgyzstan. Although Central Asia's contribution to global energy security is low, it matters in a context of energy diversification, in which China's One Belt One Road brought a more promising dynamics to the cooperation between Beijing and Central Asian countries.

The international system has experienced a shift from a Western-centric world dominated by the United States to a decentralized world. This accompanies a shift in the distribution of energy wherein China overtakes the US as the biggest oil importer. Energy is vital for the survival of countries' economic facilities. Japan is a country that needs to import the energy resources to run its industry. Thus, the security of its economic activities is dependent on the security of energy supplies. The fact that Turkey is a country with ambitions to become a regional energy center and has strong ties to the Middle East makes it vital to Japanese interests. Also, Japan's advanced technology and economy are vital to Turkey's ambitions of becoming an energy hub. These reciprocal interests make it possible for both countries to attain a high level of cooperation. Nevertheless, this depends on their self-awareness and political will in order to exhibit a more independent behavior which transcends their previously US-dominated foreign policy.

The high dependency on fossil fuels, fluctuations in prices and supply have macro/micro-economics effects for both energy exporters and importers. Therefore, understanding economic stability based on energy market changes is an important subject for policymakers and researchers. Norway, as a fossil fuel exporting country, is a good choice for the analysis of the relationships between economic robustness and fossil fuel economic fluctuations. While the country is one of the pioneers in the field of sustainable energy utilization, they have tried to provide a robust economic environment for oil export revenues. In this chapter, the impacts of energy changes on the economy are investigated in Norway. In this regard, first, the impact of oil prices on macro-economic parameters is discussed. Afterwards, the main issues related to energy economics including resilience of the energy sector, energy policies, economics analysis of the energy sector, and the electricity markets are discussed.

The discovery of energy sources in the Eastern Mediterranean region, while providing opportunities, further complicates Turkey-Israeli relationship. If Israel and Turkey can cooperate on energy, they can revitalize their relationship to the extent when the military elites were strong in Turkey and the relationship between two countries was established with the hands of generals. The 2016 Israel-Turkey agreement, which ended years of tension, provides Israel and Turkey with the opportunity to cooperate in energy areas. This collaboration, in turn, could generate the eventual emergence of the new ruling elites that would fill the vacuum created by the decline of the military's role in Turkey. It was the crisis management experience of the US that made the agreement between two countries possible in June 2016. Yet, any normalization process between Israel and Turkey will not be easy. US policy in the Mideast influences Turkey-Israel relations, and Turkey-Israel relations, in turn, affect the future of the Middle East. This obliges the US to bear a tremendous responsibility.

Preface

INTRODUCTION

In today's integrated hemisphere nothing is less important than energy resource management. Allocation of those scarce resources is one of the main issue for global players in world geopolitical order. The economic dynamics has vital role in international policy formation among countries. Thus energy geopolitics is the combination of international relations, economics and geography. The dominant powers in world energy geopolitics have radically changed in last couple of decades. The bower blocks during the cold war period disintegrated and reshaped. Industrial production with innovative technologies moved from western world to newcomers in Asia. The paradigm shifts in post-cold war period had also impacts on global energy geopolitics. This chapter devoted on the bases of global energy geopolitics.

A core definition of economics is based on the nexus between scarce natural resources and the unlimited needs of humanity. In almost every aspect of daily life the simple relation between supply and demand is the main element of home economicus. The global trade between nations is shaped by this simple law of economics which we call supply-demand relations. Thus, we understand that the price mechanism plays a crucial role not only in international transactions but also in international trade.

In physics, energy is defined as work that a force can or could do. Those forces can be gravitational, electrical, magnetic, nuclear, chemical or mechanical. Energy would be in different forms such as: kinetic, magnetic, electric, heat and potential. As long as satisfy basic human needs in daily life energy also help to assemble, extract, refine, shape and synthesize in industrial production process.

The evolution of mankind both geometrically and technologically accelerated the necessities of nations. The mode of production transformed from manpower and animal power to machinery in the manufacturing process with the industrial revolution. This transformation in the industrial production process generated some changes in world resources in the 19th century. As manpower was replaced by machinery it increased the demand for natural energy resources such as coal and oil. Prior to this, those natural resources were not classified as economic goods due to the low demand for them. The sharp increase in demand with technological improvement, soared the demand for those scarce resources.

This industrialization and technological development increased the demand for these new energy resources. Thus, the energy supply and demand increased with the new production paradigm shift, where the market mechanism determines the price structure for those scarce natural resources. Until that time, they were non-tradable goods in markets and did not have market forces for price determination.

In the early stages of the industrial revolution in the 19th century, the main source of energy transformation was coal in steam power engines. There were coal mines around the United Kingdom where the first inventions of steam power engines emerged. Both in the UK and other western European countries there was an abundant amount of coal to match the increased energy demand from new factories. Until the beginning of the 20th century, Europe, with its coal reserves was essentially energy self-sufficient. New developments in technologies such as oil-based engines made oil valuable worldwide. Contrary to coal, oil is not found in the western world, especially in Europe. The oil rich middle east countries became the focus of this new energy demand. Geographical distance between energy supply and demand had opened another dimension in energy market due to remote access in price determination, where main supplier countries differs from main demanders. So oil and other energy resources become much more expensive due to the specific nature of production process.

The two world wars ended European dominance of the world economic order. British, German and French hegemony in the peripheries of the globe had also diminished. The collapse of the Ottomans, Russians and Habsburgs created a vacuum, and power restoration by new sovereign nation states in the oil rich parts of the Middle East. Additionally, besides its international trade structure, those materials become subject to international politics where different countries are involved in price determination. But in this case, due to the strategic nature of energy resources, not only market forces but also national and international policies, started to affect prices. Increasing conflicts in energy supply zones had the inevitable outcome of interventions by hegemonic powers. So energy security becomes an important part in energy geoeconomics.

ENERGY ECONOMICS

The simple definition of energy economics in the encyclopedia is as follows:

Energy economics is the application of economics to energy issues. Central concerns in energy economics include the supply and demand for each of the main fuels in widespread use, competition among those fuels, the role of public policy, and environmental impacts. Given its worldwide importance as a fuel and the upheavals in its markets, oil economics is a particularly critical element of energy economics. Other efforts have treated natural gas, coal, and uranium. Energy transforming and distributing industries, notably electric power, also receive great attention. Energy economics addresses, simultaneously as well as separately, both the underlying market forces and public policies affecting the markets. Economic concerns differ sharply from those of natural scientists and engineers. The most critical difference is in the outlook towards supply development. Many economists argue that market forces allow smooth adjustments to whatever happens to the physical stock of resources. Potentially, these market forces can produce resources cheaper than other methods presently employed to cause adjustments. At worst, the cost rises will be gradual and manageable. In contrast, these economists stress the harmful effects of governments on energy. (Macmillan Encyclopedia of Energy, 2000)

The importance of energy resources is based on their operational support in production processes for human needs. So energy resources function to transform raw materials into final products. Energy economics is a social science that focuses on economic agents – firms, individuals, governments –supply of energy resources. It studies the roles of alternative markets and the regulatory structures of these activities, economic distributional impacts, and environmental consequences. Energy economics basically gathers raw materials from the physical environment, processes them and releases back to the physical environment. Thus, energy supply is a conversion process of chemical or mechanical energy resources for human utilization. This process either storable or non-storable while some energy resources are renewable such as solar and wind energy. The others are perishable and nonrenewable such as coal and oil. The superiority and advantage of crude oil over other forms of energy resources is that it is transferable from supply rich countries to others, whereas renewable energy sources in their form tend to be can only applicable in locally.

The demand characteristics of energy resources is different from other goods that are derived from industrial production. Thus, the higher the economic growth the higher the energy demand. But in today's modern high tech lifestyle, energy is an essential good where demand is always positive. The standard of living of various people on the planet is directly related to per capita energy consumption. It is not surprising that the higher standard of living in the western world is subject to an increased level of energy consumption. These two different forces combine aggregated demand for energy resources. The demand for energy resources has a significant pattern of substitution effect. So consumers have the potential to shift their demand among different energy sources by considering price and other variables such as new technologies.

The market conditions of energy resources were not a priority until the 1970's. The main product- oil-, was inexpensive and there weren't any bottlenecks in the production process. The OPEC restrictive supply quotas policy sharply affected global prices and created a new paradigm in the world order. Western countries with solid economic performances faced new a phenomenon where they both experienced inflation and stagnation, what is now called stagflation. Price shocks with declining economic growth and employment forced western countries to focus on those energy supply issues. The obvious western superiority in technological and industrial development had its first challenge when Middle Eastern countries increased energy prices.

Energy supply policies both at local and international levels involve the policy triangle of economic efficiency, energy security and sustainability. The following factors are the main determining factors for energy price levels (Applied Energy Partner, 2017):

1. **Supply:** Energy from nuclear, coal, gas, oil, and renewable sources reacts quickly in response to the available supply (or lack thereof). This is a key contributing factor to price fluctuations, which can occur on an hourly basis.
2. **Demand:** Demand for heating, cooling, light, and processes varies in response to demand in terms of economic, technological, and efficiency measures.
3. **Gas Storage:** This is a term for energy "inventory" (since you can't store electricity), i.e. the difference between supply and demand. Gas injections and withdrawals are announced weekly, and prices adjust accordingly.
4. **Weather Forecasts:** The predicted weather forecast, as well as actual weather events, are important considerations, affecting spot market prices and short-term contracts. Whether the forecast becomes reality is less critical to longer-term prices.
5. **Generation Changes:** While seemingly more localized, these changes can have a broad effect on the markets.
    a. **Nuclear**: Retirement of older plants as they require re-licensing can cause fluctuations.
    b. **Coal**: Coal plant conversions to natural gas to avoid scrubbing-technology costs can also cause fluctuations
    c. **Transport**: Across the U.S. there are severe constraints in gas pipeline and electrical transmission capacity, which take time and investment to reverse. With the difficulty of transportation, prices rise.

6. **Global Markets:** Despite the massive growth in shale gas production, major changes in global oil supplies can affect U.S. domestic energy costs.
7. **Imports and Exports**: Global oil and gas prices determine the relative profits suppliers can make selling fuels domestically or overseas. All energy prices are connected to some degree.
8. **Government Regulation**: Federal (FERC) and state (PUCs) regulations can change both supply and demand costs quickly and significantly, which, as noted above, affects the cost of energy.
9. **Financial Speculation**: Like most other traded commodities, energy prices can be affected significantly by financial speculation, which is the least transparent factor of all. If a market doesn't seem to be following the direction indicated by supply or demand-related factors, the cause is almost always financial speculation, which is largely invisible and causes unexpected movements.

Those factors primarily affect short term and medium-term prices. In the long term the most important factor, especially in the EU, is the increasing usage of renewable energy resources such as solar and wind energy. This factor reduces energy import dependency of those countries via conversion of those resources into electricity.

ENERGY GEOPOLITICS

Due to the increasing importance of energy security and sustainability, the energy economy has become much more entangled with geopolitics in international relations. Some geo-economic developments such as the rising power of China, a paradigm shift in the new world order and the Arab spring developments have had inevitable impacts on the economic dynamics of global energy geopolitics. In the last decades there was a tremendous increase in GDP levels and economic growth in Asian countries such as China, India and others. Meanwhile, after the foundation of Shanghai Security Organization in 1996 they became a new player in energy security. According to Kaplan (2014), "Geopolitics is the battle for space and power played out in a geographical setting. Just as there are military geopolitics, diplomatic geopolitics and economic geopolitics, there is also energy geopolitics. For natural resources and the trade routes that bring those resources to consumers is central to the study of geography."

The issue of leveraged soft power in hemispheric studies has stimulated the spatial analyses of conflict resolution with geo-economics in strategic decision-making process. The term geo-economics entered the lexicon in 1990 with an article by Edward Luttwak, which argued that following the Cold War, the importance of military power was giving way to geo-economic power. One reason the term is more commonly used now is the rise of China, which is increasingly using economic tools from trade policy to restore its global power. Two other factors are also relevant: the revival of state capitalism and state-owned enterprises means that states have more economic resources at their disposal; and the deep integration of global trade links and financial markets has made geo-economic tools more powerful (Petsinger, 2016).

Under these circumstances, the classical economics toolbox based on demand-supply relations cannot easily be applied. The economic dynamics are intertwined with political actors in an international global arena. The geopolitical dimension in natural resource economics of energy_became one of the hot topics in today's globe. Energy geo-economics is a joint research area of three main social sciences. Economics, international relations and geography. Geo-economics, a new discipline in the social sciences, has become the basic concept of political analysis with an increasing interest in the interaction between international trade and strategy. Since the 1990s, despite the attempts to integrate the concept of geo-economics in disciplines such as International Relations, Geography, Economics, Geopolitics, History and Business Administration, it seems that it is too abstract or differs according to the point of view of the researchers (Erkişi, 2017).

In Figure 1, The intersection of the three circles is the locus of energy geo-economics where elements of international relations, economics and geography interact. The relation between economic power and the geographical location of countries is the focus of Geo-economics and the new international conflict of interest zones situated in the international trade emanates from those countries.

British geopolitical strategy in the 19th century concentrated on the security and economic purposes of Empire. British academic John Mackinder wrote his book *The Geographical Pivot of History* and introduced Heartland theory. According to this theory, the world strategic center was situated in central Asia and the Volga Basin. This region is like a natural castle with its topographic and geographic location. From those times, the British experts were aware of the rich energy resources of Eurasia and the Middle East. His assumption, based on the geographic location of the land is important. Thus, the world can be divided into three geographical locations:

1. **Heartland:** Eastern Europe and Siberia
2. **Inner Crescent:** Turkey, Germany, India, and China
3. **Crescent:** United Kingdom, South Africa, and Japan

The master of Eurasia would be the world leader in geo-economics. The failure of both Napoleon and Hitler on the Russian steppe, clearly verified this argument.

After WWI, German geopolitics reinforced the Lebensraum strategy constructed on the idea of Darwin's survival of the fittest theory. Every country must be considered a living organism, and in order to keep it alive, it has to create its own living area

Figure 1. Energy geo-economics

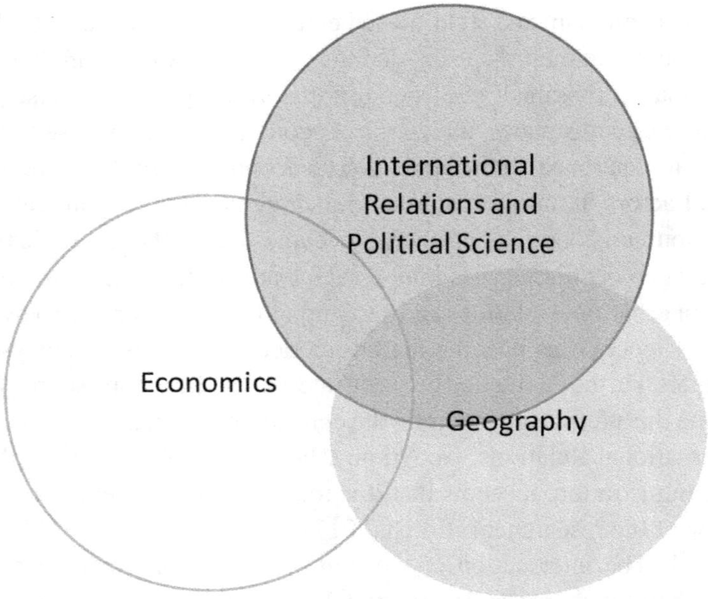

for growth. This strategy is based on controlling energy supply sources in a region. German strategy, ironically reincarnated in different forms of this theory, when German economic power seized most of the Balkan countries in the 1990's under the EU umbrella. First, the German mark, and then the Euro, circulated as hard currency in those countries as economic hegemonic power of Germany returned after a couple of decades since German tanks had been in those countries.

US hegemonic dominance in the Middle East, consolidated after the second world war by the diminishing power of continental Europe and the UK. That strategy achieved by containment policy of US. Contrary to the Heartland Theory, Rimland's approach focuses on the crescent of Eurasian countries. In order to diminish the USSR's power in Eurasia, the US must seek to improve its penetration in those countries. The North Korean, Vietnam and Afghanistan wars were the main power struggles and conflict zones during the cold war since they are vital countries under this policy.

Even though Francis Fukuyama called the disintegration of the USSR The End of History, -where the US led economic system has an undisputed victory-, during the Putin presidency, the foreign policy of the Russian Federation radically changed, shaped by Dugin's Neo Eurasian philosophy. China, the Russian Federation, Kazakhstan, Kirghizstan, Tajikistan and Uzbekistan founded the Shanghai Security Organization which has a similar function to NATO for those countries. This new paradigm possibly aims to solve the integration problems of multi ethnic Russian society. The authoritarian, state controlled economic structure and governance will create a modern society without western democratic norms.

In modern geo-economics, the variables of spatial location, identity and economics are the main determinants of strategic analyses. In today's global society the historical legacies of ancient societies become quite significant in identity formation. Preserved identities in the former USSR countries are very essential leverage for the foundation and integration of the Turkic Union in that part of the world. Hard power must always be accompanied by soft power in international relations. The historical heritage of Turkish culture in those countries will serve as an important tool for that. All these theories became much more complex in today's cyber global society, where a small minor change in any part of the world can create a butterfly effect in another remote part of the globe. We may consider that those studies explain global and regional international relations between energy-rich countries and links them to political and economic power. In geographic dimensions' new term geoenergeia classifies states according to a combination of economic, political, environmental, and cultural factors along with their states' wealth in certain energy resources, based on the view that energy is a strategic weapon. From another perspective, the analysis assesses the impact of the possession and use of energy resources which have direct implications for states' sovereignty (Vidakis,2015).

NEW DYNAMICS AND THE CHANGING NATURE OF ENERGY ECONOMICS

Global industrial production in the last couple of decades moved to the Asian continent from the western world due to its low-cost advantages. One of the main final outcomes is the sharp and sustainable high growth rates in China. In turn, the energy demand from those countries has accelerated.

Since 2003 global trends in energy shipping routes moved to Indo Pacific waters owing to the dominance of China in the global economy. Economic hegemony generally associated_with cultural and political hegemony. So the tensions between an economically and demographically stagnant, aging Europe and a troubled, autocratic but energy rich Russia would expose the decline of Greater Europe, while North America and the Indian Ocean world become the new pulsating centers of commerce. As Brzezinski says Eurasia will be next chess board for global powers. It is quite obvious that self-sufficiency is much higher in the US compared to the EU by recent investments in shale gas resources in North America.

In Figure 2 we observe the changing nature of energy demand in oil. While the US is turning to her own resources for import substitution, China, India, South Korea and Japan are confronted with increasing import dependency. This has some implications for international relations. The more their energy import dependency increases, the stronger will the need be to secure the international trade and shipping routes to minimize disruptions to their supplies. So the Asian countries are becoming more dominant not only as economic powers but also political powers at a global level. New breaking points in the international arena thus lie in Chinese borders while Asia is becoming the new global economic attraction center.

Figure 2. The demand for oil
Source: U. S. Energy Information Administration (2014)

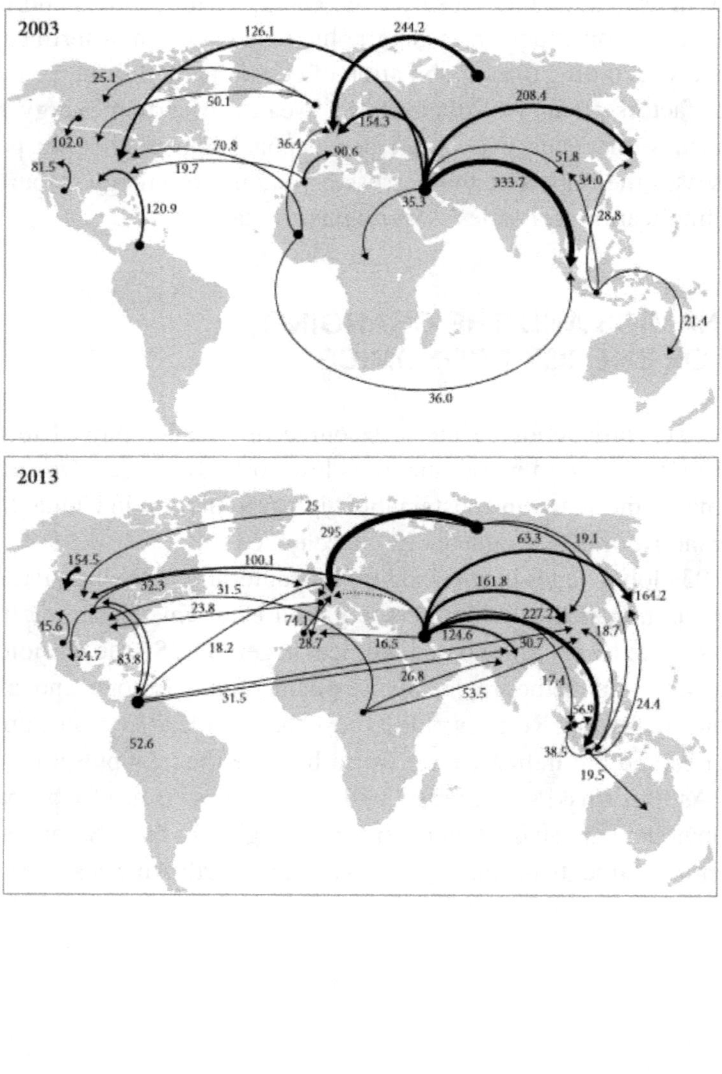

ORGANIZATION OF THE BOOK

Global energy consumption vis-a-vis production has become a subject attracting significant attention over the last couple of decades. The energy market and price determination, global energy security become an important issue in today's world. Energy economics not only focuses on local issues but also on cross border and international issues. The multinational dimension is essential for understanding further solutions. The modest purpose behind this book was to draw a cross country reference for energy geopolitics with contributors from different countries around the globe.

Chapter 1 questions whether western philosophies applicable to Muslim economies in case of energy? This question is highlighted by Ansari in energy efficiency for sustainable development. This author explores cultural and economic development, with a normative approach which embraces a Super Natural Being. Hence, the difference between the Muslim world and the West in education, political, economic, and spiritual values are searched in sustainability. Thus, she gives a cultural and religious background of energy efficiency.

Chapter 2 explores Egypt case in Muslim world. It is obvious that much of the world's energy supply comes from the Islamic world. In the first chapter Amal provides detailed information about Egypt, one of the main pillars of the Arab world. He explains the core dynamics of revolutions after 2011 with proper updated information. He carefully analyzes the weakness of the country and proposes possible solutions.

Chapter 3 devoted on foreign direct investment for oil producing countries. FDI is the crucial factor for economic growth in almost all of the developing countries. Nurseiit investigates the main drives of FDI in oil producing countries from those of the developed world. In his cross-country analyses the author compares oil producing countries. He comments that national economic policies such as tax levels and currency stability are not the priority factors which lure FDI to those countries. Rather, the discovery of new fields, a modern infrastructure for transport and processing and good public governance are the significant factors for attracting high FDI in the energy sector.

Chapter 4 investigates EU and Central Asia relations. The European Union has been an economic giant and a political dwarf in most of the global affairs in recent decades. In order to restore its global power, the EU broadened its penetration scope to Central Asia. Azalov. in her chapter on Europe's new international strategy for energy politics in this region, clarifies that regional cooperation and assistance in the region for energy politics are not sufficient and that the EU needs to promote and take further steps to ensure that democratization and development cooperation are embedded in energy politics. Thus in her article, she analyzes energy politics with its link to security and EU politics in Central Asia. In this manner, the EU's

Energy Strategy Papers for 2020, 2030 and 2050 are studied with its external relations' dimension.

Chapter 5 presents sustainable and clean energy resources for Turkey. Sustainable development is the locus of development efforts of most developing countries concerned with preserving their natural resources. The main effort is to promote renewable energy usage in economic development. Gok focused on Turkish development efforts by considering clean energy usage. She clarifies that, there is a need for alternative energy sources in Turkey. In order to eliminate the hazardous side effects of electrical energy production by thermal sources policy makers must turn to renewable energy sources, in which Turkey has potential. In that chapter, Turkish renewable energy potential with necessary technical substructure is analyzed.

Chapter 6 concludes strategic location of Turkey for EU energy needs. Turkey has been in the EU's waiting room since 1960. This long-lasting relation has had lethargic side effects on both parties. The EU o is unable to provide clear signals to Turkey due to the multilevel decision process and fragmented opinions. Turkey, on the other side, is confused by the statements of different EU members such as France and Germany on its national politics agenda in the past. On the other hand, both are dependent on each other regarding energy transportation. Turkey's strategic location, makes it a land of passage for transporting oil and gas to Europe. This geopolitical importance is an important opportunity for both sides. In her chapter Ayture, presents her research findings on Turkey's energy policy regarding EU.

Chapter 7 reviews Turkeys potential as energy corridor. In many aspects Turkey has the potential to serve as an energy transit corridor, with the Baku-Tbilisi-Ceyhan (BTC) and Kirkuk-Ceyhan pipelines and with seaborne oil trade, where large volumes are carried by tankers through its straits. Gokirmak asserts his research findings for the future prospects of Turkey becoming an energy market with new projects connecting producers in Central Asia, the Middle East and the Caucasus, and major consumers of oil and natural gas in Europe and other regions in the world. The projections for two recent mega projects, the Trans Anatolian Natural Gas Pipeline Project (TANAP) and the Turkish Stream will help Turkey come closer to fulfilling this ambition. In IEA forecasts for Turkey supporting that argument, however there are some requirements to be considered for a mature energy market such as: infrastructure, storage capacity and market reforms for well-functioning transactions. In that chapter, the framework of those requirements is briefly discussed.

Chapter 8 analyses Azerbaijan economy which is heavily dependent on energy exports. The development path of countries has many pillars, including economic growth. The developing world presents different patterns than the developed world in its path towards development. They arrive at different outcomes even while following the same policy guidelines of developed countries. Aida and Ulviyya focus on the meaning of the concept of asymmetry in a small group of developing countries and countries with economies in transition. The authors consider the geopolitical asymmetry associated with the dominance of the dollar in this market, analyze the existing political situation and offer their vision of Azerbaijan's economy, for which oil has become an obstacle to development. This is manifested primarily by the fact that developed countries are "pure" exporters of capital to developing countries, whereas developing countries are the debtors of developed countries. Thus, as a logical conclusion of the study, they comment on the subject of oil's de-dollarization, and, assessing all the shortcomings of the resource asymmetry from the perspective of a weak state, present their vision for solving this challenging problem.

Chapter 9 focus on One Belt One Road project of China. Awakening giant China is becoming a world leader regarding the volume of its GDP. Its increasing demand for energy resources has forced it to follow a much more active role in international policy formation. The strategical background of this new era is the new Silk Road Project called One Belt One Road policy. Duarte outlined the recent developments in Chinese rising energy demand. Soaring energy demand revealed the energy security issue in Central Asian countries. He concludes that although Central Asia's contribution to global energy security is low, it matters in a context of energy diversification, in which China's One Belt One Road brought a more promising dynamic to the cooperation between Beijing and Central Asian countries.

Chapter 10 addresses Turkish Japan relations. The fact that Turkey is a country with ambitions to become a regional energy center and has strong ties to the Middle East makes it vital to Japanese interests. Also, Japan's advanced technology and economy are vital to Turkey's ambitions of becoming an energy hub. These reciprocal interests make it possible for both countries to attain a high level of cooperation. Nevertheless, this depends on their self-awareness and political will in order to exhibit a more independent behavior which transcends their previously US dominated foreign policy.

Chapter 11 identifies Norwegian energy sector. Norway, as a fossil fuel exporting country, is a good choice for the analysis of the relationships between economic robustness and fossil fuel economic fluctuations. While the country is one of the pioneers in the field of sustainable energy utilization, they have tried to provide a robust economic environment for oil export revenues. In this chapter, the impacts of energy changes on the economy are investigated in Norway. In this regard, first, the impact of oil prices on macro-economic parameters is discussed. Afterwards, the main issues related to energy economics including resilience of the energy sector, energy policies, economics analysis of the energy sector, and the electricity markets are discussed.

Chapter 12 establishes sound evidence on cooperation between Turkish Republic of Northern Cyprus and mainland in energy issues. The discovery of new energy resources in the Eastern Mediterranean region presents new opportunities for neighboring countries. Turkey with its close ties with the Turkish Republic of Northern Cyprus had long term interests and benefits from those resources. Turkey and Israel are main players in the specified region. Şahin and Sozen explain the possible outcome of mutually collaborative policies. They argue that US policy in the Middle East influences Turkey-Israel relations and these, in turn, affect the future of Middle East and Eastern Mediterranean energy sources. This compels the US to bear a tremendous responsibility. US President Trump's decision in December 2017 to recognize Jerusalem as the capital of Israel can be regarded as unconditional support of Israel. The US can help Turkey and Israel to avoid future crises but not by not through this kind of asymmetrical support to Israel; but rather by applying the same level of pressure on both countries.

REFERENCES

Applied Energy Partner. (2018). Retrieved from http://www.appenergy.com/2017/01/23/10-factors-that-affect-the-cost-of-energy/

Erkişi, K. (2017). Strategic Use of Economic Power: Geo-economics. *The Journal of International Social Research Volume*, *10*(52), 1025–1035. doi:10.17719/jisr.2017.19

Kaplan, R. D. (2014). Geopolitics of Energy. *Forbes*. Retrieved in April 2018 https://www.forbes.com/sites/stratfor/2014/04/04/the-geopolitics-of-energy/#61be93a93b39

Macmillan Endcyclopedia of Energy. (2000). *Energy Economics*. Retrieved from https://www.encyclopedia.com/environment/encyclopedias-almanacs-transcripts-and-maps/energy-economics

Petsinger, M. (2016). *What is Geo-economics*. Retrieved in April 2018 from https://www.chathamhouse.org/system/files/publications/twt/WiB%20YQA%20Geoeconomics.pdf

U.S. Energy Information Administration. (2014). *Annual Energy Outlook*. Retrieved on August 22, 2014 from https://www.eia.gov/outlooks/aeo/#pet_oli

Vidakis, I., & Baltos, G. (2015). Security Aspects of "Geoenergeia" and the Significance of Energy Resources Management in International Politics. *Geopolitics of Energy*, *37*(3), 1-16.

Acknowledgment

I would like to acknowledge the help of all the people involved in this project and, more specifically, to the authors and reviewers that took part in the review process. Without their support, this book would not have become a reality.

First, as editor I would like to thank each one of the authors for their contributions. Our sincere gratitude goes to the chapter's authors who contributed their time and expertise to this book.

Second, I would like to wish to acknowledge the valuable contributions of the reviewers regarding the improvement of quality, coherence, and content presentation of chapters. Most of the authors also served as referees; we highly appreciate their double task.

Ahmet Salih İkiz
Muğla Sıtkı Koçman University, Turkey

Chapter 1
Sustainable Development in Muslim Economies: Peace at What Cost?

Aisha Ansari
Capella University, USA

ABSTRACT

This chapter addresses geopolitics and sustainability and their interrelations with peace and stable economic conditions. Historical events in energy and geopolitics often suggest economic and social patterns which aim for sustainable growth but result in impoverished living conditions. Through analyzing the downward spirals in supply and demand, behaviors, and lifestyles, economies have the potential to offer diverse strategies for management of change and development. Exploring cultural and economic development, striking a balanced and normative approach which acknowledges the existence of societal dogmas, and finding alignment in educational, political, economic, and spiritual values are keys to sustainability.

INTRODUCTION

Recognizing world power as a viable strategy, and a systematic and iterative process, scholars often agree that the traces of Islamic teachings are found fundamentally in all global government schemes. Economic progress in most Western societies is based on a riba, usury and interest-bearing profits. As these developments gain notoriety, they also fuel debates on attitudes and behaviors crucial to our global societies and real world sustainable growth. Sub-systems and networks based upon local cultures and communities are indeed catalysts in sustainability - for example

DOI: 10.4018/978-1-5225-4203-2.ch001

boycotts, strikes, riots, and social unrest. As Islam is the religious foundation in many governing nations, Emile Durkheim's model (1858-1917) suggests that the Islamic contribution to establishing and maintaining sustainable and efficient economies should not be omitted when considering a nation's past, present or potential sustainable development. Using a literature review in this chapter which supports reasoning and evidence in ethics, morals, and global patterns in sustainability, we draw upon a qualitative theoretical framework acceptable in both Eastern and Western economies.

As a part of our mission to offer real world recommendations in sustainability, an Islamic approach is discussed in terms of its potential to assist lasting economic growth. This chapter also offers evidence in the form of a literature review as well as lessons learned and best practices. An analytical timeline of violence and global socioeconomic conditions is compared to peacetime economic conditions. Devaluation of the United States dollar, which the Organization of Petroleum Exporting Countries (OPEC) used as a pricing and purchasing unit of currency, began in world financial markets, causing geopolitical rifts. Leveraging natural resources, largely oil production, has caused an imbalance in world peace, power, and economics.

Decision making in geopolitical matters calls for problem identification and peace-keeping measures that allow for economic sustainability and growth in governed societies. Based on maintaining security and stability and addressing basic human needs in society, sustainable energy is a vital economic process. A timeline dating back before the 1973-1974 Arab-Israeli conflict is presented in this chapter along with a feasible strategic research plan on energy and sustainability.

HISTORICAL EVENTS SHAPING ENERGY GEOPOLITICS AND SUSTAINABILITY

Some Islamic thinkers now consider global conditions to have been deteriorating since World War II and to be steadily progressing toward the destruction of humanity. Examining a concept of social welfare, Ibn Khaldun, and sustainability, a claim of natural resources depletion and a lack of sustainable measures exist in Islamic societies (Muhammad, 2010). Ibn Khaldun's theory of maintaining a balance between basic human needs and excessive consumption fits well with the Islamic principle of caring for the good of society. Emile Durkheim's concepts, while not labeled Islamic, also suggest a need for moderation in consumption. Durkheim's theory does, however, argue for a separation between labor and religion, or functionalism (Durkheim & Swain, 2015). From this perspective, social phenomenon in our societies are made up of individualism and self-sufficiency, and are united by the common thread found in our beliefs and social behaviors. Sustainability requires a combination of both self-sufficiency, such as nations, and unity, as in the philosophies of our societies.

Basic targets in sustainable energy revolve around pricing and supply, being described as security in resources (Bilgin, 2015). Dating back to before the 1967 Arab – Israeli conflict, energy security challenged decision makers and governing entities. With a particular focus on private sector participation in sustainable resources, there is scrutiny about the shape and process of geopolitical activities (Bilgin, 2015).

During the Arab-Israeli conflict of 1973, oil resources were commodities traded in global markets. A decision by OPEC to use the US dollar currency during the embargo sent economic shocks through the industrial world, affecting financial stability. Conditions prior to the oil embargo were described as "cheap energy"; with increased pricing and scarcity in supply, OPEC would leverage "Petro dollar recycling," increasing foreign investments and ensuring oil exporting security (Licklider, 1988).

Daniel Yergin offers a different truth about global dependency on oil and sustainable energy citing that, before Pearl Harbour, Winston Churchill's quest precedes the 1967 war, the Bush presidency, and the Iraqi – Kuwaiti conflict (1991). In his historical account of energy and conversion from coal to oil, the author declares oil as being at the root cause of war, politics, and economics (Yergin, 1991). At this stage in our global history, the roles of most key players have changed. However, nations are still in a quest for oil, power, and sustained wealth. Perhaps similar views on how quadrupled oil prices affected the United States economy during the 1973 – 1974 embargo resonate with Merrill's account of the oil crisis, making a polemic argument that the disruption in energy surpassed the Vietnam War (2007). Furthermore, other nations - for example Netherlands, the United Kingdom, Canada, and Japan, following the United States' lead - attempted political actions to conserve energy use and promote lifestyle changes (Licklider, 1988). Most experts agree little or no progress has been made in terms of sustained energy production and consumption since the beginning of these efforts.

MILESTONES/TIMELINE IN WAR AND PEACE AND OIL PRODUCTION

The following events have affected energy security and sustainable development, including geopolitics, power, economic climates, and times of war and peace in modern history.

Table 1. Timeline in war and peace; oil production

Timeline	Event
1967	1967 War - the Six Day War, began with Israeli strikes in Egypt, Jordan, Syria
1971	Richard Nixon takes United States off gold standard Dollar conversion and gold devaluation
1973	October 6, 1973 – Yom Kippur War – Egypt strikes Israel
1973 – 1974	Arab – Israeli War OPEC oil embargo
1975 – 1980	US recession and budget restructuring
1987 – 2000	Intifada Palestinian – Israeli Conflict
2005	Israel withdrew from Gaza
2006	Israel Lebanon War
2011	Al-Jazeera leaked document about Palestinian War and Peace intelligence
2012	UN votes – Palestine as non-member observer
2017	OPEC back in the news with h prospects in shale production, Oil prices below 45 dollars

GLOBAL SOCIAL AND ECONOMIC PATTERNS

Defined by our behaviors, philosophies, and lifestyles, sustainability takes on a different purpose when espousing truth, meaning, and value. Khaldun and Islam, Durkheim and Christianity, and geopolitics shape how we have chosen to define sustained economies in diverse global conditions. When comparing societies and economies, many similarities as well as differences become clear. Yet, an argument and call for discipline and appreciation with regard to natural and man-made environments is universal. According to Khaldun, economic structure within an Islamic society envisions citizen awareness of natural boundaries in our daily life. The argument defines sustainable resources as, once depleted, almost impossible to replace. Although they are referred to as "environmental sins", developed and undeveloped nations experience resistance to sustainable behaviors. This is both costly and consequential (Muhammad, 2010).

The United States consumes more energy from petroleum than any other country, 9.3 million barrels per day in 2007, according to the EIA, (US Energy Information Administration). In 2015, gasoline fuel consumption in the United States dropped slightly to at 9.2 million barrels per day. EIA reports 47% of all petroleum use in the United States is for gasoline fuel. Heating oil accounts for only 21% at 4 million barrels per day. The Wall Street Journal reports that while American demand for petroleum is slightly down, prospects in shale and energy production are now considered competitive and sustainable energy sources (Bordoff, 2016).

The Sura Khaf (28:77) calls for seeking the earth's wealth without destroying its natural resources, in line with current demands in sustainable behavior (Muhammad, 2010). Muhammad (p.b.u.h.) is revered as an excellent role model in spiritual and worldly affairs. Hart, makes the argument that Muhammad (p.b.u.h.) surpassed all men in history and when considering environment, conditions, timing, history, and economics is number one on Hart's ranking of the most influential persons in our history (1992). Muhammad (p.b.u.h.) is also credited with shura, consultation as advised in Islamic shari'ah, and key in negotiations and collaborations (Ibrahim, Basir, & Rahmàn, 2011).

Echoing the same call for conservation and adherence to fundamentals, Durkheim and functionalism define religion as a doctrine enshrouded in adherence, being unified and sustainable (Durkheim & Swain, 2015). While Durkheim's theory and positioning supported a division in labor and sacred values, anomie, also a part of his theory, highlighted a collapse in sustainable resources due to the decline of societies' and normality's support for former value systems (Durkheim & Swain, 2015). Societies and economies in a global environment are built on behaviors which recognize boundaries and adherences for the maintenance sustainable resources.

A WORD FORWARD: SOCIAL-CULTURAL CHANGE AND DEVELOPMENT

Going forward, energy sustainability requires collaborative efforts and continuous improvement while benchmarking progress in education, health, holistic values, and technological advances. World Bank geopolitics has funded innovative projects, promising to preserve resources and the environment. Although not without scrutiny in transparency and social responsibilities, Mexico has received 672 million dollars in funding for 43 projects (Godoy, 2011). Turkey is also partnering with Japan, France, Canada, and China on hydro energy needs, and tackling the construction of new facilities to ensure secure and affordable energy (Bilgin, 2015).

Geopolitics, multinational efforts grounded in resource processes and geography, has the potential to make both war and peace (Evans, 2008). Our basic concern in this environment is to monitor prices, supply, and demand for energy (Evans, 2008). Solutions targeting sustainable energy require global partnerships and policies that improve systems of security and peace. Ongoing calls in multidisciplinary research and sharing in best practices ensure a higher chance for sustainable resources and global security.

REFERENCES

Bilgin, M. (2015). Turkey's energy strategy: Synchronizing geopolitics and foreign policy with energy security. *Insight Turkey*, *17*(2), 67–81.

Bordoff, J. (2016). The New Geopolitics of Declining Oil Demand. *Wall Street Journal*. Retrieved from https://blogs.WSJ.com

Dirioz, A. O. (2013). Energy security, politics, markets, peace. *Journal of Foreign Policy and Peace*, *1*(1), 90–97.

Durkheim, E., & Swain, J. W. (2015). *The Elementary Forms of the Religious Life, a Study in Religious Sociology – Scholars Choice Edition*. Scholars Choice.

Evans, R. L. (2008). *Fueling Our Future: An Introduction to Sustainable Energy*. Cambridge University Press.

Godoy, E. (2011). *World Bank Calibrating its Measurement of Sustainability*. Global Geopolitics and Political Economy. WordPress.

Hart, M. (1992). *The 100: A Ranking of The Most Influential Persons in History*. New York, NY: Kensington Press.

Ibrahim, D., Basir, S.A., & Rahman, A. A. (2011). Sustainable Economic Development: Concept, Principles and Management from Islamic Perspective. *European Journal of Soil Science*, *24*(3).

Licklider, R. (1988). The Power of Oil: The Arab oil Weapon and the Netherlands, the United Kingdom, Canada, Japan, and the United States. *International Studies Quarterly*, *32*(5), 205–226. doi:10.2307/2600627

Merrill, K. R. (2007). *The Oil Crisis of 1973 – 1974: A Brief History with Documents*. Bedford: St. Martin's.

Muhammad, M.T.S. (2010). Principals of sustainable development in ibn Khaldoun's economic thought. *Malaysian Journal of Real Estate, 5*(1).

World Bank Archives. (n.d.). *Global Geopolitics & Political Economy*. Author.

Yergin, D. (1991). *The Prize: The Epic Quest for Oil, Money and Power*. New York, NY: Simon and Schuster.

KEY TERMS AND DEFINITIONS

Functionalism: Religion as a doctrine enshrouded in adherence, being unified and sustainable.

Intifada: Arabic for "tremor," a term used in 1987 for Palestinians revolt against Israeli West Bank occupation.

Riba: An agreement of mutual gain between two individuals, companies, etc. One party profits while the other receives little or nothing from the transaction, greed.

Shura: Consultation as advised in Islamic sharia, and key in negotiations and collaborations.

Usury: Goods and exchanges in trade are not equally matched in quality and quantity, but for the same price despite deceptive practices.

Chapter 2
Power, Character, and Leadership:
The Case of Egyptian Revolutions

Amal Nagah Elbeshbishi
United Nations Economic Commission for Africa, Morocco

ABSTRACT

On the road ahead for Egypt, the formidable obstacle is a two-headed monster: character and leadership. Now, after two revolutions, it is time to analyze this monster. Egypt must deal with this as success in development is simply not possible without first slaying this monster. As Egypt embarks on its immediate future, targeting its immediate political, economic, and social priorities, it must not overlook the crucial objective of determining the character of the nation, and the responsibility of its leadership. This is a prerequisite that if not met will delay further progress and development.

INTRODUCTION

Character and *leadership are a* formidable two-headed monster on the road ahead for Egypt. This obstacle must be tackled through a mix of soft and hard powers, employed to pursue varying strategic goals in conquering hearts and minds, to create convincing incentives and exercise influence. Now, after two revolutions, it is time to analyze this monster. Success in development is simply not possible without first dealing with this obstacle.

DOI: 10.4018/978-1-5225-4203-2.ch002

As Egypt embarks on its future, targeting in on its political, economic and social priorities, it must not overlook the crucial objective of determining the *character* of the nation, and the responsibility of its *leadership*. This is a prerequisite that, if not met, will negatively affect development.

This chapter is based on two main methodologies:

1. Literature review on the socio-economic situation of Egypt; and
2. Empirical analysis to explore the current state of Egypt's socio-economic environment.

The following hypothesis is explored in this chapter:

Widespread unrest does not arise from the long-standing political failings of the system, but rather from its sudden perceived failure to use its soft and hard power to provide essential security to the population.

This chapter will outline the various institutions and factors needed to establish sustainable levels of economic prosperity, which in turn will usher in an era of development for Egypt. It is critically important to note that these institutions and factors are time-tested, proven elements of development all over the world. A necessity for a country's productivity and ability to spread prosperity to its citizens; they allow a nation to use available resources efficiently.

The focus is then turned to the mix of soft and hard powers we refer to as an all-critical two-headed monster; It deals with how the character of a nation is formed and the indispensable signs and characteristics of viable leadership. Without a character overhaul and the emergence of genuine and passionate leadership, the ability to produce sustainable development in Egypt will be greatly hindered.

DEVELOPMENT: STAGES, FACTORS AND PILLARS

The World Economic Forum, in its annual global competitiveness report, examines the many factors enabling national economies to achieve sustained economic growth and long-term prosperity. The 2016-2017 global competitiveness report ranks countries on the state of their national economies. Egypt ranks 115 out of 138 economies in the Global Competitiveness Index (GCI); measured on a scale from 1 to 7 Egypt scored a 3.67.

The global competitiveness report addresses three stages/factors of development and their respective pillars as shown below:

- **Stage I. Basic Requirements:** Institutions; infrastructure; macroeconomic environment; health and primary education.
- **Stage II. Efficiency Enhancers:** Higher education and training; goods market efficiency; labor market efficiency; financial market development; technology readiness; market size.
- **Stage III. Innovation and Sophistication Factors:** Business sophistication; innovation.

These pillars are interdependent; while their strengths reinforce each other, a weakness in one area often has negative impacts on other areas. For example, innovation is very difficult without a well-educated and trained workforce adept at absorbing new technologies. Sufficient financing for Research and Development (R&D) or an efficient goods market makes it possible to take new innovations to market.

With regard to these three stages of development, Egypt is in stage II. To get to stage III's innovation enhancer pillars, Egypt's focus should be on stage II's efficiency enhancer pillars.

For any nation, solid character and competent, ethical leadership are prerequisites for viability and proper functioning of the twelve pillars, allowing for their expected impact on the economy to materialize. The alternative is a bleak state of paralysis and lack of development. The components of the twelve pillars that are both pertinent and problematic for Egypt are presented below.

- **Institutions:** The world economic crisis has made the importance of a sound and fair institutional environment more apparent given the increasingly direct role played by the state in the economies of many countries. For example, owners of land, corporate shares, or intellectual property are less likely to invest in the improvement and upkeep of their assets if their rights as owners are not protected. Excessive bureaucracy and red tape, over-regulation, corruption, dishonesty in dealing with public contracts, lack of transparency, and the political dependence of the judicial system impose significant economic costs on businesses and slow the process of economic development. Proper management of public finances is critical to ensuring trust in business environments. The global financial crisis, along with numerous corporate scandals, has highlighted the relevance of accounting and reporting standards and transparency. They aid in the prevention of fraud and mismanagement, ensuring good governance, and maintaining investor and consumer confidence. An economy is well served by businesses that are run honestly, where managers abide by strong ethical practices in their dealings with the government, other firms, and the public at large. Private sector transparency is indispensable and can be brought about using standards as well as auditing and accounting practices that ensure access to information in a timely manner.

- **Infrastructure:** Economic growth is significantly impacted by the quality, extensiveness and availability of infrastructure networks. Effective modes of transport, including roads, railroads, ports, and air transport, enable entrepreneurs to get their goods and services to market in a secure and timely manner and facilitate the movement of workers to their workplaces. Economies depend on an electric supply that is free of interruptions and shortages so that business and factory work can proceed unimpeded. A solid and extensive telecommunications network allows for the rapid and free flow of information, increasing overall economic efficiency by helping to ensure that businesses can communicate and that economic actors can make decisions based on all available relevant information.
- **Macroeconomic Environment:** The stability of the macroeconomic environment is important for business and is significant for the overall competitiveness of a country. Running fiscal deficits limits the government's future ability to react to business cycles. Firms cannot operate efficiently when inflation rates are out of hand. The economy cannot grow in a sustainable manner unless the macro environment is also stable.
- **Health and Primary Education:** A healthy workforce is vital to a country's competitiveness and productivity. Sick workers cannot function to their full potential. Poor health in workers leads to significant costs for businesses, as sick workers are often absent or operate at lower levels of efficiency. Workers who have received little formal education can carry out only simple manual tasks and find it much more difficult to adapt to more advanced production processes and techniques. They also contribute less to executing innovations. In other words, lack of basic education can become a constraint on business development, with firms finding it difficult to move up the value chain via more sophisticated or value-intensive products.
- **Higher Education and Training:** Quality higher education and training is crucial for economies that want to move up the value chain beyond simple production processes and products. Today's globalizing economy requires countries to nurture pools of well-educated workers who can perform complex tasks and adapt rapidly to both their changing environments and the evolving needs of the production system.
- **Goods Market Efficiency:** Healthy market competition, both domestic and foreign, is important in driving market efficiency and thus business productivity. It ensures that the most efficient firms, producing goods demanded by the market, are those that thrive. The best possible environment

for the exchange of goods requires minimization of impediments to business activity through government intervention. Market efficiency also depends on demand conditions such as customer orientation and buyer sophistication.

- **Labor Market Efficiency:** The efficiency of the labor market is critical to ensuring that workers are allocated based on their most effective use in the economy and are provided with incentives to give their best effort in their jobs. Labor markets must have the flexibility to shift workers from one economic activity to another quickly and at low cost, and to allow for wage fluctuations without much social disruption. Efficient labor markets must also ensure a clear relationship between worker incentives and their efforts, and equity in the business environment between women and men.
- **Financial Market Development:** The financial crisis has highlighted the central role of a sound and well-functioning financial sector for economic activities. The banking sector needs to be trustworthy and transparent. Financial markets need appropriate regulation to protect investors and other actors in the economy at large.
- **Technological Readiness:** The technological readiness pillar measures the agility with which an economy adopts existing technologies to enhance the productivity of its industries. Specific emphasis is placed on its capacity to fully leverage Information and Communication Technologies (ICT) in daily activities and production processes for increased efficiency and competitiveness.
- **Market Size:** The size of the market affects productivity since large markets allow firms to exploit economies of scale. Traditionally, the markets available to firms have been constrained by national borders. In the era of globalization, international markets have become a substitute for domestic markets, especially for small countries. Exports can be thought of as a substitute for domestic demand in determining the size of the market for the firms of a country.
- **Business Sophistication:** Business sophistication is conducive to higher efficiency in the production of goods and services. Business sophistication concerns the quality of a country's overall business networks as well as the quality of individual firms' operations and strategies. This leads to increased productivity, thus enhancing a nation's competitiveness.
- **Innovation:** The final pillar of competitiveness focuses on technological innovation. Substantial gains can be obtained by improving institutions, building infrastructure, reducing macroeconomic instability, or improving human capital. However, all these factors eventually run into diminishing returns. The same is true for the efficiency of labor, financial, and goods markets. In the long run, standards of living can be enhanced only by

technological innovation. Innovation is particularly important for economies as they approach the frontiers of knowledge. In doing so, the possibility of integrating and adapting exogenous technologies tends to disappear.

As previously mentioned, it is important to keep in mind that the above-mentioned pillars can reinforce one another when strong and negatively impact each other when weak. For example, a strong innovation capacity (pillar 12) will be very difficult to achieve without a healthy, well-educated and trained workforce (pillars 4 and 5) that is adept at absorbing new technologies (pillar 9), and without sufficient financing (pillar 8) for R&D or an efficient goods market as a funnel to consumers (pillar 6). Although the pillars are aggregated into a single index, measures are reported separately for the 12 pillars as such details provide a sense of the specific areas in which a country needs to improve.

EGYPT'S PROBLEMATIC SOCIO- ECONOMIC FACTORS

Establishing confidence through a credible and far-reaching reform program is vital to Egypt's future and to realizing the considerable potential of its large market size and proximity to key global markets. Continued reforms are needed to create favorable conditions for private sector growth, which will be crucial for job creation and social cohesion. Such reforms include more openness to trade and investment, including reduction in tariff duties, non-tariff barriers, and a more favorable environment for Foreign Direct Investment (FDI). Continued efforts to strengthen financial markets and investment in skills and education will further support private sector growth.

Egypt faces enormous socio-economic challenges. Persistent failure to address the socio-economic aspects of popular demands could ultimately lead to another more aggressive and desperate revolution in the future. *We can understand the appearance of social unrest in 2011 and 2013 based upon the hypothesis that widespread unrest does not arise from the long-standing political failings of the system, but rather from its sudden perceived failure to use its soft and hard powers and provide essential security to the population.* Failure to provide security undermines the very reason for the existence of a political system. This results in the system losing those very powers. Once this occurs, resulting protests can reflect a wide range of reasons for dissatisfaction, broadening the scope of the protest, and masking the immediate trigger of the unrest. Individuals depend on a political system to make adequate decisions that guarantee expected living standards. This is particularly true for marginalized populations, whose alternatives are limited and who live at the boundaries of survival even in good times.

The dependence of the population on political systems engenders its support of those systems, even when authoritarian. Indeed, a certain amount of authority is necessary in order to protect against atypical individuals or groups who would disrupt it. behavior and the challenge it represents to the existing order calls for a comprehensive approach that brings together hard and soft tools of power. When the ability of the political system to provide security for the population breaks down, popular support disappears. In such t cases, there is nothing to lose; i.e. even the threat of death does not deter actions that are taken in opposition to the political order. Any incident can then trigger death-defying protests and other actions that disrupt the existing system. Widespread actions that jeopardize the leadership of the political system, or the political system itself, take place. All support for the system and allowance for its failings disappear along with the loss in its soft and hard powers.

It is reasonable to hypothesize that when the underlying trend exceeds the threshold, the security of vulnerable populations will be broadly and persistently compromised. Such a threat to security should be a key concern to policy makers worldwide. Social unrest and the political instability of countries can be expected to spread as the loss of security persists and becomes more pervasive. Even though the underlying causes are global and are not necessarily due to specific governmental policies, desperate populations are likely to resort to violence even in democratic regimes.

According to the Global Competitiveness Report 2016-2017, Egypt ranks as follows in many important variables (rank/138):

- The overall security situation (133)
- Property rights (100)
- Intellectual property rights (124)
- Public trust in politicians (84)
- Irregular payments and bribes (64)
- Judicial independence (47)
- Favoritism in decisions of government officials (28)
- Wastefulness of government spending (122)
- Burden of government regulation (63)
- Efficiency of legal framework in settling disputes (81)
- Transparency of government policymaking (97)
- Business costs of terrorism (135)
- Business costs of crime and violence (124)
- Reliability of police services (114)
- Quality of overall infrastructure (108)
- Government budget balance, % Gross Domestic Product (GDP) (132)
- Inflation, annual % change (130)

- Government debt, % GDP (117)
- Country credit rating (98)
- Quality of primary education (134)
- Primary education enrollment, net % (28)
- Higher education and training (112)
- Secondary education enrollment, gross % (85)
- Tertiary education enrollment, gross % (81)
- Quality of the education system (135)
- Goods market efficiency (112)
- Effect of taxation on incentives to invest (84)
- Total tax rate, % profits (96)
- Number of procedures to start a business (76)
- Prevalence of non- tariff barriers (100)
- Trade tariffs, % duty (126)
- Business impact of rules on Foreign Direct Investment (FDI) (114)
- Burden of customs procedures (80)
- Imports as a percentage of GDP (120)
- Effect of taxation on incentives to work (104)
- Labor market efficiency (135)
- Pay and productivity (125)
- Reliance on professional management (133)
- Country capacity to attract talent (103)
- Female participation in the labor force, ratio to men (133)
- Financial market development (111)
- Affordability of financial services (72)
- Ease of access to loans (136)
- Soundness of banks (70)
- Availability of latest technologies (117)
- FDI and technology transfer (71)
- Market size (25)
- Exports as a percentage of GDP (132)
- Business sophistication (85)
- Production process sophistication (105)
- Capacity for innovation (135)
- Quality of scientific research institutions (128)
- University- industry collaboration in R&D (137)

According to the Global Competitiveness Report 2016-2017, Egypt's most problematic factors for doing business are

- Policy instability 21.0
- Government instability 12.5
- Access to financing 10.2
- Foreign currency regulations 8.4
- Corruption 7.7
- Inadequate infrastructure 5.5
- Poor work ethic in national labor force 5.4
- Inadequately educated workforce 5.4
- Crime and theft 4.3
- Restrictive labor regulations 4.1
- Tax rates 3.9
- Inflation 3.2
- Tax regulations 3.1
- Inefficient government bureaucracy 2.7
- Insufficient capacity to innovate 1.7
- Poor public health 0.9

The points mentioned above are directly linked to prevailing national character and leadership issues, its formidable two-headed monster. It creates an obstacle to Egypt's successful navigation of the above-outlined development factors and related pillars, in turn limiting its ability to achieve sustained economic growth and long-term prosperity.

To create growth and employment, Egypt can build on its large market size (25th); its business sector which, by some accounts, appears more sophisticated than those of neighboring countries (85th); and its geographical proximity to the large European market. To do so, Egypt needs to step up its reform efforts and address the major rigidities that plague its goods, labor, and financial markets, in which the country ranks 112th, 135th, and 111th, respectively. Other priorities include higher education and training (112th), which is below the performance of peer economies particularly in terms of quality (134th); and the overall security situation (133rd), which remains fragile and imposes significant costs for business.

EGYPT'S NATIONAL CHARACTER

Character, in the long run, is the decisive factor in the life of an individual and of nations alike. - Theodore Roosevelt (Goodreads, 2018).

The true wealth of a nation lies not in its gold or silver but in its learning, wisdom and in the uprightness of its sons. - Gibran Khalil Gibran (Montano-De La Cruz, Magallona-Flores, Angeles-Guevara, & Ponce De Leon-Ladera 2002)

It is said that nations survive and thrive but for their character, for if they lose their character they are gone. The listed components of the twelve pillars are being both essential and a challenge for Egypt is documented and well-known throughout Egypt in official and public circles. In fact, a lack of these components offends and shocks any Egyptian with love and passion for the piece of land straddling the magnificent Nile river, sitting north of Sudan, east of Libya, and buffered to the east by the Red Sea and north by the Mediterranean.

Egypt's current national character, a devastating testament to decades of neglect, is reflected in a great deal of chatter (in official reports and in the media) focused on the following issues and concerns:

* *Religious leaders, be it the Muslim Brotherhood or the Salafis, pushing for a resurgence of Islamism and having the audacity, and the temerity to state/advocate that:*

- Religion/Islam mixing with politics, willfully and in a sinister manner ignoring the merits of separation of mosque and state.
- Sharia law becoming the law of the land, when it should be but one source into Egypt's laws and regulations.
- Women do not have the same rights as men, with the idea that they need not work since their role is as mothers and homemakers.
- Tourism being curtailed and controlled to be consistent with Islam.
- Believers of other religions being seen and treated as infidels.

This is a position, an approach to leadership, that could be generally described as an unwillingness to reform, lack of acceptance of modern life, and a weak intellectual imagination with regard to alternatives for a good life. The idea is that life on earth will continue without prosperity, good health, education, and happiness but, instead, leadership offers the citizenry an opportunity at paradise where they receive all of these things upon death. This conveniently passes the buck to God while allowing leaders to do little to nothing in terms of development.

This approach has resulted in the following:

- *The failure to create, generate, spread, disseminate and synthesize knowledge and information; ensuring that people read, understand and practice/ implement what they learn in their lives.*

- *The critical status of Egypt's women, not only women of Cairo and Alexandria but of the remotest of villages. Ideas, programs and projects need to be generated for the equal participation of women in Egypt's workforce. Without this, Egypt is denied the creative and productive efforts of half of its population and thus has little chance of making speedy, effective and efficient movement forward. This implies that Egypt's collective wisdom expects the nation, in the 21st century, to fire its economy on only half of its cylinders.*

Egypt's tradition of moderate Islam recognized women's rights and encouraged women to study and work. By contrast, for Wahhabis, a woman's job is to please her husband. Wahhabism has influenced all Islamic societies and as it spread in Egypt, more women started to wear the hijab, or headscarf. But this has not created a more virtuous society; it led to the opposite.

Until the end of the 1970s, many Egyptian women still went without head scarves, wearing modern style dresses, yet incidents of sexual harassment were rare. Now, with the spread of the hijab, harassment has taken on epidemic proportions.

Why is it that men did not harass Egyptian women at such high rates when they used to wear short skirts while sexual harassment has increased against women in headscarves? When ultraconservative doctrine dehumanizes women, reducing them to objects, it legitimizes acts of sexual aggression towards them.

- *Listening, watching, and reading on Egypt, one cannot help but think that Egypt is Cairo and Alexandria. But, what about other people living elsewhere in the country and the enormous lack of information flow in their direction? Meanwhile, this information is the only way to shape, alter and influence their character toward the needs of modern society.*
- *A culture of blaming others, nationally and internationally, for Egypt's societal ills exists. Egyptians should focus mainly on their soft and hard powers in order to rebuild Egypt politically, economically and socially while letting go of the 'conspiracy theories' that are getting in the way of their progress.*

Worrying only drains the power to think, plan and execute. It creates confused individuals resulting in an incapacity to build for the future. It turns a proactive nation to a helpless and reactive nation. Authorities can help by being more transparent and responsive in a timely manner. While conspiracy theories have been around for ages and will continue to evolve, there can be a focus on work, trust and moving forward.

- *In general, Egyptians living in Egypt are made to feel mentally fatigued, financially insecure, insufficiently informed, intellectually stretched, distant from modern living, and in a constant fight with corruption, unsure of the ethics code permeating their society.*

It is not an exaggeration to say that those involved in the revolutions of January 25th, 2011 and June 30th, 2013 and people all over the world who care for Egypt, were and are deeply concerned about the damaging and harmful national character that persists in the nation. They know the history and remember better times. They recall an Egypt that was a magnet for people seeking a better, safer life, people who came from Turkey, Greece, Italy, and so on. It was an Egypt where young people from all over the Arab world came seeking college education, establishing glossy careers in arts and literature. The Egypt they now can only read about is home to Cairo, at one-time a cosmopolitan city rivaling the most cultured and sophisticated cities the world had to offer and Alexandria, once the shining star of the Mediterranean.

These young Egyptians who took to the streets in revolution, with the prideful support of many around the globe, were using their soft and hard powers to say enough is enough. They said, 'we cannot take it anymore; we do not care about arcane traditions, and we will no longer accept whatever is thrown at us politically or culturally.' If properly supported, these young revolutionary Egyptians can help move the country through the mandatory prerequisites needed to confront and overhaul today's national character for the sake of Egypt's development.

LEADERSHIP IN EGYPT

Pity the nation whose statesman is a fox, whose philosopher is a juggler, and whose art is the art of patching and mimicking.

Pity the nation that welcomes its new ruler with trumpeting, and farewells him with hooting, only to welcome another ruler with trumpeting again.

Pity the nation whose sages are dumb with years and whose strong men are yet in the cradle.

Pity the nation divided into fragments, each fragment deeming itself a nation.

Gibran Khalil Gibran (AZ Quotes, n.d.)

Thirty years of Mubarak dictatorship have in effect made the concept of leadership non-existent; today there is a dearth of leaders in all aspects of Egyptian society. Leadership that can master the perfect mix of soft and hard power is what Egypt must have if there will be forward movement.

Leadership: The What

Courage, not complacency, is our need today. Leadership not salesmanship, John F. Kennedy (C.E., 2007)

After decades of incompetent leadership across the board, Egypt is more than deserving of a better group of leaders that know how to both channel physical quantifiable resources while also building support via its cultural brand and reach. It should not take a miracle for Egypt to uncover or discover leaders of character, with passion, ethics and a strong vision of the future. Words like rich, famous or powerful should not be in their dictionary.

A strong leader is not enough; a gang leader is an example of a strong person with a bad character, while an outstanding leader is one with both strong and good characteristics. Egypt needs leaders that will guide people to the future, leaders with the capacity to instill hope and ambition, and the capability to convey confidently that change is necessary to survive and thrive. For Egypt's future leaders to be effective, their people must have trust in them. Such leaders need to display a good sense of character composed of beliefs, values, skills, and positive traits.

- **Beliefs:** Sharing with confidence their beliefs of what is good, what is bad, what is human nature, etc.
- **Values:** Being open as to what they value in life, the worth of people, and what concepts or issues hold relevance
- **Skills:** Demonstrating knowledge and abilities gained throughout life, as well as a continued devotion to learn
- **Traits:** Displaying by action, not words, honesty, competency, forward-looking vision, inspired performance on the job, fair and broad-mindedness, courage, intelligence, imagination, and straightforwardness

No one expects Egypt's future leaders to possess all or even most of these characteristics, as these are traits of an ultimate, perfect leader. However, nations that have reared men and women possessing a good deal of these beliefs, values, skills and traits have prospered. Therefore, it is crucial for Egypt's leaders - wherever they work - to establish an ethical framework, teach and coach, as well as harmoniously manage the many activities within their realm. This is what it would take for their people to trust in their vision and buy into their plans. This will be their real power.

Leadership: The Who

Egypt has an opportunity it has not seen in generations. It is a golden opportunity for Egyptians across the spectrum of professions, organizations, and institutions to rise as leaders, as persons who hope, can and strive to make a difference. Missing this opportunity would mean at best a wait for another generation or two.

The vigor of today's young Egyptians and their disillusionment with authority could mark a departure from long-standing ineffective traditions and usher in a people-led transformation. An absolute necessity is for leaders to emerge from all over Egypt - media executives, university presidents, village mayors, community leaders, youth groups, military leaders, government officials, parliament members, the judiciary, arts and entertainment celebrities, business executives, labor organizations, and older and newly forming political parties.

After a long history of indifference and fear of exposure, it is not easy for such leaders to emerge. However, it is incumbent on those who fought for and supported the revolutions, those who are the majority of the Egyptian people, to effectively organize and speak credibly to potential leaders, pleading them to stand up, take risks and be a catalyst for a vibrant Egypt.

Liberals, leftists and all political activists with a vision of a modern Egypt should work fast and hard to identify such leaders, understand their interests and fears, lobby them, partner with them, and ask them to speak up. These potential leaders can then tell the nation what they know about what it would take to make things happen. This work could highlight economic and social issues which resonate with ordinary Egyptians- schools, hospitals, jobs, labor conditions- and are the traditional focus of the left. They could also work with corporate leaders to speak up about the importance of a robust and independent corporate world. Independent and well-financed private organizations are essential to the success of democratic transitions. They are also critical to maintaining democracies, once they have emerged.

A highly relevant effort could be to put together a program for information flow specifically tailored to reach deep into Egypt's most remote communities. Essentially a "Train the Trainer" program, influential members of such communities would be invited to various centers around Egypt for periodic information-sharing sessions that they can carry back and use to train potential leaders within their communities, later reporting on progress.

Leadership: The Why

Leadership's first role must be to gradually replace the above-noted national character with a healthier one. To bring back confidence in governance structures, new leaders need to enact initiatives that will show their commitment to a new way of doing things. It is just as important for people to have confidence in their local representatives as they do in national representatives. People want to feel that their local government is there to represent their wishes and act in their interest. There are potential leaders everywhere in Egypt. We must reach out to them and mobilize them so that they can assume responsibility and act with regard to the leadership objectives and tasks to be outlined below.

Leadership: The Objectives

To spur an overhaul of the national character and develop a model to suit the modern world, Egypt's leadership would go a long way by focusing on the following two top-priority objectives.

Separation of Mosque and State

- Egypt's laws must be appropriate, just, fair and proportionate for all. Sharia law should be but one of many sources of Egypt's constitutional, civil and criminal laws and here should be no question about the merits of separation of mosque and state. Religion is a spiritual matter to be left to the individual.
- The raison d'être for religious groups is interjecting religion into all aspects of life. This is what religious organizations bring to the table; it is their profession. The need here is to work with them in reaching an understanding that it is fine to preach Islam but not to legislate life in Egypt. A main reason is because no society today can claim total homogeneity.

Women's Full Participation in All Aspects of Life in Egypt Is a Must

- If full equality of women and men is not achieved in reasonable time, Egypt will continue to be mired in all manner of political, economic and societal issues effectively preventing any chance for progress.

When the 2011 and 2013 protests began in Egypt, women accounted for 40 to 50 percent of demonstrators. Women of all ages, with and without veils, set up barricades, led debates, shouted slogans and together with men risked their lives. The idea that men and women should behave differently was set aside during the two revolutions.

- A fulfilling life for all the people in Egypt will not happen while women continue to be prevented or discouraged from living up to their duties and responsibilities for their country, in all aspects.

Leadership: The Tasks

Egypt's leadership should start off with the easier task of grasping and etching in its collective soul the twelve pillars necessary for a nation's development. The youth who spearheaded the revolutions can be a vehicle for that effort. The leadership should then begin the enormous challenge of completely reshaping and reframing the national discourse to mobilize the nation to urgently address an overhaul of its national character. This is a mandatory prerequisite for development that requires moving determinedly and expeditiously to tackle the following tasks:

Task 1: Free Men From Their Obsession on Women

Developed a national campaign to educate both men and women with the purpose of:

- Presenting the benefits of establishing equality of men's and women's rights;
- Persuading men to focus less on lust and more on self-control and their role and responsibilities toward effecting social reform; and
- Informing the populous that men do not own women and that the status of women must change; that a man is not superior to a woman, only different anatomically.

Task 2: Revolutionary Youth

Use soft power and the popular mobilization skills and network of youth groups to engage the masses, gauge their interests, incorporate their views and get them involved. The major problem which denies Egypt a demographic dividend is youth unemployment. The youth bulge could reach a 'Goldilocks Moment' in which things are 'just right': more young people = more workers = economic growth = social good. It could also lead to a 'Third Revolution' if growth is not distributed fairly and there are no jobs for these potential 'workers'.; Revolutionary youth used soft and hard powers and have already shown twice that they will not accept autocracy, that they can bend the future to the pattern of their dreams.

Task 3: Leadership to Educate and Inform

Do not ask why they do not know, but ask how we can educate and inform underprivileged Egyptians. Remember *"Destroy ignorance, or ignorance will destroy you."*

Task 4: Business Leadership to Fight Corruption and Lobby for a Civil Society

Convince business leaders to file official charges against everyone they had to bribe to start or develop their businesses. Exposing corrupt officials in the ministries, municipalities and government institutions and dismissing them from their positions would be a major benefit to the nation.

Task 5: Social Reform

Be clear that there is simply no avoiding social reform. Without it there is no significant boost to the strength and popularity of genuine progressive democratic forces in Egypt. The nation should be preoccupied less with external issues and much more with internal reform and the many fights to be waged against corruption.

Task 6: Leadership Performance Evaluation

Institute evaluation mechanisms throughout the nation's leadership infrastructure to ensure that leaders demonstrate their value-added. Evaluations should be carried out by an independent agency that reports to the prime minister, the president, or the parliament as appropriate.

Officials need to decide, before taking office, if they have the competency, the capacity and the intent to make a difference, with consequences for erroneous decisions made clear to them. Officials should always be held accountable They should also be made to understand that the only purpose in accepting a powerful public service position is to add value and affect change in a challenging situation and not to acquire wealth, fame, power and an enhanced reputation to be parlayed into lucrative positions and deals once they leave office.

Task 7: Information Campaign About Desirable Social Characteristics

Launch a nation-wide information campaign outlining the social characteristics of progressive nations as a clear example of use of soft power. Emphasis should be on the key issues of successful cultures, including what helps countries achieve a low level of corruption and a high level of transparency.

Without the above-outlined leadership structure and framework, the second (*leadership*) prerequisite is neglected. This results in the character overhaul prerequisite not being addressed and further delays development.

CONCLUSION

Knowing what it takes to produce and support development is never the issue. In a global economy, progress is possible and an improved life standard subsequently waits on the horizon. Throughout history, however, the character of a nation and its inspired and exceptional leadership are unquestionably what secure success. The fate of the 25 January 2011 and 30 June 2013 revolutions are at stake. Stellar national character and passionate leadership lead the way to development. If that fails to emerge in the near future, the Egyptian revolutions will go down in history as a mere regime change rather than the change in the tide they had the potential to bring forth.

REFERENCES

AfDB, Organization for Economic Cooperation and Development (OECD), United Nations Development Programme (UNDP), & Economic Commission for Africa (ECA). (2012). Youth Employment: Five Challenges for North Africa. Paper for the regional conference: Promoting Youth Employment in North Africa, Tunis, Tunisia.

African Development Bank (AfDB). (2012). Inequality and Arab Spring Revolutions in North Africa and the Middle East. *Africa Economic Brief*, *3*(7).

African Development Bank (AfDB). (2012). *Political Transitions and New Socioeconomic Bargains in North Africa*. Economic Brief.

African Development Bank (AfDB). (2012). *Jobs, Justice and the Arab Spring: Inclusive Growth in North Africa*. Author.

Asseburg, M. (2012). Protest, Revolt and Regime Change in the Arab World: Actors, Challenges, Implications and Policy Options. German Institute for International and Security Affairs.

C.E. (2017). *11 Leadership Quotes From JFK That Still Resonate Today*. Retrieved from https://www.cashthechecks.com/11-leadership-quotes-jfk-still-resonate-today/

Goodreads. (2018). *Theodore Roosevelt*. Retrieved from https://www.goodreads.com/quotes/63681-character-in-the-long-run-is-the-decisive-factor-in

Hankel, W., & Isaak, R. (2011, May). Megatrends of the Global Economy of Tomorrow. *The World Financial Review*.

Hobson, E. (2011). *After the Arab spring Revolutions: 7 reasons why LED offers a major solution to North Africa's economic problems*. Local Economic Development Network of Africa (LEDNA). Retrieved from www.ledna.org

Kinninmont, J. (2012). *'Bread, Dignity and Social Justice': The Political Economy of Egypt's Transition*. Chatham house briefing paper. Retrieved from www.chathamhouse.org

Kouddous, S. A. (2011). Crackdown in Cairo. *The Pulitzer Center.* Retrieved from http://pulitzercenter.org/articles/cairo-egypt-habib-al-adly-trial-postponed-tahrir-square-revolution?format=print

Lynch, D. (2008). Tension in Egypt shows potency of food crisis. *USA Today.* Retrieved from http://www.usatoday.com/money/world/2008-04-29-egypt-food-shortage_N.htm

MacKenzie, D. (2011). Can complexity theory explain Egypt's crisis? *New Scientist.* Retrieved from http://www.newscientist.com/article/7dn20082-can-complexity-theory-explain-egypts-crisis.html

Mason, P. (2011). Twenty reasons why it's kicking off everywhere. *BBC News.* Retrieved from http://www.bbc.co.uk/blogs/newsnight/paulmason/2011/02/twenty_reasons_why_its_kicking.html

Mobarak, A. (2005, May). Democracy, Volatility and Economic Development. *The Review of Economics and Statistics, 87*(2), 348–361. doi:10.1162/0034653053970302

Montano-De La Cruz, E., Magallona-Flores, M., Angeles-Guevara, E., & Ponce De Leon-Ladera, H. (2002). *Dimensions in Learning English: A Series for Philippine Secondary Schools*. Retrieved from https://books.google.co.ma/books?id=gN74sdoJSkQC&pg=PA282&lpg=PA282&dq=%22The+true+wealth+of+a+nation+lies+not+in+its+gold+or+silver+but+in+its+learning,+wisdom+and+in+the+uprightness+of+its+sons.%22+Gibran+Khalil+Gibran&source=bl&ots=EyQyBucsFI&sig=ja2oj9U0x2ANh7uj978OlL2fQFc&hl=en&sa=X&ved=0ahUKEwji9ZOFzPPaAhWCTd8KHV-5DekQ6AEIJjAB#v=onepage&q=%22The%20true%20wealth%20of%20a%20nation%20lies%20not%20in%20its%20gold%20or%20silver%20but%20in%20its%20learning%2C%20wisdom%20and%20in%20the%20uprightness%20of%20its%20sons.%22%20Gibran%20Khalil%20Gibran&f=false

National Bureau of Economic Research (NBER). (2005). *Forms of Democracy, Policy and Economic Development*. Working Paper 11171, March 2005.

Powell, F. (2008). Civil Society, Social Policy and Participatory Democracy: Past, Present and Future. *Social Policy and Society*.

QuotesA. Z. (n.d.). Retrieved from http://www.azquotes.com/quote/434274

Robert, L., & King, S. (2000). *Discovering the Leader in You- A Guide to Realizing Your Personal Leadership Potential*. Center for Creative Leadership.

Shah, A. (2011). Middle East and North African unrest. *Global Issues*. Retrieved from http://www.globalissues.org/article/792/mideast-north-africa-unrest

United Nations Economic and Social Commission for Western Asia (UNESCWA). (2011). Youth Exclusion in the ESCWA Region: Demographic, Economic, Educational and Cultural Factors. Population Report, issue no. 5. United Nations.

United Nations Economic and Social Commission for Western Asia (UNESCWA). (2012). Promoting Participatory Democracy in the ESCWA Region. *Social Development Bulletin*, *4*(2).

World Economic Forum. (2014). *The Global Competitiveness Report 2014- 2015*. Author.

World Economic Forum. (2015). *The Global Competitiveness Report 2015- 2016*. Author.

World Economic Forum. (2015). *The Africa Competitiveness Report 2015*. Author.

World Economic Forum. (2016). *The Global Competitiveness Report 2016- 2017*. Author.

Zakaria, F. (2011). How democracy can work in the Middle East. *TIME Magazine*. Retrieved from http://www.time.com/time/magazine/article/0,9171,2046038,00.html

ENDNOTES

1 A "revolution" is an uprising that forces a change of leadership or regime (as in Egypt, Tunisia, Libya and Yemen). The term "revolt" is used where regime change has failed to materialize, or an uprising has been suppressed (as in Bahrain and Syria).

2 Rank among 2014- 2015 economies was 114, GCI 2014- 2015 (out of 144) was 119, GCI 2015- 2016 (out of 140) was 116.
Source: The 2014-2015, 2015- 2016, 2016- 2017 global competitiveness reports

3 Scale from 0 to 24, from the list of factors, respondents to the World Economic Forum's Executive Opinion Survey were asked to select the five most problematic factors for doing business in their country and to rank them between 1 (most problematic) and 5. The score corresponds to the responses weighted per their rankings.

4 Countries with a youth bulge, relatively good access to regional and international news and information, growing social inequality, widespread corruption and authoritarian governments are simply not going to be stable forever. Youth bulge could give countries a major competitive advantage but in the absence of vibrant, diversified economies able to create the increasing number of jobs needed, it has proven to be more of a curse.

Chapter 3
Key Factors Affecting the Investment Attractiveness of an Oil Producing Country

Nurlan Nurseiit
Eurasian Innovation Academy of Economics and Management, Kazakhstan

Darmen Sadvakassov
Center of Strategic Initiatives, Kazakhstan

ABSTRACT

Foreign direct investment (FDI) is the engine of growth of all countries, contributing to the inflow of financial capital, technology, skills, employment, to the establishment of production of modern goods and services, which enables a national economy to become more competitive in the global market. However, the developing or transition countries often lack the capital to finance their own development. Analysis of 21 developed and developing oil-producing countries from 2008 to 2014 show that the most important factors for attracting investment in the oil and gas sector are the discovery of attractive fields, the creation of a developed and modern infrastructure, increases proven reserves of hydrocarbons, and of corruption. Less important but still factors are a stable currency, an open trade regime, favorable business conditions, as well as lower taxes on oil-producing business.

DOI: 10.4018/978-1-5225-4203-2.ch003

INTRODUCTION

Foreign direct investment (FDI) is the engine of growth in all countries, contributing to the inflow of financial capital, technology, management skills, employment opportunities, modern products and services, which makes a national economy more competitive in the global market. This is particularly true for developing and transition countries, which often lack funding to support their economic growth that is often reflected in the imbalances between internal savings and investments, as well as exports and imports. Developing countries have an insufficient amount of savings to finance their investment needs. To overcome this gap, they need an inflow of foreign capital. FDI is now an important source of capital in developing countries, which allows them to attract long-term capital flows, as well as international companies, which help to promote the growth of production and productivity. However, FDI has not always being regarded so favorably.

In the 1960s and 1970s, many countries maintained a cautious and sometimes negative attitude towards FDI, because they did not want their economies to be overly defendant on foreign capital. Therefore, the developing countries preferred to attract bank loans and the poorest ones - aid from the rich countries instead of FDI.

The situation changed dramatically in the 1980s, when FDI emerged as an important vehicle in the financing of the economies of developing countries. This was mainly due to a sharp rise in interest rates and, and as a consequence the emergence of the debt crisis, which negatively affected many developing countries. Because of the crisis, access to bank lending for these countries was closed, as interest rates rose to prohibitive levels. Moreover, the sharply worsened investments ratings of many developing countries, due to the debt problems led to their de facto withdrawal from the credit market. Thus, developing countries were forced to reconsider their previous attitude to FDI.

As a result, FDI grew rapidly afterwards, while the importance of other forms of capital dropped significantly. For example, the share of subsidized foreign capital has been steadily declining since 1960, and commercial loans, which accounted for the main capital flows in the 1970s, completely ceased after the debt crisis of the 1980s. It is now generally accepted that FDI contributes to the economic growth and modernization of developing countries.

The Asian Development Bank in its Asian Development Outlook (ADB, 2004) noted that FDI inflows greatly accelerated in recent years as a result of many factors, such as the rapid development of technical progress, the emergence of globally integrated production and marketing networks, bilateral preferential investment agreements, the recommendations of major international organizations, and the positive experiences of developing countries that opened their doors to FDI.

The FDI peak in developing countries came in 2008, followed by a period of decline, a consequence of the global economic crisis and the related decrease in demand for raw materials. New growth in FDI in developing countries began in 2009 and continued until 2015. Their share in 2016 increased to 59% of global FDI inflows, while in 2007 it was less than 40% (WIR, 2017, p. 16). In 2016, the decline in flows to developing economies was due to a sharp drop in commodity prices, which fell 14 percent to $646 billion (WIR, 2017, p. 10). As a result, developing countries again began actively competing to attract FDI.

There is competition between the developing and developed countries to attract FDI, as FDI provides a good opportunity for developing countries to enter or strengthen their position in the global system of production, trade, and investments. As mentioned by Bouoiyour (2003), many countries are actively trying to attract foreign investment by offering tax holidays, exemptions from import duties, subsidies and infrastructure development measures, preferences, and sometimes monopoly rights on production granted to foreign companies.

This study aims to identify the key factors that determine the inflows of FDI in oil-producing countries. Once these factors are identified, policy recommendations can be made i to remove obstacles to inflows. Another task is to study ways to increase the attractiveness of the oil and gas sector to FDI, which is crucial for many oil-producing countries, including Kazakhstan.

The high value of the oil and gas sector of Kazakhstan's economy, evidenced by the fact that the share of the oil and gas industry in the GDP is 25%, in foreign currency inflows - about 60% of government revenues - about 44% (Alimov, 2014; Danesh, 2010). The situation in many developing oil-producing countries is similar. For example, the share of the oil and gas sector in the GDP is 12% in Venezuela and 66% in Angola. In government revenues, oil and gas account for 45% in Iran, 60% in Nigeria and up to 75% in Saudi Arabia (Giles, 2014). Oil and gas account for 80% in Venezuela and 95% in Kuwait of all export earnings (Ubiznes.ru, 2015).

In recent years, the relevance of this study has increased, since Kazakhstan is experiencing a stagnation in oil and gas production, this has been frozen at 80 million tons of oil since 20xx. Of course, the main reason for this is the structure of ownership in the oil sector of the country, it is dominated by foreign players (they account for more than 80%). Another reason is the sharp drop in world oil prices, which has forced large oil companies to postpone the implementation of their oil production plans. However, according to international observers, there is also the problem of excessive regulation and control in the Kazakhstan oil and gas sector.

Thus, FDI, on the one hand, is desirable for developing countries, especially Kazakhstan. It has enabled the country to rapidly increase oil production, which since independence has been the main engine of growth. FDI helped to attract new technologies, managerial skills and create new jobs in the country. However, on the other hand, FDI has caused today's economics problems, such as the high degree of vulnerability of the economy to the changes in world oil prices; a high economic dependence on foreign countries; an increasing focus of the country on the production of raw materials while their processing is carried out in other countries In other words, if FDI had been involved in the processing industry, rather than solely in oil production, the country's economy would be more stable and dynamic, and the current decline in oil prices would not have affected significantly the pace of its growth and fiscal sustainability.

In this regard, the choice of adequate policies to attract FDI in the oil and gas sector is of great importance today, as it should be aimed at maximizing the benefits of the country's oil and gas resources in the long term. In this study, we focus on FDI in the oil and gas sector of the country, using data for 21 oil and gas producing countries.

BACKGROUND

Initially, FDI flows were explained by the theories of international trade and firms. They assumed that the difference in interest rates between countries as the main cause of FDI, as well as the conditions of perfect competition and the same production function. However, the theory of interest rate differences can only partly explain FDI flows between countries (Majeed & Ahmad, 2009).

More recent theories (Barrel & Pain, 1999) considered a market of imperfect competition, and FDI flows associated primarily with the actions of transnational corporations (TNCs) because they received a monopoly advantage over local competitors during the investments. For example, Agiomirgianakis et al. (2003) noted that FDI is mainly related to the behavior of transnational companies (TNCs). Factors affecting the behavior of TNCs may affect the magnitude and direction of FDI. These include such factors as economies of scale, and the benefits associated with the life cycle of products, etc. National governments also compete with each other to attract FDI by changing key parameters of their economic policies, such as labor market conditions, the amount of corporate taxes, tariff barriers, subsidies, and by the privatization and liberalization of the trade regime.

However, the underlying hypotheses of the beneficial effect of low wages, tariff barriers, taxes, etc. do not always find confirmation. For example, Tsai (1994) found support for the hypothesis of the beneficial effect of cheap labor on FDI for the period 1983 to 1986, but only weak evidence for the period from 1975 to 1978. On the other hand, Goldsbrough (1979), Schneider and Frey (1985) and Shamsuddin (1994) showed that higher wages reduce the inflow of FDI to developing countries.

Some studies (Hartman, 1994; Cassou, 1997; Kemsley, 1998) showed that the amount of corporate tax in the host country also has a negative impact on FDI flows. Others (Root & Ahmed, 1979; Jackson a&nd Markowski, 1995; Porcano & Price, 1996) believe that taxes do not have a noticeable effect on FDI and Swenson (1994) found a positive relationship between FDI and taxes. This ambiguous situation is caused by the fact that the paradigm of the global economy after the 1980s has changed significantly. In the past, FDI was directed to those countries that had relatively inexpensive resources, and currently, it goes mainly to countries where there is skilled labor.

An ODI (1997) study confirms that relative labor costs are important for FDI in labor-intensive industries and export-oriented subsidiaries. However, in the capital-intensive industries, where labor costs tend to be a small percentage of overall costs, it is the availability of appropriately skilled labor, rather than cheap labor that influences the location of FDI. In Kazakhstan, this is particularly relevant as the extractive industries are capital-intensive industries.

For example, in 2012, the direct share of the oil industry in the employment of the population of Kazakhstan was only 0.8%, nevertheless its share in the income of the population was 1.6%, in GDP - 18.9%, and in the country's tax revenues - 38.8%. Taking into account the total indirect effects on other sectors, the importance of the oil and gas industry for the economy of Kazakhstan is even higher. The share of oil and gas industry in employment was 4.8%, in labor income - 4.8%, in the GDP - 32.8% and in tax revenues – 43.3% (KazEnergy, 2013, pp. 32-40).

Another approach to explaining the movement of FDI is the "OLI paradigm" (Dunning, 1988). According to this theory, companies invest abroad in order to realize benefits from the use of their ownership (O), location (L), or internationalization (I).

1. Advantages of the property, a company receives from its ownership of the intangible assets such as technology, management skills, patents, trade networks, reputation. These assets give the company a market force that compensates for the shortcomings of production abroad.
2. Advantages of the location associated with the presence of valuable assets such as abundant natural resources, a large market size, the cost of production factors and infrastructure in the foreign local markets.
3. The benefits of internationalization associated with the ability of the firm to minimize its transaction costs, if it, for example, produces products abroad through subsidiaries and joint ventures, rather than through the issuance of licenses or subcontractors.

While the benefits of ownership and internationalization are associated with the company's activity, the advantages of its location arise due to the peculiarities of the country in which it is located.

There are three types of investments: horizontal, vertical and mixed types (Markusen & Venables, 1999).

For the horizontal type, FDI is the investment by foreign companies seeking a new market (a market-seeking investment) to obtain benefits in trading costs and improve their strategic advantage through a more intimate acquaintance with the taste and the behavior of local buyers in competition with local rivals. In this case, a foreign company goes to the local market through its subsidiary with the same products that it offers at home.

The vertical type of FDI is, on the contrary, aimed at resources (a resource-seeking investment) to enhance its own security or to reduce its production costs by using cheaper resources. Such investments, usually stimulated by the difference in production costs and prices, are made due to the difference in the trade value.

The mixed type of FDI (the knowledge-capital model) includes both vertical and horizontal type investments at the same time (Markusen, 2002).

For the oil and gas industry of Kazakhstan, all three types of FDI are typical, but the second type dominates, since FDI is mainly aimed at finding cheap hydrocarbon resources.

However, despite the fact that these factors determine the main reasons for FDI, they are not able to effectively explain the unequal distribution of investments among countries (Krifa-Schneider & Matei, 2010, p. 54). However low country risk is one of the several important reasons for attracting FDI. These risks are usually associated with cross-border investments (Nordal, 2001; Luo, 2009). Indeed, investments that cross international boundaries are not typical of domestic investments. They are caused by a variety of national differences in socio-political institutions, economic structures, politics, geographies, and currencies. For example, De Mello (1997) in his study argued that the political regime of host countries is a potentially important determinant of FDI. Foreign investors are guided primarily by the desire of rent seeking under standard assumptions of profit maximization.

The presence of developed institutions that protect property rights, political stability, and low levels of corruption contribute to creating a more favorable investment climate (Barrell & Pain, 1999; Bevan, Estrin & Meyer, 2004). Singh and Jung (1996), analyzed the impact of a composite index of two variables, political risk and business climate, and Busse and Hefeker (2007), who studied the effect of 12 individual indicators of political risk on FDI, have obtained similar results.

Sekkat and Varoufakis (2007) estimated the value of the business climate in the inflow of FDI and found that the improvement of trade regime (openness), infrastructure, economic and political conditions lead to an increase in FDI inflows.

A similar analysis carried out by Botric and Skuflic (2006) in relation to the countries of South-Eastern Europe for 1996-2002 confirms that trade openness and a high level of infrastructure development led to an increase in FDI flows to these countries. Kok and Ersoy (2009), who studied the factors affecting FDI in 24 developing countries for the period 1976-2005 also concluded that trade openness, growth of GDP per capita, gross capital formation and the availability of infrastructure (measured by access to telephone lines) positively influenced FDI.

Economic growth can also be attributed to positive business climate factors. Lim (1983) argued that the rapid growth of an economy provides more opportunity for profit. However, this hypothesis is not always confirmed. For example, Lynn (1980), Schneider and Frey (1985), Culem (1988), Ancharaz (2003) found a significant positive relationship between growth and FDI, while Nigh (1985) confirmed it only for developing countries, and found a weak negative relationship for developed countries. On the other hand, an increase in FDI, even in the manufacturing sector, does not necessarily lead to higher rates of economic growth.

FDI can accelerate growth only under certain conditions (Asiedu, 2013): education in the host country is above a certain threshold; (Borensztein et al, 1998); domestic and foreign capital are complementary (de Mello, 1999); the country has reached a certain level of per capita income (Blomstrom et al., 1994); it is open (Balasubramanyam et al., 1996), and its financial sector is sufficiently developed (Alfaro et al., 2004).

At the same time, the burden of servicing external debt and inflation had negative impacts on FDI. Demirhan and Masca (2008) investigated the factors influencing FDI flows in 38 developing countries between 2000-2004. They also found that the business climate and the level of infrastructure development in the country had a significant impact on its attractiveness to FDI. Thus, the growth rate of per capita income, open trade regimes and the number of telephone lines positively affected the inflow of FDI, while the inflation rate and the tax rate had a negative influence.

Jordaan (2004) argued that a well-developed infrastructure improves the performance of the investment potential in the country and, consequently, stimulates the flow of FDI to the country. According to Asiedu (2002), the number of telephones per 1000 inhabitants is a standard measure for assessing the level of infrastructure development. However, in accordance with Asiedu (2002), this measure is not adequate, because it detects only the presence, but not the quality of the infrastructure. In addition, it includes only the infrastructure of fixed telephony, excluding the presence of cellular (mobile) communication.

Majeed and Ahmad (2009) carried out a more detailed analysis, using panel data of 72 developing countries for 1970-2008. They found that the income per capita, GDP growth, remittances, trade openness, and a well-developed communications infrastructure have also a positive effect on FDI as a percentage of GDP, while the real exchange rate, inflation, military spending, balance of payments, and the level of non-performing investments per employee have a significant negative impact. At the same time, external debt, loans to the private sector, as well as official development assistance had virtually zero impact on FDI inflows.

Jordaan (2004) argued that the effect of opening up to FDI also depends on the type of investment. When investments are looking for a market (a market-seeking investment), in this case, trade restrictions have a positive impact on FDI, as the restrictions to the market for foreign firms force them to set up branches in the host country. By contrast, with regard to investments in natural resources, there is a positive relationship between the openness of the market and the influx of FDI as TNCs prefer to invest in a more open economy, where there are less costs related to export-import transactions.

Nevertheless, political risk, the business climate and the level of infrastructure development in a country are not always dominant factors in making decisions on FDI. According to the ODI (1997), when the host country has abundant natural resources, there is no need for extra incentives to attract FDI, as seen in politically unstable countries, such as Nigeria and Angola, where the high profits in the extractive industries more than compensate for the risks associated with political instability. For example, large mining companies overcome some of the political risks by investing in their own infrastructure and security forces. Moreover, these firms are not limited by the small size of local markets or currency risks, as they sell the products on the international market at world prices.

A study conducted by the EBRD (2000) showed that the investment climate in the CIS depends on a wide range of factors such as onerous taxation, widespread corruption, poor governance, weak legal and regulatory frameworks (including the violation of property rights). This is accompanied by a widespread practice of public participation in the economy and the need to follow a program of structural reforms.

As for Kazakhstan, the following problems were identified: high levels of corruption in the provision of basic services (health, education, internal order, law and customs services), overregulation and the rigidity of legislation, and a high degree of monopolization of markets (Clinton R. Shiells, 2003).

METHODOLOGY

Data Sources

The data panel contains observations from 21 developed and developing countries for the period from 2008 to 2013. Only those countries for which complete data sets are available for all variables have been included. The list of countries selected for the model is given in Table 4. Nevertheless, there are two countries in the sample (Angola and Iran), which do not have data on individual indicators. Angola does not have data on the effective exchange rate, and Iran - on the volume of trade with GDP.

Table 5 includes a list of variables and their sources. Almost all the variables used in the study were taken from the World Bank database, except for data on oil production, capital and operating investment, proven oil reserves and indices of corruption. They were taken from other sources, such as Rystad, GEM WoodMac, Transparency International and so on.

The Variables and Their Measurement

Dependent Variables

As a dependent variable, we used a proxy variable - the natural logarithm of the per capita capital investments in the oil and gas industry of different countries. This variable was taken as an explanatory, as the statistics on FDI in the oil and gas sector in different countries are not always collected, or available to the public in view of their commercial confidentiality. Data on capital investment were taken from the database Rystad (2010).

Foreign investment by foreign corporations that received 10% or more of a local firm was attributed to FDI. Therefore, FDI does not cover all the resources invested by transnational corporations in the host country, but includes a significant part of them. For example, investments financed through debt financing or the issuance of shares are not included in FDI.

Explanatory Variables

The names of the selected explanatory variables and sources of information about them are listed in Table 5 in appendix. Their short names and methods of calculation are given in Table 6 in appendix. Descriptive statistics of the variables used in the basic model are contained in Table 7 in appendix.

We also tested all the potential variables for inclusion in the model for stationarity, using the ADF test procedure (Table 8 in appendix). All data are stationary, including the steady-state changes in the first and second levels.

In addition, we have checked for long-term data convergence by using the Johansen cointegration test. Results are presented in Table 9 in appendix. The results show that there are 13 equations that converge at the 5% level of significance, which means that the long-term convergence of the variables and the ability to construct meaningful models exist.

The coefficients of correlation between the variables are given in Table 10 in appendix. The cross-correlation between most of the variables are in the low range, except for the explanatory variable (INV_CAP) and the independent variables (FBS and PROD_CAP), all are greater than 60%. Therefore, these variables should not be included in the model, or if included in the model, the absolute values should not be used and then be tested for the presence of residual cross-correlation with the dependent variable.

The method of least squares was used as a method of data processing.

Description of the Model

The relationship between per capita capital investment in the oil and gas sector, and the explanatory variables will have the following form:

ln_inv_cap = c*ln_inv_cap(-1) + c(2)*ln_corr + c(3)*gdpr + c(4)*ln_reer + c(5)*ln_res + c(6)*ln_fbs + c(7)*ln_tax_rate+ c(8)*defl + c(8)*trade_open, (1)

where variable ln_inv_cap shows the changes in per capita capital investments in oil and gas sector; ln_corr - the changes in the perceived level of corruption; gdpr - in country's business climate; ln_reer – the changes in the price competitiveness of products in relation to other countries, ln res - the changes in proven oil reserves; ln_fbs - the rates of modern infrastructure development, and ln_tax_rate - the changes of the overall tax burden on enterprises. The variable defl reflects the average annual net change in inflation, and the variable trade open shows the influence of the degree of openness of the trade regime on capital investments in the oil and gas sector.

Almost all the variables, except for the variable of oil rents, fixed broadband casting, age dependency ratio, real economic growth, and trade openness are taken in logarithmic form, that makes it possible to trace the influence of the rate of change of the analyzed variables. It also provides an opportunity to eliminate the problem of multicollinearity between the explanatory variables.

The inertia in the process of capital investments is reflected through the use of time lags.

We expect the independent variables to influence the explanatory variable (Table 1) in the following manner:

- inv_cap with temporal lag (-1) is expected to be positive, since the current level of per capita capital investment will encourage new capital investment in the oil and gas sector of the country in the future.

- The impact of corruption on the inflow of capital investment in the oil and gas sector (corp) is negative.
- Real GDP growth (gdpr) has a weak negative relationship between developed countries, as it improves the prospects for non-oil business development, which could lead to a decrease in the attractiveness of investments in oil production.
- An increase in the real effective exchange rate (reer) leads to a rise in the costs of production of a country compared with other countries, which reduces the profitability of local production and of new investments. In this case, we should expect a negative effect on FDI. However, the growth of the real effective exchange rate helps reduce the cost of imported equipment, which is a significant share in the cost of oil and gas projects, thereby increasing the attractiveness of oil and gas projects. In this case, a positive effect t can be expected. Thus, we can conclude, that in the initial stages of the operation of oil and gas fields the share of foreign capital will prevail in the cost of the project, as local investors feel the need for investment and do not have technologies to produce modern oil and gas equipment. Therefore, at this time we should expect real effective exchange rate to have a positive effect. However, at the stage of maturity, when all attracted foreign investment is fully paid off, we should expect a negative effect for this variable. Therefore, the impact of the real effective exchange rate variable will depend on the overall stage of development of oil and gas projects in the country.

Table 1. Expected sign of influence of independent variables on the investment in the oil and gas sector

Name of Variable	Sign of the Variable	Name of Variable	Sign of the Variable
Inv_Cap	+	Trade	+
Res or Res_Cap	+	Gdpr	–
Capex_Bar Opex_ Bar Cost_Bar	"– "- if these are lower and "+"- if these are higher than world oil prices	Doi_Buss	–
Prod Prod_Cap	+	Reer	"+" in the in the initial, and "–" in the final stage
Oil_Rent Nat_Rent	+	Defl	–
Tax_Rate	–	Corr	–
Fbs	+	Adr	–

- Proven reserves (res or res_cup) on investments in oil and gas sector is expected to be positive, since an increase in proven reserves increases the attractiveness of the country to investments in the oil and gas sector.
- The cost variables (opex_bar, capex_bar, and cost_bar) depends on the internal prices of oil. If they are lower than world oil prices, their influence will be positive. If they are higher than world oil prices, their influence will be negative.
- The expected effect for the variable tax rate will be negative, as the tax increase reduces income, which has a negative effect on the attractiveness of capital investment in the oil and gas sector.
- The expected sign of the variable local infrastructure (fbs) should be positive as its improvement reduces the indirect costs and, hence, increases the share of profit in the price of final products.
- The impact of oil rent on FDI will be positive, as this indicator shows the attractiveness of natural resource extraction and oil for the economy of a particular country. The higher it is, the more profitable will be investments in the oil industry.
- The expected effect of oil production (prod and prod_cap) will be positive, as with its increase in general and per employee, the attractiveness of investments in the oil and gas sector increases.
- Improving the business climate in the country by the Doing Business (doi_buss) ranking leads to an inflow of investments in the economy, including oil and gas sector. Therefore, the expected effect wıll be positive.
- Foreign investors prefer countries with more open economies, as in this case they are free in their actions and can easily and cost-effectively withdraw money from the invested country. Therefore, the effect of the trade regime variable is expected to be positive.
- The inflation rate (defl or cpi) has a negative impact on the inflow of FDI, as it promotes the reduction of the expected real return on investment.
- The level of corruption in a country reduces its attractiveness to investments. In this regard, the expected sign of the corruption indicator on FDI will be negative.
- The impact of the age dependency ratio (adr) on investments will be negative, as the growth of the pension load per worker increases the cost of production and reduces the attractiveness of investments.

Finally, these theoretical assumptions are confirmed by the presence of correlation between these explanatory variables and the dependent variable (see. Table 10 in appendix).

CALCULATIONS OF THE MODEL OF CAPITAL INVESTMENTS IN OIL

The following models of per capita capital investments in oil production, depending on various factors have been received based on a comprehensive analysis of the data on 21 oil-producing countries (Table 2).

The models explain the per capita capital expenditure in the oil and gas sector in the mentioned oil-producing countries. Almost all the effects of the explanatory variables correspond to the theoretical expectations.

The first model includes the following variables: accumulated investments, the size of proven oil reserves, and the costs of production. The most significant indicator is the accumulated investment in the oil and gas sector of the country (t statistic = 13.33), followed by the size of proven reserves (t statistic = 3.37) and the cost of capital expenditures for the production of one barrel of oil (t statistic = 2.83).

The growth of accumulated investment by 1% leads to an increase in investment by 0.7%. This is because investments in the oil business have a long-term nature. Wherein capital investment in one year tend to draw capital investments in the following years.

The growth of proven oil reserves by 1% leads to an increase in investments by 0.09%, and the increase in the capital cost of production for one barrel of oil by 1% leads to an increase in investment by 0.18%.

In general, this model explains the change in investment in the oil and gas sector by 65%.

The second model shows that current oil production is not a decisive factor when investors choose to invest in the oil and gas sector (t statistic = -0.59). As for the alternative variable prod cap, it shows a strong correlation with the variable invest cap - 61.6%. For this reason, this variable was excluded from the model.

The third model shows that another important factor for investors is the quality of the oil fields in terms of economic rent. Oil rent is a very significant indicator (t statistic =3.48). An increase in the value of oil rent by 1% leads to an increase in investment by 0.18%.

The fourth model takes into account the effect of a change in the tax rate. This variable is significant (t statistic = 1.97), but shows a negative impact on investment. Thus, an increase in the burden of taxes by 1% leads to a decrease in investment in the oil and gas sector by 0.28%.

The fifth model adds the impact of modern infrastructure development on investments in the oil and gas sector. For this purpose, the fixed broadband costs per 100 people is used as a proxy variable. This indicator for attracting investment is very significant (t statistic = 9.18). Improving the infrastructure by 1% leads to an increase in investment in the oil and gas industry by 0.059%.

Table 2. Impact of various factors on FDI in oil and gas sector

Variables	1st Model	2nd Model	3rd Model	4th Model
Ln_Inv_Cap(-1)	0,738* (13,33)	0,731* (12,93)	0,660* (11,49)	0,630* (10,708)
Ln_Res	0,09* (3.37)	0,111* (2.82)	0,057** (1.98)	0,038**** (1.26)
Ln_Capex_Bar	0.183* (2.83)	0.161** (2.15)	0.353* (4.48)	0.390* (4.87)
Ln_Prod		-0.07**** (-0.59)		
Oil_Rent			0.018* (3.48)	0.018* (3.51)
Ln_Tax_Rate				-0.282** (-1.97)
Ln_Fbs				
Trade				
Gdpr				
Ln_Doi_Buss				
Ln_Reer				
Ln_Defl				
Ln_Corr				
Adr				
Statistics				
Observations	2125	2125	2125	2125
Adjusted R-squared	0.647	0.646	0.685	0.685
Sum squared residual	61.662	61.662	55.98	54.21
Log likelihood	-132.634	-132.634	-126.642	-124.65
Durbin-Watson stat	1.819	1.819	1.663	1.604

Note: *Probability less than 1, ** - less than 5, *** - less than 10, **** - more than 10.
The numbers in brackets show t statistics.

The sixth model additionally shows the influence of such macroeconomic factors as the openness of trade, real GDP growth, and business climate on investments in the oil and gas industry. The most significant among these factors is real GDP growth (t statistic = 4.03), followed by business conditions (t statistic = 2.56) and openness of trade (t statistic = 2.47)

GDP growth of 1% leads to an increase in FDI in the oil and gas industry by 0.067%. This corresponds to our expectations that, given the rapid growth of real GDP, investments in non-oil sectors become more attractive.

Table 3. Impact of various factors on FDI in oil and gas sector (continues)

Variables	5th Model	6th Model	7th Model	8th Model
Ln_Inv_Cap(-1)	0,491* (9,18)	0,380* (6,90)	0,391* (6,80)	0,244* (4,16)
Ln_Res	0,013**** (0.51)	0,099* (3.16)	0,132* (3.17)	0,114* (2.30)
Ln_Capex_Bar	0.481* (6.97)	0.624* (8.69)	0.635* (5.94)	0.508* (5.13)
Ln_Prod				
Oil_Rent	0.039* (7.40)	0.046* (8.50)	0.043* (6.91)	0.056* (9.13)
Ln_Tax_Rate	-0.259** (-2.14)			
Ln_Fbs	0.059* (9.34)	0.208* (5.15)	0.249* (5.26)	0.386* (5.47)
Trade		0.003* (2.47)	0.004* (2.28)	0.006* (3.36)
Gdpr		-0.067* (-4.03)	-0.062* (-3.52)	-0.053* (-3.33)
Ln_Doi_Buss		0.119* (2.56)	-0.008**** (-1.32)	0.304* (3.22)
Ln_Reer			-0.015**** (-0.05)	0.087**** (0.303)
Ln_Cpi			-0.123**** (-0.364)	-0.054**** (-0.170)
Ln_Corr				-0.614* (-5.08)
Adr				0.021* (2.88)
Statistics				
Observations	2120	2120	2120	2120
Adjusted R-squared	0.775	0.812	0.811	0.848
Sum squared residual	38.37	31.37	30.09	23.70
Log likelihood	-103.22	-89.51	-85.59	-72.09
Durbin-Watson stat	1.371	1.347	1.310	1.399

Note: *Probability less than 1, ** - less than 5, *** - less than 10, **** - more than 10. The numbers in brackets show t statistics.

The business environment and the openness of trade have a positive impact on investment in the oil and gas sector. Their improvement by 1% led to an increase of FDI in the oil and gas industry by 0.119% and 0.03%, respectively. This is in line with our expectations for these variables, as their improvement stimulates investment in the country's economy.

The seventh model includes additional factors: the real effective exchange rate and the consumer price index. In this model, the first indicator shows the wrong, and the second shows the correct signs. Nevertheless, both indicators are statistically insignificant (t statistic for the first indicator = -0.05, and for the second = -0.36). Therefore, these indicators should be excluded from the model. Their inclusion into the model also led to an incorrect sign in the indicator of the conditions of doing business, which became negative and not significant.

In the eighth model, instead of the real effective exchange rate and inflation, new indicators were added: the change in the level of corruption and the age dependency ratios.

Corruption has a significant negative impact on the influx of FDI (t statistic = -5.08). At the same time, the age dependency ratio shows the opposite sign to the expected, but statistically significant (t statistics = 2.88). This means that the hypothesis that an increase in labor costs impedes the growth of FDI in the oil and gas sector is not confirmed in practice. This is not surprising, given that the oil business is capital-intensive, but not labor-intensive. Therefore, labor costs are not a determining factor in the direction of investments in the oil and gas industry.

Summarizing our conclusions (the eighth model), the most important factors for attracting FDI to the oil and gas industry are the growth of the oil rent (t statistic = 9.14), improving infrastructure (t statistic = 5.47), the growth of capital expenditures for (t statistic = 5.14), decrease of corruption (t statistic = -5.09, the accumulated level of investments (t statistic = 4.16) and the growth of proven mineral resources (t statistics = 2.30). This confirms our assumption that investors are attracted by countries with the most promising fields, a developed and modern infrastructure, high growth of proven hydrocarbon reserves, as well as to less corrupt countries.

Other growth factors for investment in oil and gas are the improvement of business conditions (t statistic = 3.22) and real GDP growth (t statistic = 3.36). If the growth inthe openness of the economy and the improvement of business conditions have a positive impact, the growth of real GDP has a negative impact on FDI into the oil and gas sector.

As for the other statistically significant indicator – the coefficient depending on age (t statistic = 2.88), it does not adversely impact on FDI, which was typical of other sectors of the economy. Moreover, factors such as the exchange rate and inflation did not significantly affect investment in the oil industry. The increase in the tax burden on oil companies reduces their desire to expand their investments in oil and gas production.

The obtained model (eighth model) helps to understand what needs to be done by states to improve the attractiveness of investments in the oil and gas sector. These are the following measures in the decreasing order of importance: the discovery of new more attractive deposits, improving infrastructure, the expenditures for the extraction of one barrel of oil, the reduction of corruption, the increase in proven reserves, a more open trade regime and the creation of favorable conditions for doing business. Maintaining the exchange stability of the national currency and a reduction in the inflation rate are factors that improve the investment climate.

Moreover, the eighth model allows us to quantify the impact of various factors on the attractiveness of investments in the oil and gas sector. For this purpose, it is necessary to pay attention to the value of the coefficient of the variable of interest. For example, an increase of 1% in the following variables: oil rents, the improvement of infrastructure, and the increase in capital costs leads to an increase in investments per capita by 0.056%, 0.386%, and 0.508%, respectively. A 1% growth in proven reserves, increased trade openness and improving of business environment, as well as the reduction of corruption by 1% lead to an increase of investments of 0.114%, 0.06%, 0.304% and 0.614%, respectively.

CONCLUSION

Analysis of data on 21 oil-producing countries for the period 2008-2014 showed that the most important factors for attracting investment in the oil and gas sector of oil producing countries are the discovery of more attractive fields, the creation of a developed and modern infrastructure, the increase in proven reserves of hydrocarbons, and the decline in levels of corruption. In addition, increased trade openness of the economy and improved business conditions are no less important measures to attract FDI in the oil and gas industry.

Increasing the stability of the national currency and reducing inflation also have a positive, but not a statistically significant impact on attracting FDI. On the other hand, increased taxes had a clear negative impact on the inflow of FDI into the oil and gas sector.

Although the growth of real GDP reduces the attractiveness of investments in oil, it provides an excellent opportunity to move away from dependence on oil by an appropriate choice of the economic policy.

To improve the attractiveness of investments in the oil and gas sector, the government may take the following steps in the decreasing order of importance:

- Discover more attractive new fields,
- Improve infrastructure,
- Reduce corruption,
- Increase the size of proven reserves,
- Increase trade openness and improve the conditions for doing business,
- Maintain a stable national currency,
- Lower business taxes.

REFERENCES

ADB. (2004). *Asian Development Outlook 2004*. Retrieved from http://www.adb.org/Documents/Books/ADO/2004/default.asp

Agiomirgianakis, G., Asteriou, D., & Papathoma, K. (2003). *The Determinants of Foreign Direct Investment: A Panel Data Study for the OECD Countries*. Retrieved from http: //www.city. ac.uk/economics/ dps/discussion_papers/0306.pdf

Alfaro, L., Chanda, A., Kalemli-Ozcan, S., & Sayek, S. (2004). FDI and Economic Growth: The Role of Local Financial Markets. *Journal of International Economics*, *64*(1), 89–112. doi:10.1016/S0022-1996(03)00081-3

Alimov, M. (2014). *Oil despotism: Black Gold" Kazakhstan: blessing or curse?* Central Asia Monitor. Retrieved from http: //camonitor.com/14029-neftyanaya-despotiya.html

Ancharaz, V. D. (2003). Determinants of Trade Policy Reform in Sub-Saharan Africa. *Journal of African Economies*, *12*(3), 417–443. doi:10.1093/jae/12.3.417

Asiedu, E. (2002). On the Determinants of Foreign Direct Investment to Developing Countries: Is Africa Different? *World Development*, *30*(1), 107–119. doi:10.1016/S0305-750X(01)00100-0

Asiedu, E. (2013). *Foreign Direct Investment, Natural Resources and Institutions*. Working Paper, International Growth Center, 1-47.

Balasubramanyam, V. N., Salisu, M., & Sapsford, D. (1996). FDI and growth in EP and IS countries. *Economic Journal (London)*, *106*, 92–105. doi:10.2307/2234933

Barrell, R., & Pain, N. (1999). Domestic institutions, agglomerations and foreign direct investment in Europe. *European Economic Review*, *43*(4-6), 925–934. doi:10.1016/S0014-2921(98)00105-6

Bevan, A., Estrin, S., & Meyer, K. (2004). Foreign investment location and institutional development in transition economies. *International Business Review*, *13*(1), 43–64. doi:10.1016/j.ibusrev.2003.05.005

Blomstrom, M., Lipsey, R. E., & Zejan, M. (1994). What explains developing country growth? In W. J. Baumol (Ed.), *Convergence of Productivity: Cross-National Studies and Historical Evidence* (9th ed.). New York: Oxford University Press, Incorporated.

Borensztein, E., De Gregorio, J., & Lee, J.-W. (1998). How Does Foreign Direct Investment Affect Economic Growth? *Journal of International Economics*, *45*(1), 115–135. doi:10.1016/S0022-1996(97)00033-0

Botric, V., & Skuflic, L. (2006). Main Determinants of Foreign Direct Investment in the Southeast European Countries. *Transition Studies Review*, *13*(2), 359–377. doi:10.100711300-006-0110-3

Bouoiyour, J. (2003). *The Determining Factors of Foreign Direct Investment in Morocco*. Retrieved from http: //www.erf.org.eg/tenthconf/Trade_Background/Bouoiyour.pdf

Busse, M., & Hefeker, C. (2007). Political risk, institutions and foreign direct investment. *European Journal of Political Economy*, *23*(2), 397–415. doi:10.1016/j.ejpoleco.2006.02.003

Cassou, S. P. (1997). The Link between Tax Rates and Foreign Direct Investment. *Applied Economics*, *29*(10), 1295–1301. doi:10.1080/000368497000000019

Culem, C. G. (1988). The Locational Determinants of Direct Investment among Industrialized Countries. *European Economic Review*, *32*(4), 885–904. doi:10.1016/0014-2921(88)90051-7

Danesh, A. (2010). In the oil and gas industry to strengthen the role of the state, and one of the levers - the strengthening of KMG and its mining "daughter" continues. *Panorama*. Retrieved from http: //www.nomad .su /? a = 4-201009060034

De Mello, L. R. (1997). FDI in Developing Countries and Growth: A Selective Survey. *The Journal of Development Studies*, *34*(1), 1–34. doi:10.1080/00220389708422501

De Mello, L. R. (1999). Foreign Direct Investment-Led Growth: Evidence from time series and panel data. *Oxford Economic Papers*, *51*(1), 133–151. doi:10.1093/oep/51.1.133

Demirhan, E., & Masca, M. (2008). Determinants of Foreign Direct Investment Flows to Developing Countries: A Cross-Sectional Analysis. *Prague Economic Papers*, *4*(4), 356–369. doi:10.18267/j.pep.337

Dunning, J. H. (1988). *Explaining International Production*. London: Harper Collins Academic.

EBRD. (2000). *Transition Report 2000: Employment, Skills, and Transition*. London: European Bank for Reconstruction and Development.

Giles, C. (2014). Winners and losers of oil price plunge. *The Financial Times*. Retrieved from http://inosmi.ru/world/20141217/224991616.html#ixzz3by68oQVq

Goldsbrough, D. G. (1979). The Role of Foreign Direct Investment in the External Adjustment Process. *Staff Papers*, *26*(4), 725–754. doi:10.2307/3866968

Hartman, D. G. (1994). Tax Policy and Foreign Direct Investment in the United States. *National Tax Journal*, *37*(4), 475–488.

Jackson, S., & Markowski, S. (1995). The Attractiveness of Countries to Foreign Direct Investment. *Journal of World Trade*, *29*, 159–180.

Jordaan, J. C. (2004). *Foreign Direct Investment and Neighbouring Influences* (Unpublished doctoral thesis). University of Pretoria.

KazEnergy. (2013). *Social responsibility of the oil and gas industry of the Republic of Kazakhstan in 2012-2013*. Astana: Association KAZENERGY.

Kemsley, D. (1998). The Effect of Taxes on Production Location. *Journal of Accounting Research*, *36*(2), 321–341. doi:10.2307/2491480

Kok, R., & Ersoy, B. A. (2009). Analyses of FDI Determinants in Developing Countries. *International Journal of Social Economics*, *36*(1/2), 105–123. doi:10.1108/03068290910921226

Krifa-Schneider, H., Matei, I., & Matei, I. (2010, November). Business Climate, Political Risk and FDI in Developing Countries: Evidence from Panel Data. *International Journal of Economics and Finance*, *2*(5), 54–65. doi:10.5539/ijef.v2n5p54

Lim, D. (1983). Fiscal Incentive and Direct Foreign Investment in Less Developed Countries. *The Journal of Development Studies*, *19*(2), 207–212. doi:10.1080/00220388308421859

Lunn, J. (1980). Determinants of U.S. Direct Investment in the E. E. C.: Further Evidence. *European Economic Review, 13*(1), 93–101. doi:10.1016/0014-2921(80)90048-3

Luo, Y. (2009). Political risk and country risk in international business. Concepts and measures. In The Oxford Handbook of International Business (2nd ed.). Oxford University Press.

Majeed, M. T., & Ahmad, E. (2009). An Analysis of Host Country Characteristics that Determine FDI in Developing Countries: Recent Panel Data Evidence. *The Lahore Journal of Economics, 14*(2), 71–96.

Markusen, J.R., & Venables, A. J. (1999). Foreign direct investment as a catalyst for industrial development. *European Economic Review, 43*(2), 335-356.

Markusen, J. R. (2002). *Multinational firms and the Theory of International Trade.* Cambridge, MA: MIT Press.

Nigh, D. (1985). The Effect of Political Events on United States Direct Foreign Investment: A Pooled Time-Series Cross-Sectional Analysis. *Journal of International Business Studies, 16*(1), 3–17. doi:10.1057/palgrave.jibs.8490439

Nordal, K. B. (2001). Country risk, country risk indices and valuation of FDI: A real options approach. *Emerging Markets Review, 2*(3), 197–217. doi:10.1016/S1566-0141(01)00017-6

ODI. (1997). *Foreign Direct Investment Flows to Low-Income Countries: A Review of the Evidence.* Retrieved from http: //www.odi.org.uk/publications/briefing/3_97.html

Porcano, T. M., & Price, C. E. (1996). The Effects of Government Tax and Non-Tax Incentives on Foreign Direct Investment. *Multinational Business Review, 4*, 9–20.

Root, E., & Ahmed, A. (1979). Empirical Determinants of Manufacturing: Direct Foreign Investment Developing Countries. *Economic Development and Cultural Change, 27*(4), 758–767. doi:10.1086/451139

Rugman, A. M. (2009). *The Oxford Handbook of International Business* (2nd ed.). Oxford University Press. doi:10.1093/oxfordhb/9780199234257.001.0001

Schneider, F., & Frey, B. (1985). Economic and Political Determinants of Foreign Direct Investment. *World Development, 13*(2), 161–175. doi:10.1016/0305-750X(85)90002-6

Sekkat, K., & Veganzones-Varoudakis, M. A. (2007). Openness, Investment Climate, and FDI in Developing Countries. *Review of Development Economics, 11*(4), 607–620. doi:10.1111/j.1467-9361.2007.00426.x

Shiells. (2003). *FDI and the Investment Climate in the CIS Countries*. IMF Policy Discussion Paper, PDP/05/03, 1-35.

Singh, H., & Jung, K. W. (1995). *Some new evidence on Determinants on Foreign Direct Investment in Developing Countries*. Policy Research Working Paper 1531, World Bank.

Swenson, D. L. (1994). The Impact of U. S. Tax Reform on Foreign Direct Investment in the United States. *Journal of Public Economics, 54*(2), 243–266. doi:10.1016/0047-2727(94)90062-0

Tsai, P. (1994). Determinants of Foreign Direct Investment and its Impact on Economic Growth. *Journal of Economic Development, 19*(1), 137–163.

Ubiznes.ru. (2015). *Top 5 countries for oil production in the world*. Retrieved from http: //ubiznes.ru/top/reyting-stran-po-dobyche-nefti-v-mire.html

WIR. (2017), *World Investment Reports*. Division on Investment and Enterprise United Nations Conference on Trade and Development.

ADDITIONAL READING

Antill, N., & Arnott, R. (2002). *Valuing oil and gas companies*. Camridge, UK: Woodhead Publishing Ltd.

Bryan, R. D. (2012) Company Valuation: Oil and Gas vs. Other Sectors, *Economist's Corner*, 8(3).

Iskakov, S., & Yilmaz, N. (2015). Performance Evaluation Of Major Integrated Oil & Gas Companies. *International Journal of Economics. Commerce and Management, 3*(6), 332–361.

KEY TERMS AND DEFINITIONS

Capital-Intensive Industries: The industries which require relatively large amounts of money for machinery and infrastructure to produce a good or service.

Cross-Border Investments: Investments that cross national borders.

Foreign Direct Investments (FDI): The investments made by a foreign company or individual to undertake ownership or gain managerial control on a domestic company.

Investment Attractiveness: Attractiveness of business or assets for investors.

Proven Reserves: Reserves of petroleum which by analysis of geological and engineering data can be estimated with reasonable certainty to be commercially recoverable.

APPENDIX

Table 4. List of countries used in the construction of the model

USA	Saudi Arabia	Argentina
China	Australia	United Arab Emirates
Russia	Mexico	Malaysia
Canada	Angola	Venezuela
Norway	Brazil	Azerbaijan
Nigeria	United Kingdom	Algeria
Kazakhstan	Indonesia	Iran

Table 5. Variables and data sources

Variables	Data Sources
Operating costs of firms in the oil and gas industry	Rystad (2015)
Capital expenditures of firms in the oil and gas industry	Rystad (2015)
Proven recoverable oil reserves	Woodmac (2015)
GDP	World Bank (2015)
Real GDP growth	World Bank (2015)
Population	World Bank (2015)
Index of corruption feeling	Transparency International (2015)
Doing Business Ranking	World Bank (2015)
The real effective exchange rate	World Bank (2015), NBK(2015), Central Bank of Azerbaijan (2015), http://www.bis.org/statistics/eer/
Oil production	Rystad 92015)
Fixed broadband costs	World Bank (2015)
Age dependency ratio	World Bank (2015)
Total tax rate	World Bank (2015)
Natural rent	World Bank (2015)
Oil rents	World Bank (2015)
The share of foreign trade in GDP	World Bank (2015)
Inflation	World Bank (2015)

Table 6. Brief names and methods of variable's calculation

Name	Methods of Variable's Calculation
INV_CAP	Capital expenditures of firms in the oil and gas industry per capita (in constant $ 2010)
OPEX_BAR	Operating costs of firms in the oil and gas industry for 1 barrel of production (in prices of 2010 $)
CAPEX_BAR	Capital expenditures of firms for 1 barrel of production (in constant $ 2010)
COST_BAR	Total costs of firms in the oil and gas industry for 1 barrel of production (in prices of 2010 $)
CORR	Index of corruption feeling (place in the world), calculated according to the International Transparency Agency
GDPR	The rate of real GDP growth (annual in %)
REER	The real effective exchange rate is taken from the National Bank of Kazakhstan Database
RES	Proved oil reserves (trillion barrels)
RES_CAP	Proven recoverable reserves of oil per capita (mln. barrels per capita)
FBS	Fixed broadband casting (per 100 people)
TAX_RATE	Total tax rate (% of commercial profits)
NAT_RENT	Natural rents (% of GDP)
OIL_RENT	Oil rents (% of GDP)
PROD	Oil production, mln. barrels per year
PROD_CAP	Per capita oil population, barrels per person per year
ADR	Age dependency ratio (% of working population)
DOI_BUSS	Business climate in the country by the Doing Business ranking (the country's place in the world)
Defl	Inflation calculated by the GDP deflator (annual average,%)
CPI	Consumer price index (CPI), 2010=100
Trade_open	(Exports + Imports of goods and services (in current $) / GDP current $) x100

Table 7. Descriptive statistics of the variables used in the mode

	Inv_Cap	**Opex_Bar**	**Capex_Bar**	**Cost_Bar**	**Corr**	**Gdpr**	**Reer**		**Res**	**Res_Cap**
Mean	697.8649	19.98246	37.97193	57.95439	78.07018	3.450848	98.62202		57043.32	2192.188
Median	333.0000	15.95000	23.60000	39.10000	79.50000	3.387127	100.0000		14648.11	229.7500
Maximum	6631.200	58.80000	505.0000	563.8000	172.0000	10.77242	134.1506		298350.0	14385.20
Minimum	24.00000	5.700000	3.200000	9.600000	5.000000	-7.820890	70.42078		2354.209	11.80000
Std. Dev.	1178.823	11.74125	60.21804	68.76549	51.27209	3.430302	8.864041		84021.93	3700.276
Skewness	3.264016	0.905362	5.487407	4.757052	0.009717	-0.388309	-0.550866		1.746320	1.701153
Kurtosis	13.64501	3.199778	38.11136	31.27901	1.662585	3.485852	6.660812		4.784625	4.429012
Jarque-Bera	740.6739	15.76351	6427.956	4228.548	8.498013	3.986146	69.42296		73.07123	64.68436
Probability	0.000000	0.000378	0.000000	0.000000	0.014278	0.136276	0.000000		0.000000	0.000000
Observations	114	114	114	114	114	114	114		114	114
	Inv_Cap	Opex_Bar	Capex_Bar	Cost_Bar	Corr	Gdpr	Reer	Res	Cp	Res_Cap
Mean	697.8649	19.98246	37.97	57.954	78.0701	3.4508	98.622	57043.32	104.15	2192.188
Median	333.0000	15.95000	23.60	39.100	79.500	3.3871	100.00	14648.11	100.0	229.7500
Maximum	6631.200	58.80000	505.0000	563.80	172.000	10.772	134.15	298350.0	214.7	14385.20
Minimum	24.00000	5.700000	3.200000	9.6000	5.0000	-7.8208	70.420	2354.209	61.387	11.80000
Std. Dev.	1178.823	11.74125	60.21804	68.765	51.272	3.4303	8.8640	84021.93	16.3	3700.276
Skewness	3.264016	0.905362	5.487407	4.7570	0.00971	-0.3883	-0.5508	1.746320	2.807	1.701153
Kurtosis	13.64501	3.199778	38.11136	31.279	1.66258	3.48585	6.6608	4.784625	19.194	4.429012
Jarque-Bera	740.6739	15.76351	6427.956	4228.54	8.49801	3.98614	69.422	73.07123	1530.07	64.68436
Probability	0.000000	0.000378	0.000000	0.00000	0.01427	0.13627	0.00000	0.000000	0.00000	0.000000
Observations	114	114	114	114	114	114	114	114	125	114

Table 8. Results of testing by ADF method

	Level	1st Diff	2nd Diff	Level	1st Diff	2nd Diff
Inv_Cap	-4.22389	-4.913042	-5.215418	***	***	***
Corr	-3.858983	-4.453464	-7.725159	***	***	***
Gdpr	-5.382387	-7.751817	-10.81759	***	***	***
Reer	-4.462514	-7.365722	-9.349910	***	***	***
Res	-4.644021	-5.072250	-5.427023	***	***	***
Res_Cap	-4.674319	-4.875888	-5.336816	***	***	***
Fbs	-3.795886	-5.180074	-5.180074	***	***	***
Tax_Rate	-4.713597	-4.647650	-5.466039	***	***	***
Nat_Rent	-3.388626	-6.189159	-9.357876	**	***	***
Oil_Rent	-3.918155	-6.236794	-9.380185	***	***	***
Opex_Bar	-3.890742	-5.288185	-7.838809	***	***	***
Prod	-4.223890	-4.913042	-5.215418	***	***	***
Prod_Cap	-4.454439	-4.510203	-5.484746	***	***	***
Adr	-3.871684	-4.871044	-4.871044	***	***	***
Doi_Buss	-4.256027	-4.807868	-6.493099	***	***	***
Defl	-4.435573	-8.146244	-11.520000	***	***	***
Cpi	-5.011485	-5.973601	-8.416251	***	***	***
Trade_Open	-3.763260	-4.507889	-6.696004	***	***	***

Note: *, **, *** represent respectively 10%, 5% and 1% significance level.

Table 9. Data Validation on the convergence according to the Johansen Cointegration test

Hypothesized No. of CE(s)	Eigenvalue	Likelihood Ratio	5 Percent Critical Value	1 Percent Critical Value
None **	0.970330	1739.437	114.90	124.75
At most 1 **	0.942729	1373.604	87.31	96.58
At most 2 **	0.897318	1076.167	62.99	70.05
At most 3 **	0.815308	839.4505	42.44	48.45
At most 4 **	0.742125	663.7876	25.32	30.45
At most 5 **	0.680600	522.8383	12.25	16.26
At most 6 **	0.552486	404.1419	233.13	247.18
At most 7 **	0.512696	320.5209	192.89	205.95
At most 8 **	0.452891	245.7587	156.00	168.36
At most 9 **	0.422748	183.0356	124.24	133.57
At most 10 **	0.353731	125.8900	94.15	103.18
At most 11 **	0.255454	80.48992	68.52	76.07
At most 12 *	0.210658	49.81194	47.21	54.46
At most 13	0.157738	25.21013	29.68	35.65
At most 14	0.063091	7.357054	15.41	20.04
At most 15	0.005557	0.579497	3.76	6.65

*(**) denotes rejection of the hypothesis at 5%(1%) significance level. L.R. test indicates 13 cointegrating equation(s) at 5% significance level

Table 10. The coefficients of correlation between the variables

	Inv_Cap	**Opex_Bar**	**Capex_Bar**	**Cost_Bar**	**Corr**	**Gdpr**	**Reer**	**Res**	**Res_Cap**	**Fbs**
Inv_Cap	1.000	0.161	0.261	0.256	-0.477	-0.220	0.062	-0.010	0.097	0.622
Opex_Bar	0.161	1.000	0.682	0.768	-0.591	-0.235	-0.034	-0.287	-0.395	0.600
Capex_Bar	0.261	0.682	1.000	0.992	-0.411	-0.078	0.103	-0.198	-0.214	0.375
Cost_Bar	0.256	0.768	0.992	1.000	-0.461	-0.108	0.084	-0.222	-0.254	0.430
Corr	-0.477	-0.591	-0.411	-0.461	1.000	0.276	-0.137	0.109	-0.064	-0.753
Gdpr	-0.220	-0.235	-0.078	-0.108	0.276	1.000	0.021	-0.039	-0.063	-0.416
Reer	0.062	-0.034	0.103	0.084	-0.137	0.021	1.000	-0.171	-0.090	0.122
Res	-0.010	-0.287	-0.198	-0.222	0.109	-0.039	-0.171	1.000	0.820	-0.069
Res_Cap	0.097	-0.395	-0.214	-0.254	-0.064	-0.063	-0.090	0.820	1.000	-0.082
Fbs	0.622	0.600	0.375	0.430	-0.753	-0.416	0.122	-0.069	-0.082	1.000
Tax_Rate	-0.196	0.011	0.023	0.022	0.267	0.000	-0.152	-0.299	-0.428	-0.105
Nat_Rent	-0.049	-0.651	-0.313	-0.385	0.454	0.281	-0.002	0.370	0.416	-0.495
Oil_Rent	0.288	-0.166	-0.180	-0.186	-0.045	-0.041	-0.027	0.046	0.081	0.004
Prod	-0.058	-0.223	-0.191	-0.206	0.004	-0.015	0.073	0.541	0.312	0.035
Prod_Cap	0.616	-0.378	-0.188	-0.230	-0.292	-0.093	0.051	0.426	0.682	0.191
Adr	-0.062	0.086	0.037	0.047	0.265	0.061	-0.038	-0.077	-0.371	-0.135
Doi_Buss	-0.430	-0.387	-0.306	-0.334	0.807	0.182	-0.260	0.071	-0.062	-0.674
Defl	-0.125	-0.248	-0.149	-0.173	0.413	0.202	-0.123	0.199	0.129	-0.281
Cpi	-0.052	0.020	0.012	0.014	0.177	0.030	-0.114	0.223	0.128	0.001
Trade_Open	0.077	-0.221	-0.180	-0.195	-0.178	0.079	0.144	0.079	0.425	-0.162

	Tax_Rate	Nat_Rent	Oil_Rent	Prod	Prod_Cap	Adr	Doi_Buss	Defl	Cpi	Trade
Inv_Cap	-0.196	-0.049	0.288	-0.058	0.616	-0.062	-0.430	-0.125	-0.052	0.077
Opex_Bar	0.011	-0.651	-0.166	-0.223	-0.378	0.086	-0.387	-0.248	0.020	-0.221
Capex_Bar	0.023	-0.313	-0.180	-0.191	-0.188	0.037	-0.306	-0.149	0.012	-0.180
Cost_Bar	0.022	-0.385	-0.186	-0.206	-0.230	0.047	-0.334	-0.173	0.014	-0.195
Corr	0.267	0.454	-0.045	0.004	-0.292	0.265	0.807	0.413	0.177	-0.178
Gdpr	0.000	0.281	-0.041	-0.015	-0.093	0.061	0.182	0.202	0.030	0.079
Reer	-0.152	-0.002	-0.027	0.073	0.051	-0.038	-0.260	-0.123	-0.114	0.144
Res	-0.299	0.370	0.046	0.541	0.426	-0.077	0.071	0.199	0.223	0.079
Res_Cap	-0.428	0.416	0.081	0.312	0.682	-0.371	-0.062	0.129	0.128	0.425
Fbs	-0.105	-0.495	0.004	0.035	0.191	-0.135	-0.674	-0.281	0.001	-0.162
Tax_Rate	1.000	-0.294	-0.021	-0.239	-0.466	0.108	0.490	0.157	-0.501	-0.501
Nat_Rent	-0.294	1.000	0.096	0.217	0.440	-0.085	0.160	0.201	0.345	0.345
Oil_Rent	-0.021	0.096	1.000	-0.086	0.245	-0.009	-0.140	-0.059	0.253	0.253
Prod	-0.239	0.217	-0.086	1.000	0.312	-0.144	-0.119	-0.053	-0.087	-0.087
Prod_Cap	-0.466	0.440	0.245	0.312	1.000	-0.333	-0.360	-0.062	0.444	0.444

continued on following page

Table 10. Continued

	Inv_Cap	Opex_ Bar	Capex_Bar	Cost_Bar	Corr	Gdpr	Reer	Res	Res_Cap	Fbs
Adr	0.108	-0.085	-0.009	-0.144	-0.333	1.000	0.240	0.241	-0.437	-0.437
Doi_Buss	0.490	0.160	-0.140	-0.119	-0.360	0.240	1.000	0.444	-0.334	-0.334
Defl	0.157	0.201	-0.059	-0.053	-0.062	0.241	0.444	1.000	-0.124	-0.124
Cpi	0.123	-0.073	-0.023	-0.018	-0.074	0.068	0.216	0.211	1.000	-0.110

[1]For example, the share of foreign companies in the ownership structure of in oil and gas industry of Kazakhstan is about 80%.

[2]This leads to a high loss of value added.

Chapter 4
EU and Central Asia:
The Politics Behind Energy Supply

Tuğba Aydın Azalov
Artvin Coruh University, Turkey

ABSTRACT

Energy politics are one of the effective ways to interact with other states' political systems. When external relations are important for internal energy security, then energy politics are covered with other political strategies. As the case for European Union, energy strategies structures several sectors and one of them is energy supply security and promotion of sustainable use of energy worldwide. Thus, security of supply is an important part in EU energy politics based on competitiveness and sustainability. In the case of regional cooperation with Central Asia, EU is trying to be self-sufficient for its internal market to build its future safe in energy. Thus, security on energy supply becomes on the fore. So, EU, as a prominent actor in global political system, has growing interest on Central Asian states because of their high energy potential. But, regional cooperation and assistance in the region for energy politics are not enough, and EU needs to promote and take further steps toward democratization and development cooperation embedded in energy politics.

INTRODUCTION AND BACKGROUND

Energy is a word that includes political, economic and cultural ideas beyond its definition in the natural and applied sciences. Its effect is more than just shaping people's daily lives. So, it is possible to state that energy meets an individual's needs such as heating, lighting, transportation and the like, but also performing in several fields such as diversification, usage and operation, allocation and security of other forms of energy sources.

DOI: 10.4018/978-1-5225-4203-2.ch004

Energy sources have been shaped and changed differently in the historical process. In the recent past, there was a transformation from coal to oil and natural gas. This transformation directs itself to renewable energy sources, together with nuclear energy as an essential part of future energy demands. Also, shale gas production in the USA is seen as a promising energy source and may shape and change future energy politics.

But currently, there are countries and regions which possess rich sources of oil and natural gas. North Africa, the Middle East and Central Asia are some of these. After independence, the Central Asian states have emerged as superpower states. But, Central Asia is not alone regarding Russian politics and plans on the region, along with the USA and China. The European Union is the leading power aiming to consider interests in the region, specifically from energy. As the European Union is shaping its energy policy, it considers not only intra-European parameters, but extra-European actors depending on its policy. And while the EU is shaping its energy policy towards Central Asia, several factors such as consumption, transit countries and access to sources are considered in a broad sense. Given the EU's dependence on external energy sources and the crises it has faced in the past, the necessity of building a stronger energy policy is obvious.

In this paper, Central Asian energy resources and the EU's energy policies towards Central Asia will be analyzed. Thus, energy supply and security will be the main topics that will be studied in this article. In the first part, a definition of energy and its effects on politics and security will be studied. The second part will include a statistical analysis of current Central Asian energy resources and the EU's energy demands. Finally, the European Union's policies as presented in the Energy Strategy Papers for 2020, 2030 and 2050 will be analyzed, primarily focusing on energy supply, energy security and international partnerships.

ENERGY

Concept and Importance

Energy has a wider concept more than what people frequently hear or mention about it in their daily lives. The term 'energy' mostly includes rising security, improvement of energy sources, foreseen of future energy sources or researches for new technologies. But according to Sovacool cited by Smith, it is a mistake to define energy without asking what kind of values and ethics it includes and who benefits (Smith & High,2017; pg:4). So, when describing energy, culture, geography and ethics should be taken into account and policies should be shaped with these factors in mind.

But, to give a general and simple definition, energy is defined as 'power as a form of light, heat or so embedding in elements" (TDK,2017) Another definition put forward by the U.S Energy Information Administration takes us one step further and provides an approach defining our daily lives. So, energy is defined as "the ability to do work, coming in different forms as Heat (thermal), Light (radiant), Motion (kinetic), Electrical, Chemical, Nuclear energy and Gravitational" (EIA, 2017). Here, we can say that energy is one of indispensable component for the continuity of current civilization (Sevim, 2012; pg:4378).

As seen above, energy rooting originating from seven different sources has a vital importance in shaping our environmental, cultural, economic and political life. So, energy is kind of fundamental rights of health, development and education centering states to reach energy sources and services (LaBelle, 2017; pg: 615). That right puts energy justice on the agenda and forces us to discuss the issue of global and particular justice embedded in politics and economics. Citing from Sovacool and Dworkin, global and particular energy justice can be described as below:

'A Global energy system that fairly disseminates both benefits and costs of energy services, and one that has representative and impartial energy decision-making' (LaBelle, 2017, pg:616).

'Particular energy justice is defined as 'a nuanced accounting of localized issues and interpretations of social, political and economic actions affecting energy resources and services.' (LaBelle, 2017, pg:616).

On a global dimension, energy is related to how people understand it. If energy is seen as the ability to work, then *energy improves lives by allowing people transportation and mobility and opportunities to light homes, schools and businesses.* (Smith & High, 2017, pg:2). This definition describes *how people live with energy in their daily life, but does not consider how energy contributes to a good life.* (Smith & High, 2017, pg:2). Thus, ethical judgments on what people consider *good lives* and *good societies*, vary widely. For example, some member states of the European Union are against nuclear energy because of its potential risks, especially after the accident in Fukushima. But for some member states, nuclear energy is accepted as a secure, reliable and affordable source of low-carbon electricity generation. (Energy Roadmap 2050, 2011; pg:13)

Other aspects of energy discussed by anthropologists are energyscape and energypower. While energyscape extends from the local to the transnational, energypower shows how political power is exercised_for_energy infrastructure. (Smith & High, 2017; pg:4)

Energy Security

The other essential concept on energy is energy security. Providing energy security is markedly mentioned in mostly in European Union politics given its increasing energy consumption.

Energy security is defined by the European Commission as '*saving strategical stocks or ability to provide compassable and sustainable external sources in case of energy need in the future because of insufficient national energy sources*' (Sevim, 2012;pg: 4386). Compassable and sustainable external sources become essential for supplier countries in that definition. Considering the risk posed by terrorist attacks to energy production, transportation and distribution infrastructure, civil war, occupation, cutbacks by investment gaps or embargoes, it is necessary to analyze deeply the energy security of external sources to ensure that they are both sustainable and compatible (Sevim, 2012, pg:4385). In this case, the costs of energy sources, transportation routes, an increasing demand tendency and the geographical distribution should be taken into account while assessing energy security, energy politics and energy supply (Sevim, 2012; pg:4385).

Definitions of energy security are structured according to its source and security dimension. If energy security is considered according to its source dimension, then traceability, accessibility and admissibility are highlighted. This is to say, it considers a-) providing energy from clean and diversified sources at an affordable price; b-) using energy rationally and economically; c-) and accessing sufficient sources with a fair distribution (Sevim, 2012; pg: 4386).

But if energy is considered in its security dimension, then physical protection toward any attacks on transportation, consumption, marketing or research and development (Sevim, 2012; pg: 4385). Energy Strategy of USA's definition on security concept also includes priority for security in foreign policy and economics; fixing global energy policies as securing global economic development; transportation routes' security and protecting dependency balance on energy sources (Sevim, 2012; pg: 4386).

Energy security is also shaped differently by importer and exporters countries' approaches. While diversification of energy supplies becomes important for importer countries, diversification of energy demands puts on the fore for exporter countries. In both cases, diversification remains essential. Because, any physical problem and physical attacks on current energy sources or unstable political systems will not cause lack of energy by diversified energy supply. Together with, diversified energy demands will reflect positively on gross incomes of energy producer countries and also, they will not depend on sole consumer (Sevim, 2012; pg: 4386).

In case of the European Union, it is acceptable that EU is one of biggest energy consumer in the world and also has *vulnerable energy market to increasing supply of some commodities by global oligopolies creating internal and external imbalances*

(Commission Staff Working Paper, 2011, pg:10). Because of that, EU has energy dependency and must provide stable and long-term accessible energy policies for its internal market including international partnerships.

ENERGY STRUCTURE AND NEEDS

Central Asia

Since their independence, Central Asia has been described by many academicians as a flower of the super powers. The main reason for this attraction is its energy potential. Russia, together with the EU, China and the USA compete to benefit from the potential energy sources that region base. If 21th century is accepted as a period of energy wars, then Central Asian region must be counted more important (Çınar, 2008; pg: 21).

There are several geopolitical theories, analyzing how to gain ascendancy over Eurasia. Heartland Theory by Mackinder and The Grand Chessboard Theory by Brzezinski are one of them theorizing power wars on Central Asia. According to Mackinder, one of suitable place for a state is center and Eurasia is the center for earth power. This center is named as 'Heartland'. According to Mackinder, Asia, Africa and Europe form a world island as a whole and Asia and Europe form Eurasia as a whole that is named as Heartland (İşcan, 2004; pg: 60). For Mackinder, there are two crescents controlling the center. Any power sections on those crescents could prevent distribution of any power that holding whips in the region. Those crescents are:

- **Inner Crescent:** Germany, Austria, Balkans, Turkey, Iran, Pakistan and India
- **Insular Crescent:** England, Australia, USA and Canada (İşcan, 2014; pg:61)

If Mackinder's theory is taken into account, then we can say that Central Asia becomes a dominant region for global competition between states. And also if oil and natural gas are considered as principal energy sources for today and if wars are result by the tendency of controlling energy sources, energy wars over Central Asia can be accepted as an essential for international politics (Çınar, 2008; pg: 25). Because, controlling strategical points for trade and strategical fields for energy sources are making a point for global powers (Sevim, 2012; pg: 4387).

The graph below shows current crude oil reserves in the world:

As seen in the graph, cumulative production is 114.4 billion barrels for OPEC member states. Net additions to reserves are highest as 279 billion barrels by OPEC. The result for non-OPEC states are disappointed. Cumulative production rate is 149

Figure 1. World Proven Crude Oil Reserves 2007-2016 (OPEC, 2017)
Source: OPEC,2017

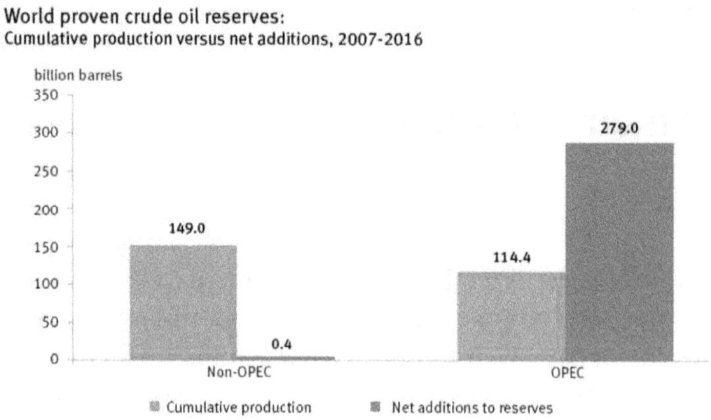

billion barrels and net additions to reserves are just 0.4. This clearly shows main reasons of energy wars in the world.

For natural gas reserves, Central Intelligence Agency's (CIA) report named The World Factbook shows the numbers of estimated reserves that the world have. This table below just shows the ranking rates in order to show the relative positions Central Asian states in the world order.

As seen on the table, Russia and Iran have the biggest shares in the list as first and second. For Central Asian states, Turkmenistan holds 6th in ranking with the amount of 7,504,000,000,000 cu m and Kazakhstan holds 15th with the amount of 2,407,000,000,000 CUM. Uzbekistan holds 19th in the order with the amount of 1,841,000,000,000 cu m, Kyrgyzstan holds 93th and Tajikistan holds 95th with the amount of 5,663,000,000 cu m.

The Central Intelligence Agency also published a report analyzing all countries reserves by crude oil, refined petroleum products, electricity, natural gas and carbon dioxide emissions from the consumptions of energy. In this article, oil and natural gas are put on emphasis, so below you will find only natural gas and crude oil production, consumption, import and export rates regarding to its ranking in world order.

Kazakhstan

See Table 2.

Kyrgyzstan

See Table 3.

Table 1. Country comparison by natural gas proven reserves

RANK	COUNTRY	(CUM)	DATE OF INFORMATION
1	RUSSIA	47,800,000,000,000	1 JANUARY 2016 EST.
2	IRAN	34,020,000,000,000	1 JANUARY 2016 EST.
3	QATAR	24,530,000,000,000	1 JANUARY 2016 EST.
4	UNITED STATES	10,440,000,000,000	1 JANUARY 2015 EST.
5	SAUDI ARABIA	8,489,000,000,000	1 JANUARY 2016 EST.
6	TURKMENISTAN	7,504,000,000,000	1 JANUARY 2016 EST.
14	AZERBAIJAN	2,550,000,000,000	
15	KAZAKHSTAN	2,407,000,000,000	1 JANUARY 2016 EST.
16	EGYPT	2,186,000,000,000	1 JANUARY 2016 EST.
17	CANADA	1,996,000,000,000	1 JANUARY 2016 EST.
18	NORWAY	1,922,000,000,000	1 JANUARY 2016 EST.
19	UZBEKISTAN	1,841,000,000,000	1 JANUARY 2016 EST.
20	KUWAIT	1,784,000,000,000	1 JANUARY 2016 EST.
21	LIBYA	1,505,000,000,000	1 JANUARY 2016 EST.
22	INDIA	1,489,000,000,000	1 JANUARY 2016 EST.
23	EUROPEAN UNION	1,300,000,000,000	1 JANUARY 2015 EST.
93	KYRGYZSTAN	5,663,000,000	1 JANUARY 2016 EST.
94	SOMALIA	5,663,000,000	1 JANUARY 2016 EST.
95	TAJIKISTAN	5,663,000,000	1 JANUARY 2016 EST.

Source: *(The World Factbook, 2017)*

Tajikistan

See Table 4.

Turkmenistan

See Table 5.

Uzbekistan

See Table 6.

As seen in the tables, the Central Asian states are important for energy sources. But, Russian dominance in the region, together with its political and historical ties on Central Asian states, force other powerful states to organize new policies on

Table 2. Crude oil and natural gas information for Kazakhstan

Type	Amount	Country Comparison to the World
Crude oil – production	1.621 million bbl/day (2016 est.)	15
Crude oil – exports	1.292 million bbl/day (2016 est.)	10
Crude oil – imports	145,800 bbl/day (2014 est.)	38
Crude Oil-proven Reserves	30 billion bbl (1 January 2016 est.)	12
Natural Gas – production	21.38 billion cu m (2016 est.)	32
Natural Gas – consumption	13.1 billion cu m (2016 est.)	43
Natural Gas – exports	13.7 billion cu m (2016 est.)	19
Natural Gas – imports	2.2 billion cu m (2014 est.)	31
Natural Gas proven reserves	2.407 trillion cu m (1 January 2016 es)	15

Source: (CIA-Kazakhstan, 2017)

Table 3. Crude oil and natural gas information for Kyrgyzstan

Type	Amount	Country Comparison to the World
Crude oil – production	1,000 bbl/day (2015 est.)	92
Crude oil – exports	19.65 bbl/day (2013 est.)	89
Crude oil – imports	0 bbl/day (2013 est.)	211
Crude oil - proven reserves:	40 million bbl (1 January 2016 est.)	81
Natural Gas – production	34 million cu m (2014 est.)	85
Natural Gas – consumption	429 million cu m (2014 est.)	100
Natural Gas – exports	0 cu m (2013 est.)	127
Natural Gas – imports	395 million cu m (2014 est.)	68
Natural Gas - proven reserves:	5.663 billion cu m (1 January 2016 est)	93

Source: (CIA-Kyrgyzstan, 2017)

region. For natural gas imports for example, Turkmenistan and Uzbekistan share the same amount of 0 cu m because of resulting from their own political preference on energy policy. Also, Tajikistan and Kyrgyzstan have 0 cu m for natural gas export. An inadequate technology infrastructure affects these results.

Table 4. Crude oil and natural gas information for Tajikistan

Type	Amount	Country Comparison to the World
Crude oil – production	181.6 bbl/day (2015 est.)	97
Crude oil – exports	78.6 bbl/day (2013 est.)	87
Crude oil – imports	0 bbl/day (2013 est.)	133
Crude oil - proven reserves:	12 million bbl (1 January 2016 est.)	90
Natural Gas – production	12 million cu m (2014 est.)	91
Natural Gas – consumption	224 million cu m (2014 est.)	103
Natural Gas – exports	0 cu m (2013 est.)	190
Natural Gas – imports	212 million cu m (2014 est.)	73
Natural Gas - proven reserves:	5.663 billion cu m (1 January 2016 es)	95

Source: (CIA-Tajikistan, 2017)

Table 5. Crude oil and natural gas information for Turkmenistan

Type	Amount	Country Comparison to the World
Crude oil – production	243,100 bbl/day (2015 est.)	36
Crude oil – exports	70,740 bbl/day (2013 est.)	40
Crude oil – imports	0 bbl/day (2013 est.)	140
Crude oil - proven reserves:	600 million bbl (1 January 2016 est.)	49
Natural Gas - production	76 billion cu m (2014 est.)	11
Natural Gas - consumption	30.2 billion cu m (2014 est.)	31
Natural Gas – exports	45.79 billion cu m (2014 est.)	7
Natural Gas – imports	0 cu m (2014 est.)	141
Natural Gas - proven reserves:	7.504 trillion cu m (1 January 2016 es)	6

Source: (CIA-Turkmenistan, 2017)

European Union

The European Union is one of biggest energy consumers, and is heavily energy dependent on external sources. Especially after the 1973 Oil Crisis, the European Union decided on several measures to reduce this dependency. Investments in nuclear energy, renewable energy sources and energy efficiency projects are just those that the European Union has implemented since then. In spite of these investments, the

Table 6. Crude oil and natural gas information for Uzbekistan

Type	Amount	Country Comparison to the World
Crude oil – production	60,130 bbl/day (2015 est.)	52
Crude oil – exports	27,000 bbl/day (2013 est.)	53
Crude oil – imports	360 bbl/day (2013 est.)	80
Crude oil - proven reserves:	600 million bbl (1 January 2016 est.)	48
Natural Gas - production	61.74 billion cu m (2014 est.)	15
Natural Gas - consumption	47.04 billion cu m (2014 est.)	20
Natural Gas - exports	14.7 billion cu m (2014 est.)	17
Natural Gas - imports	0 cu m (2013 est.)	145
Natural Gas - proven reserves:	1.841 trillion cu m (1 January 2016 es)	19

Source: (CIA-Uzbekistan, 2017)

European Union is still dependent on external suppliers for xx percent of its energy needs.

The EU's energy consumption and energy dependency are presented in the following graphs:

In this graph, the light colored proportion of each column shows net imports in gross energy consumption. The green column is for solid fossil fuels, dark blue is for total petroleum products and light blue is for gas. According to this graph, since 1990, the was observed in total petroleum products. In 2015, the amount was 358 mtoe for natural gas and 602 Mtoe for petroleum products. Solid fossil fuels use declined visibly over this period. According to Eurostat data, net import dependency was between 52.1% - 54% between 2005 to 2015 (Eurostat, Energy Dependency, 2017).

This graph presents the structure of gross energy consumption both in the EU-28and at a national level. In 2015, petroleum products (34.4%) had the biggest share, followed by gas (22.0%) and solid fossil fuels (16.1%), nuclear heat (13.6%) and renewables accounted for 13.0%. On the other hand, it is possible to say that renewable energy sources have increased from 4.3% in 1990, to 5.7% in 2000, to 13.0% in 2015, while gas has risen from 17.9% in 1990, to 22.9% in 2000 and to 22.0% in 2015 (Eurostat, Gross Inland Consumption, 2017).

There are only 3 EU countries where fossil fuel consumption is below 50% (Sweden 27.6%, France 47.3% and Finland 45.3%). In 2015, over half of the gross consumption was meet by solid fossil fuels in Estonia (60.8%) and Poland (50.5%). The average in the EU-28 was 16.1%. The smallest shares of solid fossil fuels in

Figure 2. Energy dependency by fuel, EU-28, 1990-2015, ktoe (Eurostat, Energy Dependency, 2017)
Source: Eurostat, Energy Dependency, 2017

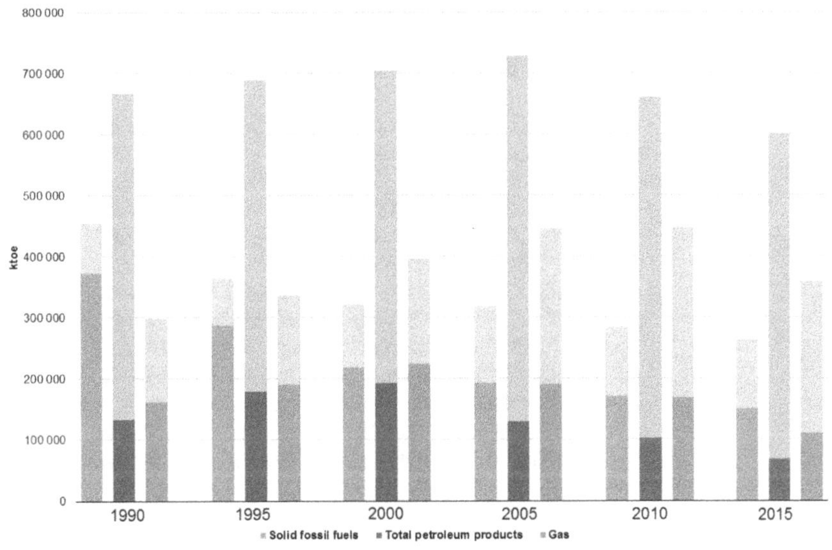

Note: the light coloured proportion of the column shows net imports with respect to gross inland energy consumption (including international maritime bunkers), which is represented by total column height

gross inland energy consumption (under 2%) in 2015 were observed in Latvia, Luxembourg, Cyprus and Malta (Eurostat, Gross Inland Consumption, 2017).

The biggest shares of total petroleum products in gross inland energy consumption were observed in: Cyprus 92.8%, Malta 85.4% and Luxembourg 63.1%.

Natural gas accounted for shares varying from 37.6% in the Netherlands to under 2% in Sweden, Cyprus and Malta. Natural gas was also an important energy source in Italy and the United Kingdom with a share of over 30% and Lithuania and Hungary reaching nearly the 30% mark (Eurostat, Gross Inland Consumption, 2017).

In 2015, there were 14 Member States with nuclear power plants. The highest nuclear share was in France (a 43.7% share of nuclear heat in gross inland energy consumption), followed by Sweden (30.7%), Slovakia (24.1%), Slovenia (22.1%) and Bulgaria (20.5%) (Eurostat, Gross Inland Consumption, 2017).

In this graph, Norway and Russia are the main natural gas suppliers for EU countries. But in 2016, the need for natural gas imports have risen from Russia (18.3%), Ukraine (17.1%), Tunisia (5.2%) and Nigeria (2.2%)

Figure 3.National shares of fuels in gross energy consumption, 2015, percentage (Eurostat, Gross Inland Consumption, 2017)
Source: Eurostat, Gross Inland Consumption, 2017

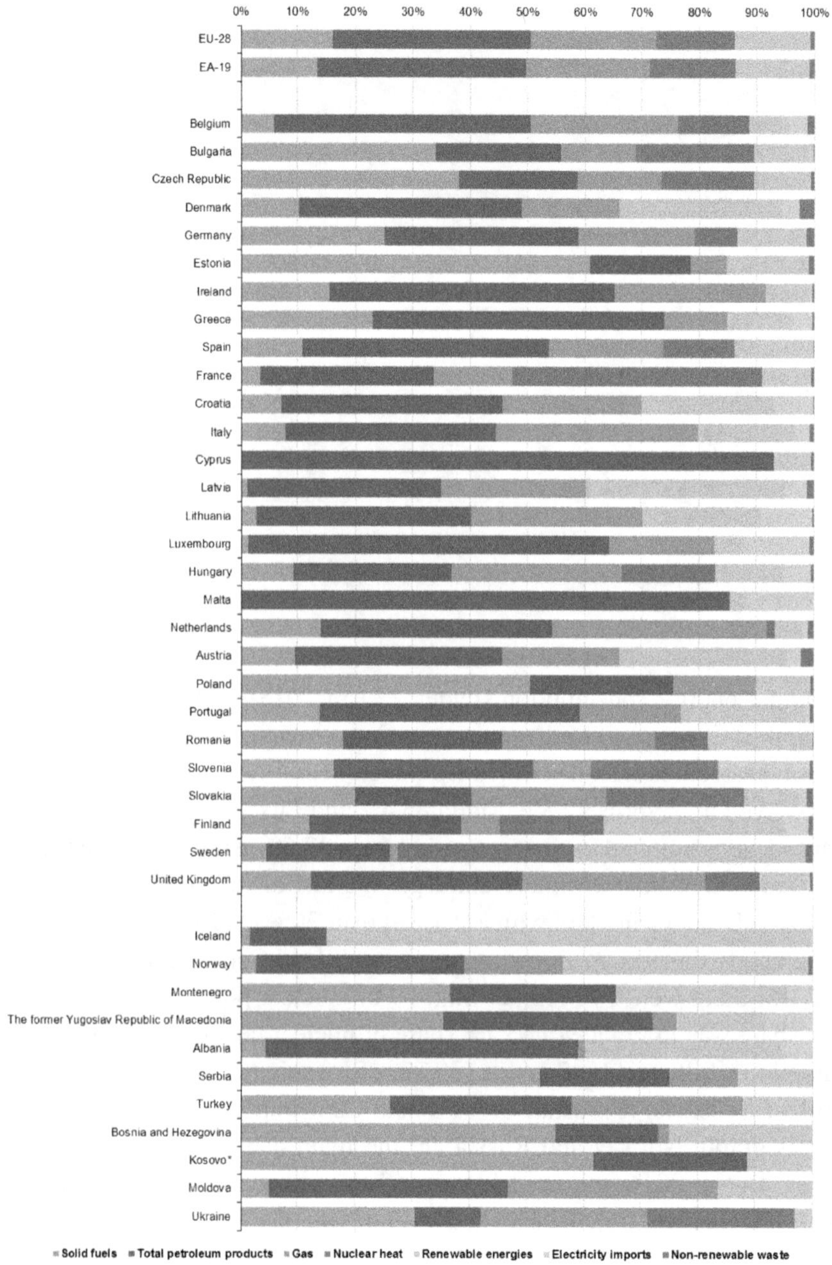

(*) This designation is without prejudice to positions on status, and is in line with UNSCR 1244 and the ICJ Opinion on the Kosovo Declaration of Independence.

Figure 4. Percentage of extra-EU imports of natural gas by country of origin, 2015-2016 (Eurostat, Extra-EU Imports, 2017)
Source: Eurostat, Extra-EU Imports, 2017

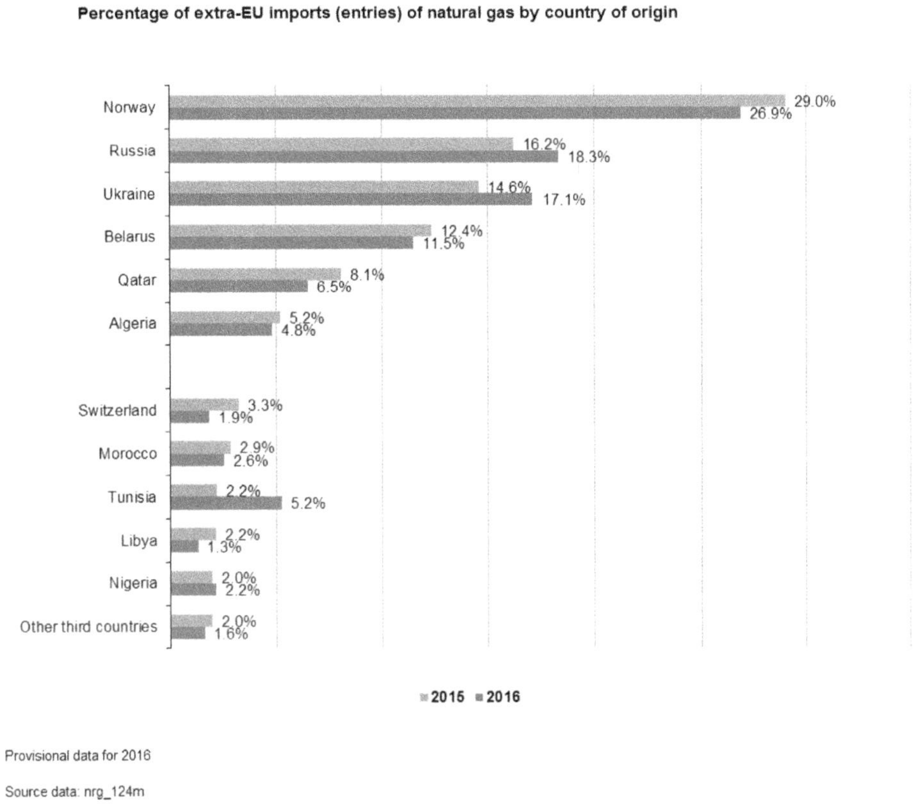

In Table 7, the main origin of solid fuels, crude oil and natural gas are counted as primary energy sources. For this listing, Russia is the main supplier from 2005 to 2015. For crude oil, Kazakhstan is 6th in the list with the rising amount of 4.1 in 2005 to 6.2% in 2015. So, Kazakhstan as one of Central Asian state has significant effect on EU-28 natural gas supply.

In this table, the main origin of solid fuels, crude oil and natural gas are counted as primary energy sources. For this listing, Russia is the main supplier from 2005 to 2015. For crude oil, Kazakhstan is 6th in the list with the rising amount of 4.1 in 2005 to 6.2% in 2015. So, Kazakhstan as one of Central Asian state has significant effect on EU-28 natural gas supply.

Table 7. Main origin of primary energy products, 2005-2015 (Eurostat, Primary Energy Imports, 2017)

	Solid fuels										
	2005	2006	2007	2008	2009	2010	2011	2012	2013	2014	2015
Russia	20.2	21.4	21.5	22.7	26.2	22.8	22.9	23.0	25.9	25.9	25.8
Colombia	10.0	9.7	11.0	10.7	15.2	16.9	20.7	21.7	19.6	18.8	21.3
United States	6.5	6.7	7.9	12.2	11.8	14.2	15.7	20.6	19.5	18.3	14.0
Australia	11.2	10.3	11.3	10.2	6.5	8.9	7.7	6.6	6.7	5.6	8.1
South Africa	21.4	19.8	17.4	14.4	13.7	8.2	6.9	5.7	6.0	8.7	6.9
Indonesia	6.2	7.9	6.8	6.3	6.1	4.7	4.4	4.0	2.8	3.0	3.1
Canada	2.7	2.4	2.6	2.3	1.2	1.7	1.9	1.5	1.6	2.2	1.4
Mozambique	0.0	0.0	0.0	0.0	0.0	0.0	0.0	0.0	0.3	0.3	0.4
Ukraine	1.9	1.4	1.5	2.0	1.5	1.6	2.1	1.5	1.4	1.3	0.4
Others	19.9	20.4	20.0	19.2	17.8	20.9	17.6	15.5	16.2	15.8	18.6
	Crude oil										
	2005	2006	2007	2008	2009	2010	2011	2012	2013	2014	2015
Russia	30.5	31.2	31.2	29.8	31.5	32.4	32.8	31.8	31.9	28.9	27.7
Norway	15.6	14.2	13.8	14.0	14.2	12.8	11.7	10.6	11.1	12.4	11.4
Nigeria	3.0	3.3	2.5	3.7	4.2	3.9	5.7	7.7	7.7	8.7	8.0
Saudi Arabia	9.8	8.3	6.6	6.4	5.3	5.5	7.6	8.3	8.2	8.5	7.5
Iraq	2.0	2.7	3.2	3.1	3.5	3.0	3.4	3.9	3.5	4.3	7.2
Kazakhstan	4.1	4.2	4.3	4.5	5.0	5.1	5.4	4.8	5.4	6.1	6.2
Azerbaijan	1.2	2.1	2.7	3.0	3.8	4.1	4.6	3.6	4.5	4.2	4.9
Algeria	3.2	2.3	1.7	2.4	1.5	1.2	2.4	2.7	3.7	4.0	4.0
Angola	1.1	0.7	1.9	2.4	2.5	1.5	2.0	1.9	2.8	3.2	4.0
Others	29.5	30.9	32.0	30.8	28.4	30.5	24.4	24.5	21.1	19.8	18.9
	Natural gas										
	2005	2006	2007	2008	2009	2010	2011	2012	2013	2014	2015
Russia	34.6	33.0	32.1	31.2	27.6	26.8	28.3	27.8	32.4	29.7	29.4
Norway	20.2	21.7	23.3	23.7	24.5	22.9	22.1	24.9	23.6	25.0	25.9
Algeria	15.0	13.6	12.7	12.3	11.9	11.7	10.7	10.8	10.1	9.7	8.8
Qatar	1.3	1.5	1.8	1.9	4.6	8.1	9.6	6.8	5.2	5.5	6.1
Libya	1.4	2.1	2.5	2.4	2.4	2.2	0.6	1.5	1.4	1.7	1.7
Nigeria	2.9	3.6	3.8	3.3	2.0	3.4	3.6	2.8	1.4	1.2	1.6
Trinidad and Tobago	0.2	1.0	0.7	1.4	1.9	1.2	0.9	0.7	0.6	0.7	0.5
Peru	0.0	0.0	0.0	0.0	0.0	0.0	0.0	0.6	0.4	0.3	0.2
Turkey	0.0	0.0	0.0	0.1	0.2	0.1	0.2	0.1	0.2	0.2	0.2
Others	24.5	23.5	23.1	23.5	24.8	23.5	24.2	23.9	24.8	26.0	25.5

Source: Eurostat (online data codes: nrg_122a, nrg_123a and nrg_124a)

Source: (Eurostat, Primary Energy Imports, 2017)

EU STRATEGIES ON CENTRAL ASIA: POLICY PAPER ANALYSIS

As mentioned before in this article, the EU has increasing demand on Central Asian natural gas and oil. Thus, to dominate in the region for securing its energy supply, EU has long term policies considering 2020, 2030 and 2050 energy strategies and roadmaps. In this section, the general frameworks of these strategies and their implications for relations with Central Asia will be analyzed.

Energy 2020: A Strategy for Competitive, Sustainable and Secure Energy

This paper starts with an introduction defining energy as the life blood of a society and the this energy needs to be safe, secure, sustainable and affordable. (Energy 2020, 2010, pg: 2). It is described that Europe is challenging tests on energy and there is still pathways for EU for a more secure and sustainable energy market. (Energy 2020, 2010, pg: 2).

In The Directorate General for Energy of European Commission website, the objectives of 2020 Energy Strategy is divided into three:

- To reduce its greenhouse gas emissions by at least 20%,
- To increase the share of renewable energy to at least 20% of consumption,
- To achieve energy savings of 20% or more (European Commission, Energy DG, 2017).

To achieve these objectives above, the EU sets out priorities to reach those goals such as energy efficiency, Pan-European energy market, empowering consumers and achieving the highest level of safety and security; extending Europe's leadership in energy technology and innovation; strengthening the external dimension of the EU energy market (European Commission, Energy DG, 2017).

Main concerns about energy policy is visible in internal market and energy security in international level. According to the 2020 Energy Strategy Paper, a fragmented internal market and different national rules on energy policy create barriers to fair competition. Also, at the international level, it is expressed that Europe is vulnerable on energy supply and complains about the lack of common approaches towards partners, suppliers and transit countries. This seems a big gap for Europe's future energy policies including all partners* (Energy 2020, 2017, pg:3). Additionally, the emphasis is given to Europe's less influence in international market is emphasized given that EU is the world's fifth largest energy consumer. Europe, as the largest energy importer, is still vulnerable to supply risks and thus, needs to make more efforts to be more effective in international affairs*(Energy 2020, 2017, pg:4).

Energy policy is seen as a key to sustainable growth and the EU is seen as a strong geopolitical partner in the energy market. So, on the supply side, Europe gives priority to the development of secure and competitive sources of energy. Thus, for oil and gas, diversified and secure new supply routes are needed to meet rising import requirements and increasing demands from emerging and developing countries (Energy 2020, 2017, pg:5).

Apart from energy supply, energy technology is also a main strategy for a sustainable energy policy for the future. In the report, it is stated that Europe is facing competition in international technology market and thus, Europe should improve cooperation with third countries in specific technologies (Energy 2020, 2017, pg:15).

Under the subheading named as 'Strong International Partnership, notably with Our Neighbors', it is expressed that with over 500 million consumers, the European energy market is the world's largest regional market and energy importer. Within nation states, there is collaboration on common voice in third countries. So, security of supply, competitiveness and sustainability remain common goal in international energy policy. Additionally, relations with producing and transit countries remain important. The significance of additional energy supply sources and routes by 2020 is stated in the report (Energy 2020, 2017, pg:17).

For energy provisions, Europe has several agreements and treaties with third countries such as Free Trade Agreements, Partnership and Cooperation Agreements, Association Agreements, Memoranda of Understanding, Energy Charters Treaty and Energy Community Treaty. Thus, in the process of building a legal basis for the EU's external energy policy, EU member states should act with more coordination reflecting EU interest and internal energy market (Energy 2020, 2017, pg:17).

Under the same heading, emphasis was also given to an articulated synergy between EU's foreign and security policy and energy security. Thus, energy objectives and other policies should be harmonized. For the EU's energy security, diversification of fuels, supply sources and transit routes was defined as essential (Energy 2020, 2017, pg:18).

Under the fifth priority of the EU 2020 Energy Strategy named as Strengthening the External Dimension of the EU Energy Market, four actions were planned:

- **Action 1:** Integrating energy markets and regulatory frameworks with our neighbors, including
    - Extended The Energy Community Treaty to all EU neighbors, who are willing to adopt the EU market model like as in Mediterranean region and transit countries such as Ukraine and Turkey
    - Setting regulatory framework between EU and third countries to develop strategic routes from new suppliers
    - Mobilizing EU technical assistance to provide effective implementation of internal market and the modernization of energy sector in neighboring countries
- **Action 2:** Establishing privileged partnerships with key partners, including:
    - Reinforced energy partnership with key suppliers and transit countries aiming at promoting key principles in Energy Charter Treaty such as freedom of transit, transparency, safety and investment opportunities
- **Action 3:** Promoting the global role of the EU for a future of low-carbon energy, including:
    - Integration of energy efficiency, clean technologies and safe and sustainable low-carbon energy should into EU and bilateral cooperation activities, particularly with major consumer and emerging economies and with global partnerships.
- **Action 4:** Promoting legally binding nuclear-safety, security and non-proliferation standards worldwide, including:
    - Developing initiatives aiming at encouraging partner States to make international nuclear safety, security and non-proliferation standards

and procedures legally binding and effectively implemented around the globe, in particular through reinforced cooperation with the International Atomic Energy Agency and the conclusion of EurAtom agreements with key nuclear suppliers and user countries (Energy 2020, 2017, pg:18-19).

A Policy Framework for Climate and Energy in the Period From 2020 to 2030

Towards 2050, the European Union has launched several measures including 2020 and 2030. The reason for those measures is to build a more competitive, secure and sustainable energy system. The Directorate General for Energy of European Commission has defined those targets as below:

- A 40% cut in greenhouse gas emissions compared to 1990 levels
- At least a 27% share of renewable energy consumption
- At least 27% energy savings compared with the business-as-usual scenario. (Energy Strategy-2030, 2017)

Compared to the 2020 Energy strategy, those targets as percentages are more. This is to say, while it was foreseen as 20-20-20 targets for 2020 Energy strategy, 40-27-27 is foreseen for 2030 energy strategy. And what kind of policies does EU have to reach these goals? Again, the answer can be found on The Directorate General for Energy of European Commission website as below:

- A reformed EU emissions trading scheme (ETS)
- New indicators for the competitiveness and security of the energy system, such as price differences with major trading partners, diversification of supply, and interconnection capacity between EU countries
- First ideas for a new governance system based on national plans for competitive, secure, and sustainable energy. These plans will follow a common EU approach. They will ensure stronger investor certainty, greater transparency, enhanced policy coherence and improved coordination across the EU (Energy Strategy-2030, 2017)

If the targets of building a Pan-European market and providing energy efficiency can be realized, then a new three steps will be taken. These new steps will include again diversification of energy supply and energy security. So, it is foreseen that energy security for the medium term may continue as developing policies and implementations towards trade partners and transit countries.

The analyze these targets, the documentation named as 'A policy framework for climate and energy in the period from 2020 to 2030' would be enough. This document starts with an introduction describing that economical and financial crisis have changed lots of things inside the EU since 2008. Analyzing current developments at Union level* afterwards, new targets have been mentioned for 2030. Regarding to current developments analyzed before, providing greater security of energy supplies and reduced import dependence for the Union as a whole (Policy Framework for 2020-2030, 2014; pg:2) put significance in the paper.

Regarding reaching to 20-20-20 targets in 2020 Energy Strategy Paper, nine new targets are mentioned in this document. The first of those targets is defined as 'An ambitious commitment to reduce greenhouse gas emissions in line with the 2050 roadmaps. Delivery of this commitment should follow a cost-efficient approach which responds to the challenges of affordability, competitiveness, security of supply and sustainability, and which takes account of current economic and political circumstances.*' (Policy Framework for 2020-2030, 2014; pg:3). As seen here, affordability, supply security and sustainability put an emphasis with a cost-efficiency The other target is defined as 'Improving energy security, while delivering a low-carbon and competitive energy system, through common action, integrated markets, import diversification, sustainable development of indigenous energy sources, investment in the necessary infrastructure, end-use energy savings and supporting research and innovation' (Policy Framework for 2020-2030, 2014; pg:3). In this report, seven key elements framing energy policy were mentioned. Those are:

- Greenhouse gas emissions target
- A Renewable energy target at EU level
- Energy efficiency
- Reform of the Emissions Trading System
- Ensuring competition in integrated markets
- Competitive and affordable energy for all consumers
- Promoting security of energy supply

Of course, none of those elements can be studied separately because of their interconnected concept. But, third and seventh element including energy efficiency and supply security will especially be analyzed regarding to their closeness in this article. So, third element is related to energy savings and contributes to improved competitiveness; security of supply; sustainability; and the transition to a low carbon economy for EU energy and climate policy (Policy Framework for 2020-2030, 2014; pg:7). Seventh element has determining concept for developing relations with supplier and transit countries. In this concept, meaning of security of energy supply is described as ensuring continuous and adequate supplies of energy from all sources to all users (Policy Framework for 2020-2030, 2014; pg:11).

Energy dependency of EU strongly emphasized and reported that EU has dependency of 80% for fossil fuel for now, but this will increase to more than 90% till 2035. For natural gas import, it is stated that dependency rate will increase to 80%. Insufficient competition of EU market and rising demand in global level reflects negatively at EU energy market. This process causes vulnerability and price shocks in market. To increase energy security, three approaches have been drawn in the report. First includes renewable energy sources and is stated declining oil and gas production in EU market can contribute for domestic reserves of conventional and unconventional fossil fuels (primarily natural gas) and nuclear according to Member State preferences over their energy mix and within the framework of an integrated market with undistorted competition. Second is about common act of member states for diversification of supply routes. Because, it is believed that competition in energy market can create liberalization. (Policy Framework for 2020-2030, 2014; pg:11). And third is stated as "*greater efforts are required to improve the energy intensity of the economy cost effectively and to generate energy savings from the improved energy performance of buildings, products and processes*" (Policy Framework for 2020-2030, 2014; pg:11).

And how will the European governance be provided? This is planned as:

- National plans for competitive, secure and sustainable energy
- Indicators and objectives for competitive, secure and sustainable energy

This is to say, national plans and indicators form the governance pillar for 2030 energy strategy.

In spite of current separate processes for reporting on renewable energy, energy efficiency and greenhouse gas reduction for the period after 2020, consolidated governance process with Member States is suggested in national plan part. (Policy Framework for 2020-2030, 2014; pg:12).

In indicators and objectives part, it is stated that '*While higher shares of renewables and a more efficient energy system will contribute to both competitiveness and security of energy supply (in addition to the positive impact on GHG and pollutant emissions), they are not in themselves enough to ensure sufficient progress towards all aspects of these objectives in a 2030 perspective. Systematic monitoring with key indicators is needed to assess progress over time and to inform any future policy intervention*' and offers six subheading explaining what these indicators should include. Second (Diversification of energy imports and the share of indigenous energy sources used in energy consumption over the period up to 2030 should also be monitored) and fourth (Intra-EU coupling of energy markets, building on the liberalization of gas and electricity markets achieved already by EU legislation) are mostly related in energy security (Policy Framework for 2020-2030, 2014; pg:12).

The last part of this report explains international concept and how global changes affect European Union. According to the last section of this report, it is stated as energy demand on a global scale will rise till 2030 by the effect of Asia. It is also expected that hydrocarbon imports will sharply increase in China and India (Policy Framework for 2020-2030, 2014; pg:16) Those developments on a global scale will cause important results for EU, highly depended on energy imports (Policy Framework for 2020-2030, 2014; pg:17).

Energy Roadmap for 2050

Energy Strategy for 2050 is a long-term EU strategy in energy. Compared with previous strategies, the targets are ambitious with a special emphasis on competition and security.

For 2050 energy strategy, new targets have been designed as energy efficiency, renewable energy, nuclear energy, and carbon capture and storage (Energy 2050, 2017). A common energy market is also planned for compassable and cheap energy when needed. (Energy 2050, 2017).

In the Road Map for 2050, long-term energy politics are described in details. According to it, the target for 2050 is defined as *reducing greenhouse gas emissions to 80-95% below 1990 levels by 2050* (Energy Roadmap 2050, 2011, pg:2). In introduction part, it is expressed as this roadmap does not replace the local, regional or national efforts and aims just rising security and collaboration in long-term expected (Energy Roadmap 2050, 2011, pg:3). This is to say, a secure, competitive and decarbonized energy system in 2050 is possible ((Energy Roadmap 2050, 2011, pg:3).

In this report, opportunities and challenges were also described. One of them is on gas reported as a critical point for a transformation of the energy system. According to this approach, "Substitution of coal (and oil) with gas in the short to medium term could help to reduce emissions with existing technologies until at least 2030 or 2035. Although gas demand in the residential sector, for example, might drop by a quarter until 2030 due to several energy efficiency measures in the housing sector17, it will stay high in other sectors such as the power sector over a longer period." (Energy Roadmap 2050, 2011, pg:11). Regarding the development of a LNG system, transportation continues more independent from pipelines. So, global gas markets are also changing and integration, liquidity and diversification or supply routes becomes on the fore. It is also important that "shale gas and other unconventional gas sources have become potential important new sources of supply in or around Europe." So, Europe could relax on the point of gas importation (Energy Roadmap 2050, 2011, pg:12).

About transforming of fossil fuels, current coal stock inside the Europe's enhanced the energy portfolio and contributes to energy security. So, with the development of CCS and other emerging clean technologies, coal may play an important role in a secured energy supply. (Energy Roadmap 2050, 2011, pg:12). For oil, it is expected that oil will remain in the energy mix till 2050. One of the main challenge is this sector is "to adapt to changes in oil demand resulting from the switch to renewable and alternative fuels and uncertainties surrounding future supplies and prices" ((Energy Roadmap 2050, 2011, pg:12).

Regarding the international dimension, it is expected to improve international partnerships till 2050. This is explained in detail as:

'It will be important to manage the transition in close partnership with the EU's energy partners, notably our neighbors, such as Norway, the Russian Federation, Ukraine, Azerbaijan and Turkmenistan, the Maghreb and the Gulf countries while gradually establishing new energy and industrial partnerships. The EU needs to expand and diversify links between the European network and neighbouring countries with a particular focus on North Africa (with a view to best harness the solar energy potential of the Sahara) (Energy Roadmap 2050, 2011, pg:18).

CONCLUSION

As discussed in this paper, the EU has a high dependency on energy and is aware of its increasing future energy demands. One of the EU's main energy assets are its coal reserves. As reported in the energy strategy papers by EU, coal is being kept as an important stock for possible future energy problems. Thus for efficient EU energy market, coal reserves remain important.

But today, one of the main objectives of EU is to diversify its energy supply and to develop successful international collaboration agreements with transit and energy supplier countries. As seen many agreements made by Central Asian states separately, economics and energy are based as key elements. Because for today, Central Asia has an impact on energy with its current energy reserves and this makes the region more attractive for countries who are dependent on energy imports. When Russia is count as a principal actor in the region, it is not easy for European Union to act without difficulty. Thus, democratization and liberalization are put to the fore in many agreements and common projects between Central Asian states and the European Union. Because, it is believed that any acts only would be possible with Central Asian states if they have broken away Russian effect. At least in the short and medium term, this expectation may not realistic. Because the newly independent Central Asian states have historical and cultural ties with Russia and most of their politics are articulated with each other. Even if the Central Asian states desire and implement their own national politics and any technical support from external sources, still Russia has dominant effect during that process.

If the European Union cannot reach any effectiveness in the region, it may face serious energy problems, shaped by Russian politics. Thus, the European Union directs itself to alternative energy sources with an approach of 'diversification' and 'security'. Because, for security and diversification could only be possible if competition is easily provided in a liberalized arena.

In short, the European Union takes cautious steps taking into account Russian interests in the region. It seems that even international partnerships remain important for the short and medium term, the EU could divert its politics from the Central Asian states if EU reaches its targets to provide and to improve alternative energy sources like renewable ones. Of course, the Central Asian states energy sources and relation with these countries will remain important, but the reality of the depletability of current energy sources, will direct countries to find new sources to relax their energy needs. This may contribute significantly to reduce future possible energy conflicts.

REFERENCES

CIA-Kazakhstan, Crude Oil and Natural Gas Information for Kazakhstan. (n.d.). Retrieved from https://www.cia.gov/library/publications/the-world-factbook/geos/kz.html

CIA-Tajikistan, Crude Oil and Natural Gas Information for Tajikistan. (n.d.). Retrieved from https://www.cia.gov/library/publications/the-world-factbook/geos/ti.html

CIA-Turkmenistan, Crude Oil and Natural Gas Information for Turkmenistan. (n.d.). Retrieved from https://www.cia.gov/library/publications/the-world-factbook/geos/tx.html

CIA-Uzbekistan, Crude Oil and Natural Gas Information for Uzbekistan. (n.d.). Retrieved from https://www.cia.gov/library/publications/the-world-factbook/geos/uz.html

Çınar, B. (2008). Tarihte Üçüncü Güç ve Orta Asya Enerji Savaşları. *Güvenlik Stratejileri Dergisi, 8*.

Commission Staff Working Paper: Energy Roadmap 2050: Impact Assesment Part 1, European Commission. (n.d.). Retrieved from https://ec.europa.eu/energy/sites/ener/files/documents/sec_2011_1565_part1.pdf

EIA. (n.d.). *U.S Energy Information Administration, What is Energy?* Retrieved from https://www.eia.gov/energyexplained/index.cfm/index.cfm?page=about_home

Energy 2020, A Strategy for Competetive, Sustainable and Secure Energy, European Commission, COM (2010), 639 Final, Brussels.

Energy 2050. (n.d.). Retrieved from http://ec.europa.eu/energy/en/topics/energy-strategy-and-energy-union/2050-energy-strategy

Energy Roadmap 2050, European Commission, COM (2011). 855 Final, Brussels. Retrieved from http://eur-lex.europa.eu/legal content/EN/TXT/PDF/?uri=CELEX:52011DC0885&from=EN

Energy Strategy-2030, European Commission. (n.d.). Retrieved from http://ec.europa.eu/energy/en/topics/energy-strategy-and-energy-union/2030-energy-strategy

European Commission Energy DG. (n.d.). *2020 Energy Strategy*. Retrieved from https://ec.europa.eu/energy/en/topics/energy-strategy-and-energy-union/2020-energy-strategy

Eurostat-Energy Dependency. (n.d.). Retrieved from http://ec.europa.eu/eurostat/statisticsexplained/index.php/Energy_trends#Energy_dependency

Eurostat. (n.d.). *Extra-EU Imports*. Retrieved from http://ec.europa.eu/eurostat/statistics-explained/images/7/71/Percentage_of_extra-EU_imports_%28entries%29_of_natural_gas_by_country_of_origin.png

Eurostat. (n.d.). *Gross Inland Consumption*. Retrieved from http://ec.europa.eu/eurostat/statisticsexplained/images/e/e3/National_shares_of_fuels_in_gross_inland_energy_consumption%2C_2015%2C_percentage_F6_update.png

Eurostat. (n.d.). *Primary Energy Imports*. Retrieved from http://ec.europa.eu/eurostat/statisticsexplained/images/9/92/Main_origin_of_primary_energy_imports%2C_EU-28%2C_2005-2015_%28%25_of_extra_EU28_imports%29_YB17.png

İşcan, İ.H. (2004). Uluslararası İlişkilerde Klasik Jeopolitik Teoriler ve Çağdaş Yansımaları. *Uluslararası İlişkiler, 1*(2).

LaBelle, M. C. (2017). In pursuit of Energy Justice. *Energy Policy*, *107*, 615–620. doi:10.1016/j.enpol.2017.03.054

OPEC. (n.d.). *World Proven Oil Reserves, 2007-2016*. Retrieved from http://www.opec.org/opec_web/en/data_graphs/331.htm

Policy Framework for 2020-2030, European Commission, COM (2014) 15 Final, Brussels.

Sevim, C. (2012). Küresel Enerji Politiği ve Enerji Güvenliği. *Journal of Yasar University*, *26*(7).

Smith, J., & High, M. (2017). Exploring the Anthropology of Energy: Ethnography, Energy and Ethics. *Energy Research & Social Science*, *30*, 1–6. doi:10.1016/j.erss.2017.06.027

TDK. (n.d.). *Enerji*. Retrieved from http://www.tdk.gov.tr/index.php?option=com_gts&kelime=ENERJ%C4%B0

The World Factbook, Central Intelligence Agency (CIA). (n.d.). *Country Comparison by Natural Gas Proved Reserves*. Retrieved from https://www.cia.gov/library/publications/the-world-factbook/rankorder/2253rank.html

KEY TERMS AND DEFINITIONS

Central Asia: BEYOND geographical order, the region is important because of rich energy sources and high tendency of superpower states to control the region.

Energy Security: One of the main concern of European Union because of its vulnerable energy stocks and high dependency on external sources. It is mostly linked with accessibility and sustainability of energy routes.

Energy Strategy Papers: Policy papers published by European Union to portray current energy demands and to adopt solutions for future energy needs.

Heartland Theory: A geopolitical theory by H. Mackinder putting Central Asia into the center for earth power. This theory also helps to explain the roots of secret energy wars in the region.

Neighborhood Policy: The policy implemented by European Union, including agreements, financial assistance, and regulatory relations with non-EU member states, which have importance for the Unions' policy areas.

Sustainability of Energy: One of the European Union principles on energy policy with external states, including democratization, physical protection and international cooperation.

Chapter 5
Turkey's Renewable Energy Potential:
Policies and Practice

Suudan Gökçe Gök
University of Gazi, Turkey

ABSTRACT

The energy sector holds a crucial strategic importance for development and its sustainability. However, the energy reserves in Turkey are insufficient for the increasing energy demand. For this reason, the need for alternative energy sources has emerged. The fact that fossil fuels used in energy production will be exhausted and their damaging effects on the environment has made it inevitable for the world to use renewable energy. In Turkey, the following are the main energy sources used in electricity production: hydraulic sources, brown coal, natural gas, coal and fuel oil. The use of renewable energy sources in electricity production has shown a considerable development in the world over the past few decades. However, these energy resources have yet to be exploited to any large extent in Turkey. In this chapter, the range of potential renewable energy resources available for electricity production in Turkey will be analyzed in order to determine the necessary technical substructure to make these renewable energy resources more attractive.

DOI: 10.4018/978-1-5225-4203-2.ch005

INTRODUCTION

Technological developments and economic growth has accelerated energy demand for alternative resources. The period that started with the 1974 oil crisis showed that oil is a finite resource, and cheaper and easier electricity generation alternatives were extensively discussed in those years.

Faced with the prospect of diminishing fossil-based energy sources, the increases in petrol prices, and environmental problems, the public is paying more attention to sustainable energy generation and sustainable development issues. Energy policies are updated according to renewable energy source usages, increasing efficiency, and environmentally friendly technologies. Many countries prioritize renewable resources and design and implement policies which make them less economically dependent on foreign countries. Ocean energy, geothermal energy, biomass energy and hydro electrical energy are among the numerous renewable energy resources available. These are the energies whose applications vary according to their specific characteristics. Solar power, being the most important among the renewable energy resources is a natural resource which underlies many other energy resources. Energy resources such as wind, solar, and geothermal are used in many fields such as power generation, the operation of industrial plants, and illumination. Although many renewable energy resources require certain costs and labour in our present day, these are the resources which are endless and can sustain the power generation requirements of humanity.

All research conducted conclude that there are ideal energy resources which are environment-friendly. They will respond to the energy demand at a broad perspective, from domestic electrical instruments to space craft, along with technological development. On the other hand, it will remove the unavoidable damage caused by limited energy resources and various contaminant impacts on the environment and thus should be used in all application areas as soon as possible. Parallel to this thought, it should be focused on renewable energy resources under the heading of clean energy.

Renewable energy sources are important for energy deficient and dependent countries. Therefore, there are several favourable legal regulations in place and governmental incentives offered for research and innovative entrepreneurs. However, at this point, the renewable energy sources cannot compete with fossil fuels in terms of costs and requirements and it is not expected that the renewable energy sources will address this issue in the short term. This situation makes the the search for increased use efficiency of conventional sources as important as renewable energy sources.

Regarding renewable energies, Turkey is ahead of the game when compared to many other countries worldwide. When wind energy -the preferred energy among the renewable energy branches- is taken as a basis, Turkey ranks 3rd for wind potential in Europe nowadays. However, despite this position our country fell behind world standards in regard to usage. In fact, although one third of the country is suitable for wind energy usage, currently there are10 completed projects and there are ongoing studies related to h 12 projects. When all these projects are completed, it is expected that wind energy will cover 3-3.5% of total power consumption. Although Turkey has much more potential than European countries, these realistic figures are a proof that it has fallen behind.

This study aims to analyse the energy sources of Turkey such as hydraulic, biomass, solar, geothermal, and tidal waves besides their potential, usage rates, and alternative positions to conventional sources. Furthermore, their role in meeting the energy requirement and decreasing the dependency on external energy sources are discussed in this chapter.

THE RENEWABLE ENERGY SOURCES OF TURKEY

Renewable energy means "the energy source that is available on the next day according to natural circulation itself" (Uyar, 2006). Given the limited reserves and destructive effects of fossil fuels that meet almost all the energy requirement of the world, alternative energy sources that are renewable, more secure, and environmentally friendly are extensively researched.

The main difference between renewable and fossil-based sources is that the fossil fuels cannot be recycled after their usage while renewable energy sources are sustainable and can be converted in the next cycle of nature. Most countries pay attention to renewable sources to combat climate change and reduce greenhouse gases (Kuban & Uyar, 2007). The International Energy Agency predicts that CO2 emissions will increase at 6% according to a 2020 scenario. In order to prevent this increase, an additional $430 billion investment should be made in energy efficient and low carbon technology (IEA, 2009). The globalised world requires alternative energy sources to sustain its development. Renewable energy sources are seen as a way to meet an increased population's energy demand in an ecological and secure way. In 2000, renewable energy sources met 11% percent of all the demand in Turkey. This share decreased to 7% in 2010 because of the applied regulations. While Turkey meets 36% of its consumed energy with its own sources, it is assumed that this will decrease to 20% by 2023 .

Figure 1 illustrates the potential and installed power sources in Turkey based on hydroelectric, renewable, geothermal and biogas energy sources. The potential of renewable energy sources has been estimated at 48.000 MW, and the total capacity of installed renewable energy plant is around 1.320 MW. The highest share belongs to hydroelectric plants with 15.729 MW as seen in Figure 1.

Table 1. Turkey's renewable energy resources potentials

Renewable Energy Type		Type of Energy Use	Natural Potential	Technical Potential	Economic Potential
Solar Energy		Elec. En. (billion Wh)	977000	6105	305
Hydraulic Energy		Heat (MTEP)	80000	500	25
Wind Energy	**Terrestrial**	Elec. En. (billion Wh)	430	215	124.5
	Marine	Elec. En. (billion Wh)	400	110	50
	Sea wave energy	Elec. En. (billion Wh)	-	180	-
		(billion Wh)	150	18	-
Geothermal Energy		Elec. En. (billion Wh)	-	-	1.4
		Heat (MTEP)	31500	7500	2843
		Fuel Modern (MTEP)	90	40	25

Turkey's renewable energy resources potentials. Resource: EİE

Figure 1. Example of a figure caption

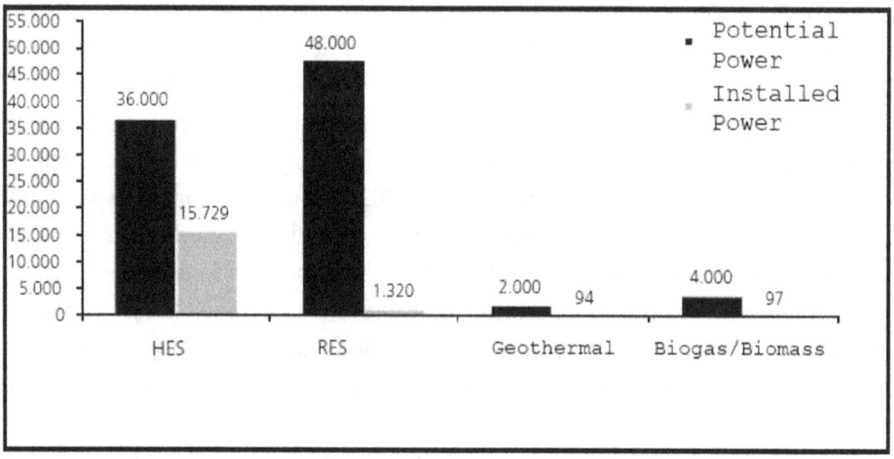

Turkey's potential sources of renewable energy are presented in Table 2.

Geothermal Energy

Geothermal is the heat energy generated by earth and includes mineral, salt, and gases that are above the average temperature of natural water. Turkey is the 7th ranked country in the world according to geothermal sources; 8% of the global geothermal energy potential is located in Turkey (Özçep & Karabulut, 2008).

Since the unit cost of geothermal energy is cheaper than others and causes minor environmental effects, it is assumed to be a more advantageous and cleaner energy source. The latest analysis shows that there is a geothermal potential of 650 MWh in Turkey. It is predicted that all the identified potential may be connected to the electricity grid by 2020. The current installed electricity potential of geothermal stations in Turkey is around 942 MWh. It is also predicted that the potential geothermal energy used in thermal tourism and heating energy is around 31500 MWh.

This potential is equal to the heating energy requirement of 5 million houses. The cost of this heating is 100 times lower than electricity, 50 times less than fuel-oil, 40 times less than natural gas, and 32 times less than coal. In addition to cost, geothermal energy systems are more secure and more flexible compared to other energy systems. The efficiency of geothermal systems may reach 97%. Another important advantage of the geothermal stations is the construction time period (Özçep and Karabulut, 2008).

The main reasons for the lack of development of Turkey's geothermal energy resources can be explained by technical, financial, and management problems. The technical problems are related to low enthalpy, calcite sedimentation in reservoirs, and environmental problems. Another issue is the absence of any long term investments in this technology in Turkey. The projects and applications of residential heating,

Table 2. Turkey's renewable energy resources potentials

Resource Type	Development Targets
Wind	20,000 MW of wind energy installed capacity by 2023
Geothermal	All known geothermal potential of 600 MW to be operational by the year 2023
Solar	Promote the use of solar energy for electricity generation, assessment of the potential of the country to ensure the maximum extent, encouraging result of studies initiated for this purpose.
Other	Technological developments and potential developments in the use of other renewable energy, legislation, regulations, depending on the preparation of production plans, taken into account
Hydroelectric	All hydroelectric potential to be evaluated technically and economically by 2023

greenhouses, and industrial plants with low enthalpy geothermal sources are realized with the support of local authorities or individuals instead of governmental policies (Solmaz & Solmaz, 2001).

Wind Power

Modern wind energy technology is known for being clean, and producing electric energy economically. Wind energy has become the preferred choice for developing countries due to the features of this technology. However, developed countries want to reduce carbon emissions and to decrease the effects of greenhouse emissions.

The potential in wind energy announced by the Ministry of Energy and Natural Resources in light of the 2008 data amounted to a total of 48GW, of which 8GW is effective and 40GW is moderately effective. The Ministry of Energy and Natural Resources had the aim of increasing installed wind energy power to 10000 MTW by 2015 in its strategic plan of 2010-2014 and to reach 20 GW installed wind energy power in the Electric Energy Submission Security Strategy Document by 2023. Turkey, which started to benefit from its wind potential lately, had almost 900 MW RES installed power at the end of 2009 and in 2010 it was a total 1300 MW .

Comparing wind turbines with fossil fuel plants, it is more economically productive because its installation costs are relatively lower. According to OECD resources, there is the possibility to obtain around twice more of the produced electricity in Turkey per year from wind.

The use of renewable energy sources and technologies is increasing in Turkey. The renewable energy law of 2007 encouraged the use of renewable energy sources in electricity generation. The positive effect of the mentioned law increased the installed wind energy potential of Turkey from 18 MW to 1300 MW in seven years. The licence applications made to the Energy Market Regulatory Authority (EMRA) show that the installed capacity may increase up to a few times of accurate potential. This situation shows that Turkey has passed an important turning point in the use of renewable energy sources. There is a similar expectation for solar energy in the projections for the next 10 years.

Solar Energy

Solar energy is a technology that harvests energy from solar irradiation. The conversion of solar irradiation to electricity is performed with solar cells that are collected in a solar panel. Solar cells are dependent on semiconductor technology. Furthermore, the solar energy can be stored as electricity with batteries or as heat with solar collectors. The direct and diffused radiation that arrives at the earth from the sun varies according to atmospheric conditions such as humidity, dust particles, and cloudiness. The sunniest places of the world are located around 35° north of the Equator. Turkey is a country that is geographically located on the solar line with a mean annual insulation value of 2906 hours, around 30% of a year (Atılgan, 2000).

It is expected that the solar cell usage will increase in Turkey with the fall seen in prices. The solar energy potential atlas of Turkey shows that there are 380 billion kWh/year of solar energy available in Turkey.

Solar energy will probably be important in the energy future of Turkey. This is not only true for Turkey but also for many countries, although solar energy is widely used to heat water in Turkey, currently it does not cover t 5% of its potential. The use of solar energy for electricity generation is one of the most important targets of Turkish energy policy. The minor applications of solar energy such as illumination, traffic lights, and traffic warning boards are increasing rapidly. During the last few years more attention has been paid to energy efficiency and renewable energies, especially solar and wind energy.

The solar energy applications are used in most of the systems in the 2010s. It is predicted that over the next 10-15 years' solar energy will spread all over the country with solar roofs, solar buildings, industrial applications and their integration into the national grid.

Biomass

Biomass is matter which does not possess a biological fossil origin. The energy derived from plant matter with carbohydrate components, is defined as biomass energy. Biomass is also defined as all organic matter which can regenerate within a period of 100 years, plants which grow on land and water, animal wastes and forestry products. In other words, plants, which absorb carbon dioxide during photosynthesis, form biomass and oxygen. It is possible to burn these plants and release the captured carbon dioxide back into the atmosphere (Taşyürek & Acaroğlu, 2012).

The use of biomass energy can be grouped into two uses, traditional and contemporary. Traditionally, biomass energy is obtained from the wooden forests, it consists of plants and animal wastes which are used for fuel. The basic specification of biomass energy is the fact that energy is obtained directly by burning the biomass product. It is very common in industrial rural regions. Contemporary biomass sources are: forestry products and wood industry wastes, agriculture products, plant-animal wastes from the agriculture sector, urban wastes and agricultural industrial wastes. The biomass products in question are processed and are converted to solid, liquid and gas fuels. The variety of biomass fuels vary from wooden briquettes to synthetic raw petrol. In order to produce biomass fuel pyrolysis, hydro gasification, hydrogenation, fractional distillation, and acid hydrolysis techniques are used. It is also possible to use a mixture of biomass fuels and fossil fuels (Dikmen, 2009).

It is possible to obtain electrical energy from cultivating fast growing plants and burning them from forests which are known as energy forests. Pilot applications have begun in our country. In addition, regarding energy obtained from urban wastes; it is estimated that the methane gas naturally formed in the waste disposal sites of Turkey, which number more than 2000, is about 650 million m3. This is approximately equivalent to 8 billion kWh. The methane gas accumulated in the waste is pumped directly to the energy production centres where t the methane gas is separated and burned. This method of electricity production is used in the waste disposal sites in Istanbul (Kemerburgaz) and Ankara (Mamak and Sincan). On the other hand, it should also be mentioned that it is more expensive to provide biomass sources than fossil sources. However, because biomass is a recyclable source, it is an important element in a sustainable global energy mix.

The Ministry of Natural Resources' data claims that the energy potential from waste disposal sites is approximately 8.6 million TEP and 70% of this is used for heating. At the same time, it was announced that biogas production capacity was measured as 1.5- 2 million TEP. What's more, it is known that bioethanol power is equal to 0.73% of total liquid fuel in Turkey, equivalent to 160.000 tones. With the Renewable Energy Law, biogas was included under the category of biomass and a guaranteed price of $0.133 /kWh was given to all electricity generation facilities which use biomass, including landfill gas.

Figure 2. Quarterly biomass and biogas based electricity production (MW) for the period 2009-2010

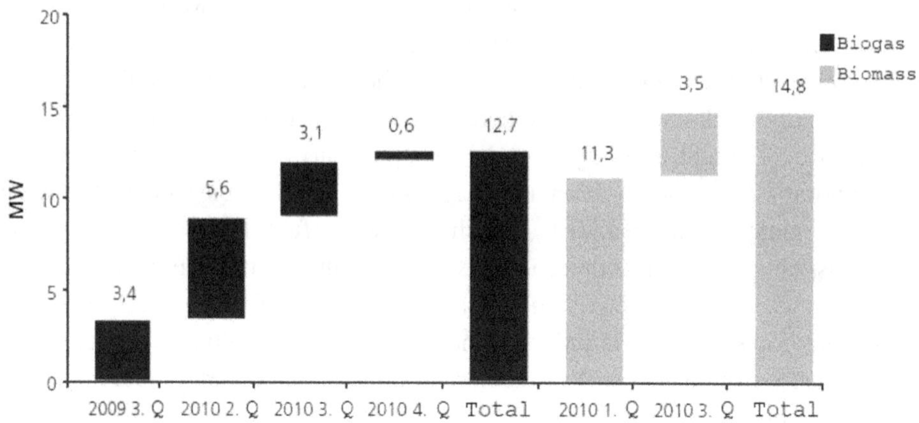

Hydraulic Energy

Hydroelectric energy is obtained by turning the static energy of water into kinetic energy. The hydroelectric power plants turn the power of the flowing water into electricity. Producing energy from hydroelectric power is a clean, productive and effective method. The initial investment in hydraulic power plants is very high. However, areas within range, even some residential areas and historical sites, will be flooded and that during periods of drought, electricity production will decrease are other disadvantages.

However, hydroelectric power plants, just like thermal power plants do not cause air and environmental pollution. Even developing half of the world's economically feasible hydroelectric production potential would result in a 13% decrease in greenhouse gas emission. Also, the lakes which are formed with time behind dams will cause the climate of the surrounding areas to soften. Hydroelectric power plants also have the advantages of preventing floods, regulating irrigation works, developing fishing, providing forestation and easing transport.

The electricity generated from hydroelectric plants is the most important source of usable renewable energy in Turkey. The theoretical hydro electrical potential in Turkey can is 433 billion kWh, technically the potential is 216 billion kWh;, as for the technical and economic potential, it was calculated at 127 billion kWh. As of today, there are 135 functioning hydro electrical power plants in Turkey. These plants have 12631 MW installed power and an annual production capacity of 45325 GW which accounts for 36% of the total potential. Approximately 45% of this is produced within the scope of the GAP project.

In the past few years the share of the electricity produced by hydroelectric plants has decreased due to the increasing use of natural gas in Turkey. With the spread of natural gas more natural gas cycle power plants have been established. Even though the initial investment is higher than for natural gas power plants, the fact that they are domestic resources and that they have no fuel cost is very important for Turkey (Altun, 1996).

Tidal Energy

The production of electricity from tidal movements is based on taking advantage of the difference between a high tide and a low tide (ideally 5 meters), by building a dam at the back of gulfs or bays the rising water goes directly into the dam; then, the power which is generated when the tide pulls back is turned into electrical energy by means of a turbine system. Another method simply utilizes the flow speed during low tide and high tide electrical energy is produced by placing a turbine(s) in the areas where flow occurs. The method in question has a similar process to the electrical energy production from sea/ocean current (flows) (Olah, Goeppert & Prakash, 2006).

Although currently there is no electric generation from the power of waves, tides, and currents in Turkey, it is known that there are investors who want to generate electricity especially from massive currents of the Bosphorus and Dardanelles straits. In a statement made by the Turkish Grand National Assembly Energy Commission in June 2009, it is claimed that the currents in the Bosphorus and Dardanelles straits have a 450 MW energy potential. To date, there is no evidence that any developments have occurred to exploit this resource.

On the other hand, it is possible to benefit from wave energies which occur in big water masses such as oceans and seas. It is one of the renewable energy forms.

The studies which have been undertaken regarding the production of energy from waves generally concentrate on electrical energy. The basic systems used in this field are divided into two: "fixed devices" and "floating devices".

Fixed devices are installed on a solid foundation, in front of breakwaters which are installed through the shores or also on a fixed sea based off shore. The most developed wave energy plants are designed according to this system. In spite of this, the floating systems are systems which move on the surface of the water just like a boat and are connected to the shore with high voltage cables. Electricity production from these systems, whose research and development are still ongoing, is made through water chambers. As the waves are taken into the water chambers, the air in the chamber is compressed and mechanical energy is obtained. This system is similar to a classical pump system (suction and force) and the transformation from mechanical energy into electrical energy is made through turbines (IEA, 2006).

Obtaining electrical energy by benefiting from oceans or seas is a relatively new resource when compared to the other renewable energy methods and its share within all of the renewable energies is below 0.1%. In forecasts made by the IEA, it has been noted that by the year 2030 it will not pass 1%. In the event that the cost level, both economic and environmental, of the technology falls, it may be possible to obtain positive results in the long run (IEA, 2006).

CONCLUSION

The biggest obstacle to the widespread adoption of renewable energy sources seems to be high costs. However, when we consider the fact that the main sources of energy in Turkey are met through imports, the investments made in renewable energy sources will bring positive results in the long run. Also, as the costs of renewable energy are decreasing daily, it seems to be inevitable that the share of energy produced from renewable sources will continue to increase. However, the increasing use of renewable energy sources also needs to be accompanied by an increasing focus on the effective use of the energy produced. It is not sufficient to produce policies which will just meet energy demands According to the energy density, which is one of the most important indicators regarding energy efficiency, Turkey is below the world average. Turkey needs to develop policies which incentivize energy productivity. Such policies, if successfully implemented, will decrease energy demand and thus decrease energy imports and decrease the investment costs required for new energy investments.

When creating the energy policies, it is very important that one does not ignore the environmental factor. The use of energy and its relation with the environment has gained a lot of importance throughout the world as well as in Turkey. While the environmental policies in Turkey ensure sustainable development, a basic objective should also be to protect nature and leave the world to the generations to come with the least damage possible. Development targets and policies should be considered within the scope of environmental issues.

Table 3. Primary energy production targets of Turkey from 2005 to 2030

	2005	2010	2015	2020	2025	2030
Hard coal and lignite	21259	28522	31820	39385	42732	45954
Oil and natural gas	2127	1735	1516	1604	1505	1465
Central heating	495	884	1336	2018	2427	2758
Hydropower	5845	7520	8873	9454	10002	10465
Wood and waste	6760	6446	6029	5681	5498	5413
Geothermal	1380	3760	4860	4860	5400	5430
Nuclear	0	3657	9143	18286	26988	29600
Solar	459	907	1508	2294	2845	3268
Wind	250	620	980	1440	1786	2154

In the case of renewable energy potential, the EU offers an important opportunity to increase the use of renewable energy sources in Turkey, however, the fact that the issue is not "internalized" by the political powers and related branches of the government is also a problematic approach to the issue. To solve the many problems which confront the development of renewable energy in Turkey and to accelerate the frozen investments for renewable energy sources, the basic starting point is for the related governmental departments and political powers to have a systematic and scientific approach to the issue. The corresponding studies performed with contributions from experts, institutions, governments, and organizations will result in successful renewable energy policies, if a renewable energy policy is designed considering the national gains and international developments and develops new power networks that integrate renewable energy sources within a framework of environmental sustainability and efficient energy use. This will provide multiple benefits for the public, the economy, society, and politicians.

REFERENCES

Altun, İ. H. (1996). Turkey's hydropower potential and development status. In *TMMOB 1.Energy Symposium Proceeding Book* (pp. 9-14). Ankara: EMO Publications.

Atilgan, İ. (2000). Overview of Turkey's Energy Potential. *Gazi University Journal of Engineering and Architecture, 15*, 1.

Çukurçayır, M., & Sağır, H. (2007). Energy Problem, Environment and Alternative Energy Sources. *Selcuk University Journal of the Institute of Social Sciences, 20*, 257.

Dikmen, A. Ç. (2009). *Locating in Turkey's Future Renewable Energy Sources for Sustainable Development Framework* (Doctoral Thesis). University of Ankara, Ankara, Turkey.

DTM. (2010). *Energy Production and Consumption in Turkey.* Undersecretariat of Foreign Trade. Retrieved from http://www.dtm.gov.tr/dtmadmin/upload/EAD/KonjonkturIzlemeDb/teut.do

EC. (2009). *The impact of renewable energy policy on economic growth and employment in the European Union.* European Commission Directorate General, Energy and Transport. Retrieved from http://ec.europa.eu/energy/renewables/studies/doc/renewables/2009_employ_res_summary.pdf

EMRA. (2011). *Electricity Market Regulatory Agency*. Retrieved from www.epdk.gov.tr

EMRA. (2013). *Electricity Market Regulatory Agency, Electricity Market Licences.* Retrieved from http://www.emra.org.tr/tr/web/electricity-market/licenses

IEA. (2006). *Renewable Energy: RD&D Priorities*. Paris: IEA Publications.

International Energy Agency. (2009). *How The Energy Sector Can Deliver On A Climate Agreement*. Copenhagen: IEA Publications.

Kuban & Uyar. (2007). Renewable Energy Regulations and Turkey. *The European Union and Turkey Renewable Energy Regulatory Workshop*, 283-290.

OECD-FAO. (2011). *OECD-FAO Agricultural Outlook 2011-2020.* Retrieved from http://www.agrioutlook.org/dataoecd/23/56/48178823.pdf

Olah, Goeppertve, & Prakash. (2006). *Beyond Oil and Gas: The Methanol Economy.* Wiley-VCH Publications.

Solmaz & Solmaz. (2001). Concept of Geothermal Energy Potential and Environmental Effects in Bursa. *Symposium on Renewable Energy Resources*, 283-290.

Taşyürek & Acaroğlu. (2012). Emission Reduction in Biofuels (Biomotor) and Global Warming Effect. *Productivity Review*, 162.

TEIAS. (2011a). *Turkish Electricity Production and Transmission Statistics 2011.* Retrieved from www.teias.gov.tr

TEIAS. (2011b). *Electricity Power Production Projection for Turkish Electrcity Energy 2011-2012*. Retrieved from www.teias.gov.tr

Chapter 6
European Energy Alternatives and Turkey's Compatibility

Selma Aytüre
Yıldırım Beyazıt University, Turkey

ABSTRACT

Energy is an issue of strategic importance to the European Union and Turkey. Both are dependent to the outside. EU is the world's largest energy importer and second largest energy consumer after the United States. Turkey's alignment with the EU's energy policy is extremely important for EU in terms of increasing the diversity and quality of its energy resources. Turkey's strategic location makes Turkey a land of passage for transporting oil and gas to Europe. This geopolitical importance is an important opportunity for both sides. In this chapter, EU's position on energy in the world has been explained first. Then the energy situation and energy policy in EU has been examined. Secondly, Turkey's energy policy and compatibility to EU together with complementary role to EU on Energy has been presented.

INTRODUCTION

Energy is an issue of strategic importance for the European Union. According to data for the year 2000, 15% of the world's energy consumption belongs to the EU. The EU is the world's largest energy importer and second largest energy consumer after the United States. After the last enlargement of the EU, this dependency has increased little more and forced to develop new initiatives. This lead the EU to diversify resources in imports in terms of energy supply security. The energy demand of the EU for the period 2010-2020 is expected to increase by 3.5% and 9.7%. In this scenario, the sources of the energy to be consumed are of critical importance.

DOI: 10.4018/978-1-5225-4203-2.ch006

The EU's energy policy is based on the Paris Treaty, which established the European Coal and Steel Community on Coal and the EURATOM Treaty on Nuclear Issues. The "EU 2020 Energy Strategy" sets priorities for the new policy and in addition, 2050 road maps are being prepared.

Turkey's alignment with the EU's Energy Policy is extremely important in terms of increasing the diversity and quality of its energy resources. Turkey is a major hydropower producer with a key role in energy, and its strategic location, makes it a transit area for transporting oil and gas to Europe.

In the ongoing negotiation process with the EU "15. Energy" Chapter, the draft screening report has not yet been approved by the EU Council and continues to be debated. The EU 2020 Energy Strategy includes priorities that need to be considered and implemented with Turkey and other regional countries in mind.

From this point of view, Turkey is not only becoming a natural energy bridge for the consumer markets in Europe and countries with large energy reserves such as the Caspian Region, Central Asia and the Middle East, but is also taking important steps to become an energy center (Yıldız, 2011, p.271). The European Union wants to take advantage of Turkey's geopolitical position in order to ensure diversity_of the resources. This geopolitical advantage is also an important opportunity for both sides because of the serious contribution that the EU can make to the security of energy supplies and the possibility of facilitating Turkey's accession negotiations with the EU (Yıldız, 2011, p.273). Turkey has already become a strategic energy partner of Germany and Europe (Hauser, 2010, p. 70).

Turkey is well positioned to become the East-West and North-South corridor with the existing pipelines, as well as including the new projects involved. This potential strengthens Turkey's position vis-a-vis the EU. It must be borne in mind, though, that energy is not the only determinant of geopolitical relations, and that the pursuit of other vital interests will also impact energy relations around the world (Amineh & Guang, 2012, p.18).

In this chapter, firstly. EU's position on energy in the world has been explained, and its energy situation and energy policy have been examined. Secondly, Turkey's energy policy and it's compatibility to the EU's, as well as its complementary role to the EU on Energy have been presented.

EUROPEAN UNION'S PLACE IN THE ENERGY WORLD

The European Union is one of the top players in the world. In terms of production, the biggest country is China (18.1%), and the EU ranks 7th (5.6%), below the United States, the Middle East, Asia, Russia and Africa (Table 1). The EU's production decreased from 1995 to 2015, while world production was on the rise. This amount was 966 Mtoe in 1995, and 771 Mtoe in 2015.

The EU's share in the world's consumption is more important than production. Final energy consumption was 11.9% in 2015. The biggest consumption region was China with 20.4%, followed by Asia (18.3%) and the United States (16.2%). The EU is the 4th biggest energy consumption region in the world (Table 2).

As one of the world's largest importers of oil, gas and coal, the EU is a major player on the international energy market. However, it remains a dwarf on the political stage as member states keep the upper hand on foreign policy. It has been argued that the energy challenges facing Europe need a coherent external policy to enable

Table 1. World energy production by region (Mtoe)

	1995	2000	2005	2010	2015	2015 (%)
EU-28	966	950	909	840	771	5.6%
China	1 064	1 129	1 707	2 316	2 496	18.1%
United States	1 659	1 667	1 631	1 723	2 019	14.6%
Middle East	1 137	1 324	1 516	1 619	1 884	13.7%
Asia*	815	922	1 105	1 343	1 568	11.4%
Russia	968	978	1 203	1 279	1 334	9.7%
Africa	774	885	1 087	1 173	1 118	8.1%
Rest of the World	1 880	2 176	2 430	2 575	2 601	18.9%
World	9 263	10 032	11 588	12 869	13 790	100.0%

Source: EU Energy, Statistical Pocketbook, 2016, 2017

Table 2. World final energy consumption by region (Mtoe)

	1995	2000	2005	2010	2015	2015 (%)
EU-28	1 133	1 180	1 242	1 208	1 114	11.9%
China	788	796	1 227	1 667	1 915	20.4%
United States	1 378	1 546	1 563	1 512	1 520	16.2%
Asia*	643	744	877	1 073	1 716	18.3%
Africa	324	369	437	508	573	6.1%
Russia	458	418	412	447	457	4.9%
Middle East	202	241	313	414	4756	5.1%
Rest of the World	1 613	1 747	1 905	2 037	1 614	17.1%
World	6 539	7 041	7 977	8 866	9 384	100.0%

Source: EU Energy, Statistical Pocketbook, 2016, 2017

Europe to play a more effective international role in tackling common problems with energy partners worldwide. It would allow the EU to speak with one voice in their external relations towards common foreign policy energy security (Amineh & Crijns-Graus, 2014, p.816).

Among the OECD group of economies, growth in GDP (expressed in real purchasing power parity [PPP] terms) was even associated with a slight decline in primary energy demand for the period 2000-2014. This is a noteworthy turn of events, but not necessarily a surprising one given that structural economic shifts, saturation effects and efficiency gains produced a peak in primary energy demand in the European Union (in 2006), since then demand has fallen by more than 10% (IEA, 2016, p.40).

Energy security is a diffuse concept and its legal definition is still in process (Sigot, 2013, p.4). Energy security is multi-faceted and differs between countries and the sector of energy considered at a given time (ibid. p.6). To some extent, energy security can be addressed directly through energy-specific policies, but ultimately energy security is inextricable from broader economic and foreign policy challenges and solutions (Gault, 2002, p.3).

Twenty-first century access to energy sources depends on a complex system of global markets, vast cross-border infrastructure networks, a small group of primary energy suppliers, and interdependencies with financial markets and technology. This is the context in which energy security has risen high on the policy agenda of governments around the world and the term 'energy security' has quietly slipped into the energy lexicon. The limited discourse about the nature of the term or its underlying assumptions has been totally eclipsed by an almost overwhelming focus on securing supplies of primary energy sources and geopolitics (Chester, 2010).

THE EU'S ENERGY SITUATION

The EU's total energy production is decreasing. Net imports are increasing. Gross inland consumption is slightly decreasing. Primary energy consumption and final energy consumption are slightly decreasing too. Import dependency is increasing. Total import dependency was 43.1% in 1995 and 54.0% in 2015. For the same year, this figure goes up in the case of hard coal (64.1%), petroleum fuels (88.8%), crude and NGL (88.4%) and natural gas (69.1%). In the case of solid fuels, it decreases (42.8%) (2015), compared to 2013 (Table 3).

Table 3. European Union 28, Energy Balance

Mtoe, unless otherwise stated	1995	2000	2005	2010	2013	2015
Production	969.1	951.9	914.3	849.6	804.7	781.9
Solid Fuels	279.8	214.6	196.0	164.8	156.5	145.4
of which Hard Coal	174.9	120.7	100.1	74.7	64.8	58.2
Petroleum and Products	180.8	180.0	142.5	109.6	85.1	89.3
of which Crude and NGL	172.3	169.9	129.9	94.3	69.7	72.7
Gases	191.5	209.4	190.8	159.8	132.0	107.9
of which Natural Gas	191.0	209.2	190.7	159.8	132.0	107.9
Nuclear	227.3	243.8	257.5	236.6	226.3	221.2
Renewables	84.1	97.9	119.6	167.9	192.8	205.0
Wastes, Non-Renewable	5.7	6.1	7.8	11.0	12.0	13.0
Net Imports	736.6	827.2	980.6	954.2	908.5	902.1
Solid Fuels	78.4	98.4	125.4	111.7	126.3	112.4
of which Hard Coal	76.6	94.3	122.7	111.0	126.0	112.6
Petroleum and Products	510.6	533.0	598.4	558.1	523.2	534.9
of which Crude and NGL	471.2	501.1	564.1	522.9	499.7	528.0
Gases	145.6	193.5	254.1	278.0	252.6	247.3
of which Natural Gas	145.6	193.5	254.1	278.0	252.6	247.3
Renewables	0.3	0.3	1.5	5.8	5.1	6.0
Electricity	1.8	2.0	1.4	0.6	1.1	1.2
Gross Inland Consumption	1 674.7	1 730.0	1 831.0	1 763.7	1 666.7	1 627.5
Primary Energy Consumption	1 567.4	1 617.7	1 712.6	1 656.6	1 569.2	1 530.7
Final Energy Consumption	1 082.7	1 132.8	1 191.3	1 163.8	1 106.6	1 084.0
Import Dependency (%)	43.1%	46.7%	52.2%	52.6%	53.1%	54.0%
of Solid Fuels	21.5%	30.6%	39.4%	39.5%	44.1%	42.8%
of Hard Coal	29.7%	42.6%	55.7%	57.9%	64.5%	64.1%
of Petroleum Fuels	74.1%	75.7%	82.1%	84.5%	87.4%	88.8%
of Crude and NGL	73.0%	74.4%	81.3%	84.6%	88.0%	88.4%
of Natural Gas	43.4%	48.9%	57.1%	62.2%	65.2%	69.1%

Source: EU Energy, Statistical Pocketbook, 2016, 2017

The EU imports more than half of all the energy it consumes. Its import dependency is particularly high for crude oil (more than 90%) and natural gas (66%). The total import bill is more than €1 billion per day. Many countries are also heavily reliant on a single supplier, including some that rely entirely on Russia for their natural gas (Buşe, 2017, p.170).

The EU's energy consumption and importation profiles have evolved since the end of the Cold War. While the immediate post-Cold War boost to neo-liberal globalization dovetailed with a gradual widening of supplier bases, which reduced its collective importation-relative (but not consumption-denominated) dependency on Russian natural gas, the EU's absolute consumption of gas from Russia and elsewhere continued to grow. Moreover, the 2004 and 2007 enlargements of the Union, which presupposed the unanimous approval of the EU 15, incorporated 12 new member states that added less to the overall size of the EU economy than to its dependence on Russian energy (Tekin & Williams, 2011, p.3).

There has been a general shift in the European Union's energy structure from solid fuels towards natural gas and renewable energy sources, which reflects the growing importance of the climate policy in the EU. Therefore, despite the overall global trend emphasizing energy efficiency and saving, the EU's natural gas demand has been growing and a similar trend is expected to continue in the future. In 2030, natural gas is expected to account for 30% of the EU's primary energy consumption, reaching nearly 700 billion cubic meters. At the same time, the EU's domestic natural gas production, even if Norway is included, is expected to decrease steadily. As a consequence, these will emerge a substantial gap between European natural gas supplies and demand, and thus the EU will be forced to increasingly rely on imported natural gas. Russia, the EU's most important natural gas supplier, possesses the largest natural gas reserves in the world. However, the country's capability to export natural gas in the future is in doubt due to several internal and external factors. Thus the EU may be forced to diversify its natural gas imports away from Russia in the future and find alternative natural gas sources, suppliers and routes (Makinen, 2010, p.58). Turkey's importance is more understood at this point._

Expectations are that the EU's dependence on imports will continue to grow. The European Commission has repeatedly warned since 2000 (Green Paper) that the EU's net energy import dependency will rise from 50.5% in 2005 to 70% of the EU's total energy requirements by 2030. More precisely, currently 45% of the EU's oil is imported from the Middle East, by 2030 this figure will reach 90%. The situation for gas is similar. Currently, 40% of the EU's gas is imported from Russia, 30% from Algeria and 25% from Norway. By 2030, this will be 60% from Russia and an 80% overall import dependency (Euractiv, 2009). It is a particular cause for alarm that the EU's import dependency on gas from Russia is likely to grow and that gas is increasingly replacing oil as the main energy source (Amineh & Guang, 2012, p.13).

Main Supplier Countries

The EU receives supplies of energy from a variety of countries around the world. It works actively with these countries to get the best deal possible, to increasingly diversify its energy sources, and to prevent disruptions to supply. The main supplier countries are: Norway, Russia, OPEC countries, Central Asia and Caucasus countries, OPEC countries.

- **Norway:** Norway is the world's third largest exporter of oil and gas after Saudi Arabia and Russia. In 2012, it accounted for about 31% of all the EU's natural gas imports and 11% of its crude oil imports. Norway also produces a large amount of hydroelectric power which it will be possible to export to the EU in greater quantities if new grid connections are built. As a member of the European Economic Area (EEA), Norway fully participates in the EU's internal energy market and cooperates closely with the EU on reducing greenhouse gas emissions.
- **Russia:** Russia is one of the EU's largest suppliers of energy. In 2013, it accounted for 39% of its natural gas imports. A number of individual EU countries are also heavily dependent on Russian supplies for certain energy resources, in particular natural gas. Natural gas supplies from Russia often go through transit countries such as Ukraine and Belarus. In 2009, the EU and Russia established an Early Warning Mechanism. This instrument aims to prevent supply interruptions in gas, oil, or electricity and to ensure rapid communication.
- **Central Asia and the Caucasus Countries:** Countries in Central Asia and the Caucasus are often rich in natural resources including oil and gas, which could help the EU diversify its energy supply. Recognizing this potential, the EU has been participating in the development of their energy sectors. To date, the EU has signed Memorandums of Understanding (MoUs) with Azerbaijan, Kazakhstan, Turkmenistan, and Uzbekistan. These MoUs outline steps to further energy cooperation with these countries. The EU also cooperates with 11 partner countries from Eastern Europe, Central Asia, and the Caucasus in the INOGATE program. The program aims to help all partners reduce their dependence on fossil fuels and energy imports, improve security of supply, and fight climate change.

- **OPEC Countries:** The 12 countries that make up the Organization of Petroleum Exporting Countries (OPEC) provide the EU with around 40% of its total crude oil imports. Of these countries, Saudi Arabia, Libya, and Nigeria are the biggest individual suppliers, each having delivered over 8% of the EU's total oil imports in 2012. The EU meets with OPEC annually at ministerial level to discuss a variety of issues including the promotion of more stable oil prices and transparent markets (EU, 2017).

Gas and Oil Supply Routes

A key part of ensuring secure and affordable supplies of energy to Europeans involves diversifying supply routes. This includes identifying and building new routes that decrease the dependence of EU countries on a single supplier of natural gas and other energy resources (EU, 2017).

Opening Up the Southern Gas Corridor

Many countries in Central and South East Europe are dependent on a single supplier for most or all of their natural gas. To help these countries diversify their supplies, the Southern Gas Corridor aims to expand infrastructure that can bring gas to the EU from the Caspian Basin, Central Asia, the Middle East, and the Eastern Mediterranean Basin. Initially, approximately 10 billion cubic meters (bcm) of gas will flow along this route when it opens in 2019-2020. Considering the potential supplies from the Caspian Region, the Middle East and the East Mediterranean, however, the EU aims to increase this to between 80 and 100 bcm of gas per year in the future.

EU actions for expanding the Southern Gas Corridor include:

- Keeping the infrastructure projects needed for the Corridor on the EU's list of Projects of Common Interest. These are projects which can benefit from streamlined permitting process, receive preferential regulatory treatment, and are eligible to apply for EU funding from the Connecting Europe Facility,
- Cooperating closely with gas suppliers in the region including Azerbaijan, Iraq and Turkmenistan,
- Cooperating closely with transit countries including Azerbaijan, Georgia and Turkey,
- Negotiating with Azerbaijan and Turkmenistan on a Trans-Caspian pipeline to transport gas across the Caspian Sea.

The other routes are: Developing the Mediterranean hub and new LNG supplies from North America, Australia, Qatar, and East Africa (EU, 2017).

Turkey's importance can be seen especially on Southern gas corridor.

The progress of Russian backed large scale gas pipeline projects such as Nord Stream and South Stream demonstrate the prominence of member state preferences within the EU, and the influence of their security and economic interests relative to the EU's objectives of diversification of supply sources which both of the Russian backed projects undermine (Maltby, 2013, p.442).

The Western Balkans are a key point for Europe to achieve supply diversity. The Western Balkan states could also provide a route for an extension of the TAP pipeline into their own states and on into Central Europe. Unfortunately, despite accession - or the prospect of accession - to the EU, the Western Balkan states are proving profoundly resistant to EU rules and values. It is argued that the only way to deal with this problem is to launch a multi-prong strategy of reform in addition to the accession process. This would include a wide range of rule of law measures, working particularly with those states who show the greatest willingness to reform. In parallel, it is argued, it would be necessary to upgrade significantly the operation of the European Energy Community. Full 'Europeanization' of the Energy Community would extend the EU energy *acquis* to the West Balkan markets. Applying the EU energy *acquis* in full would open up those markets and provide for EU levels of surveillance, thereby making them more attractive to investors. A more stable and secure legal framework which attracted investment would also ensure that investments, which added to the Union's energy security, actually took place (Riley, 2014).

The Western Balkans could play a significant role in enhancing the Union's energy security at a time of increasing insecurity. However, the EU has to deal with the problems of legal and energy market dysfunctionality seriously and develop a credible, and not just a reactive, response. The Union potentially has the tools in EU law, the accession process and the Energy Community Treaty. It needs to develop them and then be prepared to apply them. Meanwhile, the local elites have to recognize that, by entering the EU or the accession process, they have made a decisive choice to which they must now follow through on (Riley, 2014).

The interlinkage between globally designed traditional energy security concepts – that rely just on economic factors and "market-strategies" – and domestic as well as regional political stability demands new thinking with regard to both energy supply security and foreign policies. While the Russian–Ukrainian gas conflict in January 2006, has forced energy security up the European energy and foreign policy agendas, the EU-27 member states have largely failed to forge a coherent European energy security and energy foreign policy strategy after their Spring summit of 2007 because its declared political solidarity was still lacking. But the 2nd Strategic Energy Review of November 2008 has recommended new initiatives to overcome this deficiency by promoting concrete infrastructure and other projects for enhancing Europe's supply security and its political solidarity as part of a common energy (foreign) policy. If the EU is able to implement the March 2007 and November 2008 decisions, then EU oil and gas demand will drastically reduce and freeze at current levels. In this case, Putin's energy policies by using Russia's energy resources and pipeline monopolies as a political instrument to enforce its economic and geopolitical interests will be proved to be self-defeating in Russia's long-term strategic interests. It will reduce Gazprom's gas exports to a much smaller EU gas market than originally forecasted as the result of a deliberate EU policy of decreasing its overall gas demand and by diversifying its gas imports (Umbach, 2010).

The Commission told the Guardian newspaper that energy efficiency would be addressed separately, potentially with a specific target on efficiency savings. Connie Hedegaard, the EU climate commissioner, said: "Energy security and the fight against climate change are inseparable: without climate policies there can't be energy security. This is why energy efficiency and renewable will continue to be two key ingredients as they are good both for the climate and energy security. In other words, we invest the money here in Europe instead of sending it to Putin's Russia and other fossil fuel providers outside Europe''. Reducing the over dependency on Russia and sourcing new gas supplies were cited as part of a "long list of homework" for the EU by Oettinger. Increasing indigenous energy production was also listed as a priority by the commission. But as well as including renewable energy, which has been the main focus in the past, this would now explicitly include "sustainable production of fossil fuels", which would be expected to include shale gas (Guardian, 2014). But, according to Makinen, after shale gas, EU will still need more energy.

The asymmetry in import dependency among the member states, the preference of member states for a certain energy mix, "the member state" competitive position in world markets and different foreign and security approaches, will make "speaking with one voice" on energy policies a lot harder to achieve than on climate change matters (Van der Linde, 2007).

THE EUROPEAN UNION ENERGY POLICY

The vitality of the energy sector as well as its effects on economic and political decisions bring the energy issue to the center of EU policies (Sigot, 2013, p.10).

In the beginning, energy policy was at the heart of the institutions out of which the EU eventually evolved. Indeed, it could be said that the EU began with a common energy policy. With the passage of time, however, these initial institutional arrangements became less and less relevant to the energy needs and concerns of the Member States, and for many years they were never replaced or supplemented by more relevant arrangements. This marginalization of energy policy stands in stark contrast to the considerable progress that was made toward European integration in other policy areas, especially the closely related common market. Until and through much of the 1990's, energy policy remained largely an orphan of the integration process (Duffield & Birchfield, 2011, p.2).

Energy security is one of the main targets of energy policy. However, the term has not been clearly defined, which makes it hard to measure and difficult to balance against other policy objectives. This can be avoided by more clearly distinguishing between security of supply and other policy objectives. This leads us to the definition of energy security as the continuity of energy supplies relative to demand (Winzer, 2012).

The concept of energy security is widely used, yet there is no consensus on its precise interpretation. There is no one ideal indicator, as the notion of energy security is highly context dependent. Rather, applying multiple indicators leads to a broader understanding (Kruyt et al., 2009, p.2166).

A discussion on the issue of security of supply is inevitably linked to an analysis of the development of an energy policy for Europe (Haghighi, 2007, p.37). The period between 1973 and 1986 marked the beginning of the establishment of an energy policy that took the externalities of energy security into account (Haghighi, 2007, p.61). The Commission believed that the energy market is global and each player can affect the others. It was therefore accepted that energy supply could only be effectively handled if it were seen and treated as an international affair and not merely a national one (Haghighi, 2007, p.50).

EU energy security challenges have changed dramatically in the past 15 years. On the one hand, the tensions between Russia and the EU, aggravated by the annexation of the Crimea, are undermining their historical partnership on energy; on the other hand, deep transformations in the structure of global energy supply and demand, triggered by technological advances and geopolitical and economic dynamics -e.g. the US shale gas revolution and the development of Liquefied Natural Gas worldwide- are leading the EU to rethink its energy security strategy. Distinctive and new politics of energy security is definitively emerging in the EU (Prontera, 2017).

The EU aims to have a balance between objectives while establishing energy policies. Energy poverty has become a rising issue in the European Union (EU), especially in new member states (Lenz-Grgurev, 2017). New market design, security of supply and environment policy are more equilibrated (Sencar, et al., 2014, p.117).

Energy has been important since the EU was first established. The importance of coal as an energy source on ECSC which is considered as the beginning of EU, and EURATOM which is based on nuclear energy are the indications of this subject. Energy is a matter that does not take place until the 2000s in the founding agreements of the EU, although the EU is always among the main agenda items with the two oil crises of the 1970s. The issue of energy, for the first time at the end of 2009, which entered into force with the Treaty of Lisbon entered into both the TEU (Treaty on European Union) and TFEU_(Treaty on Functioning of the European Union) determination of energy policy to the first as the put into_a new title. The volatility in energy prices, especially since the beginning of the 2000s, and the supply insecurity of Russia, the largest energy importer in the EU, have been important factors. For the EU member states, energy security was considered to be the least important issue.

However, the increasing pressure of new members of the EU for a common energy policy and the continuing risk of supply interruptions in the gas coming from Russia have increased the importance of energy security in recent years (Eris, 2016, p.636).

In this respect, it is aimed to provide safe and sustainable energy at a more affordable price by preventing member countries from solely hacking. While this approach was believed to give EU member states an advantage in price bargaining, it was also thought to provide a basis for cooperation based on trust and continuity with producers and transit country partners. The most important aspect of the new energy strategy is to link energy with security and foreign policy (Çelikpala, 2013, p.397). The EU legislation lays out the basis for a competitive, qualified, diverse and cost effective energy market. This would pave the way for a more organized and efficient way of reaching various sectors for the use of energy. According to EU legislation, common rules apply to market access, organization, operation, tender procedures and authorization methods. Liberalization in the electricity and gas sectors provides opportunities for investments by private sector actors.

The energy sector is responsible for about 80% of the greenhouse gas emissions within the EU. In order to address the challenges of climate change, developing sustainable energy policies is one of the main components of the EU's energy policies.

Most European countries have concerns about the integration of large amounts of renewable energy sources (RES) into their electric power systems, and this is currently a topic of growing interest. In January 2008, the European Commission published the 2020 package, which proposes committing the European Union to a 20% reduction in greenhouse gas emissions, to achieve a target of deriving 20% of the European Union's final energy consumption from renewable sources, and to achieve 20% improvement in energy efficiency both by the year 2020. Member states have different individual goals to meet these overall objectives, and they each need to provide a detailed roadmap describing how they will meet these legally binding targets (Molina-Garcia, et al., 2017).

The European Council endorsed greenhouse gas reduction, renewable energy and energy efficiency targets for 2030 on the 23-24 October 2014 *(2030 framework for climate and energy policies)*. The framework requires the EU to collectively reduce its emissions by at least 40% compared to 1990 levels. It also sets the goals to achieve at least 27% renewable energy in the EU's final energy consumption and a 27% or greater improvement in energy efficiency by 2030.

Regarding the priority of resetting the energy policy of the Union as a new European Energy Union by reforms and reorganization, firstly "A Framework Strategy for a Resilient Energy Union with a Forward-Looking Climate Change Policy" was published in 25 February 2015 and a Vice-President of the European Commission for Energy Union was appointed by the European Commission.

The "Energy Union Framework Strategy" has five mutually-reinforcing and closely interrelated dimensions designed to bring greater energy security, sustainability and competitiveness:

- Energy security,
- A fully integrated European energy market,
- Energy efficiency contributing to moderation of demand,
- Decarbonizing the economy and
- Research, innovation and competitiveness.

There are 15 actions under the "European Energy Union Framework Strategy" and there are time schedules for regarding actions in the appendix of the Strategy. The EU shows quite strong political will and determination to form a "common energy policy" by closely following the implementation of the strategy and the action plan, also by designating a vice president in charge of Energy Union for this issue (EU, 2017).

European energy policy aims, in a more integrated way, to secure affordable energy supplies, respect market mechanisms, promote energy efficiency and protect the environment (Sigot, 2013, p.12).

The European Union's "Road to the Paris Agreement" did not include unilateral legislative instruments such as those it adopted before previous conferences pursuant to its exemplarity policy in climate negotiations. Rather, the Member States agreed in principle on new targets, some binding on both the European Union and Member States, such as those concerning greenhouse gas emissions from both emissions trading scheme and non-emissions trading scheme sectors, and others binding only on the European Union, concerning renewable energy and energy efficiency. In order to allow the European Union to meet its targets in the absence of corresponding national commitments, the Commission proposes an Energy Union Governance as a substitute for an obliterated energy policy which the Treaty does not allow to interfere with a Member State's energy mix (Thieffry, 2017).

Although the internal energy market can be seen as a convergence of interests among Member States, this integration still does not provide a sufficient basis to clinch a parallel movement on external energy policy aspects (Metais, 2013, p.23). But the EU is increasing its role in coordinating and financing the development of renewable energy and the storage and use of emergency energy supplies. Although

Member States and their energy industries appear likely to retain absolute authority in determining which energy mix makes the most sense for individual countries, the EU has set binding EU-wide targets in some areas and may continue this trend (Belkin, 2008, p.100).

The European Union (EU) is at a crossroads; it can create a competitive Europe-wide energy system that responds to energy security and climate concerns whilst stimulating job creation and growth, or it can become increasingly un-competitive, strangled by high energy imports (the EU imports 53% of the energy it consumes) and an underinvested, underperforming energy system. The EU's negative trade balance is largely due to its very high imports of fossil fuels rather than a lack of industrial competitiveness in its manufacturing sectors. In fact, despite the economic crisis, Member States have shown that energy independence and investments in renewables can reduce national trade deficits. A reduction in energy dependency and competitive, integrated energy supplies offers Europe the political and economic opportunity to unite Member States around a more integrated single energy market and to solve its energy tri-lemma of costs, security and decarbonization. It presents Europe with the chance to transform and finance critical energy infrastructure renovations, including energy efficiency improvement of the installed base, as well as proper grid development such as distribution, transmission and storage capacity upgrades (GGP, 2014, p.1).

The Energy Union is positioned to support the energy security of Europe if proper investments are made. Increasing the energy infrastructure connections between EU countries will ensure that the Energy Union is supported during times of disruption (GSSR, 2017).

New gas supplies for Europe will come primarily from North Africa, from Russia, and from the Caspian region (by pipeline via Russia and probably via Turkey, Greece and the Balkans) (Gault,2002, p.5). In general, the periphery countries from which Europe needs to draw additional volumes of oil and gas, and upon which Europe will become increasingly energy dependent over the next two decades, have significantly lower incomes per capita than do European countries. Many of these countries – especially in North Africa – have high rates of population growth. Large amounts of investment will be required to raise productivity and reduce unemployment in these countries. Yet, in the eight-year period from 1992 through 1999, Russia and North Africa together received only 1.5% of EU-15 outward direct investment – a miniscule amount considering the important roles these countries will play in Europe's energy future. EU policies designed to make these and other energy periphery countries more attractive to investors will be an important element in ensuring energy security over the long run (Gault, 2002, p.12).

Hydrogen and fuel cell technologies could form an integral part of future sustainable energy systems. This will contribute to improving Europe's energy security and air quality, while lessening climate change. The European Union prefers to use renewables, mainly hydrogen rather than fossil fuels, because of the decreasing supply of fossil fuels and the increasing demand for renewable energy sources, especially hydrogen. Hydrogen is a clean energy vector. It can be produced from a wide variety of primary energy sources. It is possible to decarbonize fossil fuels by carbon capture, allowing for the production of hydrogen from these traditional fuels with negligible carbon emissions. But, more importantly, hydrogen produced through a range of renewable primary energy sources such as wind, biomass, and solar energy is ideal for gradually replacing fossil fuels. Some hydrogen platforms can introduce a coherent European Union strategy in order to develop the use of hydrogen and hydrogen production technologies while gaining worldwide leadership (Yolcular, 2009). Much of EU's external energy policy remains to be created and coordinated (Sigot, 2013, p.25).

TURKEY'S ENERGY POLICY

The main goal of Turkey's energy policy is to supply energy while realizing economic growth and supporting social development by maintaining a steady flow of reliable, cost effective and environmentally friendly forms of energy resources. Turkey has been restructuring its energy market in order to transform it into a competitive and transparent market structure. These restructuring efforts also take into account the use of potential domestic and renewable energy resources, the incorporation of nuclear energy and energy efficiency in the legislative and technical studies. Within this scope, the main priorities of Turkey's energy policy are:

- Availability of energy for consumers in terms of cost, time and amount,
- To mobilize public and private means within the framework of free market practices,
- To reduce dependence on outside sources,
- To increase the regional and global impact of Turkey in the energy field,
- To ensure a diversity of routes, technology and resources,
- To ensure the maximum use of renewable energy resources in sustainable and efficient ways,
- To increase energy efficiency,
- To minimize any environmental impact in the production and use of energy.

The EU plays a very important role in the global energy sector. Today, oil and natural gas are the most important sources of energy in the EU member states. Oil and gas reserves are finite sources, which are concentrated in certain geographic regions of the world. EU Member States rely heavily on imports of oil and natural gas. It is expected that the demand for natural gas will increase 70 - 80% by the year 2030. The Southern Gas Corridor, which provides cooperation opportunities for Turkey and the EU, plays a vital role for energy security in these regions. Completion of the gas transmission and interconnection projects will help integrate both Turkey and the EU and contribute to the diversification of energy sources.

In this regard, natural gas interconnection projects are carried out between Turkey and the EU and studies on completion of legal and regulatory framework necessary for trade of energy sources. Turkey's geographic position as a bridge between the energy consuming countries to its West and natural gas producing countries to its East enhances Turkey's importance in terms of security of supply.

Turkey has realized projects such as the Turkey-Greece natural gas interconnection, Baku-Tiflis-Ceyhan, Kerkuk-Yumurtalik petroleum pipelines which have been completed and there are currently ongoing projects including TAP, TANAP (Trans Anatolian Natural Gas Pipeline), Baku-Tiflis-Erzurum natural gas pipeline, Samsun-Ceyhan petroleum pipelines.

The most important component of the Southern Gas Corridor "Trans-Anatolian Natural Gas Pipeline Project" transmitting Caspian gas to Europe through Turkey, is the biggest pipeline project undertaken in Turkey. The construction work of the TANAP project is ongoing. The project is expected to bring natural gas produced from Azerbaijan's Shah Deniz-2 gas field to Turkey in 2018 and to Europe in 2020. After the decisions on the TANAP project, Nabucco Pipeline project, which was planned to transit gas from the Caspian to Europe and Austria, has been cancelled and transformed into a small-scale project called Nabucco West which is expected to transit gas from Turkey to Europe. Within this context, Nabucco West and Trans Adriatic Pipeline (TAP) were two alternative routes to transit gas from TANAP to Europe. As a result, at a press conference held on 28 June 2013 the Shah Deniz Consortium (SDC) confirmed that TAP has been selected to transport gas from TANAP to Europe (MEU, 2017).

Turkey is a candidate for EU membership and, given that it is a major energy importing country, an energy dialogue with Turkey is particularly important. It is also a potential transit country for natural gas to flow from the Caucasus and Central Asia into the EU. The EU cooperates with Turkey to enable the integration of the Turkish gas and electricity markets into its internal energy market.

In addition to being a candidate for EU accession, Turkey is also a major energy importer and potentially significant transit country for natural gas supplies flowing from the Caucasus, the Eastern Mediterranean, the Middle East, and Central Asia to

Europe. To this end, Turkey's role is vital in the development of the Southern Gas Corridor through the construction of the Trans-Anatolian pipeline. The EU cooperates with Turkey to enable the integration of Turkish gas and electricity markets with the internal energy market of the Union. Additionally, Turkey is an observer in the Energy Community and the European Network of Transmission System Operators for Electricity (ENTSO-E) (EU, 2017).

According to Progress Report 2016, Chapter 15: Turkey is moderately prepared in this chapter. Good progress continued in the area of security of supply, electricity and the renewable energy sector. Turkey is at an advanced stage of alignment in the electricity sector. There has been significant progress regarding security of supply. Progress continues on the Trans-Anatolian Pipeline project (TANAP): all major procurement contracts have already been awarded with a view to finalizing the project in 2018, well ahead of the original schedule. Efforts towards expanding TANAP to bring new gas sources from the Caspian and/or the Middle East to Turkey and Europe continue. The feasibility and technical studies are ongoing for the project of a bidirectional interconnector between Turkey and Bulgaria. On the other hand, stalled negotiations on the 'Turkish Stream' were re-launched in August 2016 following the normalization of bilateral relations between Turkey and Russia. An intergovernmental agreement on the 'Turkish Stream' was signed on 10 October 2016 in the presence of the Turkish and Russian Presidents on the margins of the 23rd World Energy Congress in Istanbul. No progress was made on a transparent gas transit regime. Good progress continued on electricity networks (MEU, 2017).

Some speeches were made which are important from point of the subject concerned by some speakers in the congress titled "Energy Platform 2010: The Role of Turkey in Energy Security of Europe" organized by Konrad Adenauer Stiftung in Turkey in 2010:

- Turkey has developed "energy corridor" and "energy terminal" concepts within the context of "global energy supply security" and has been a reliable partner in international pipeline projects compatible with these concepts. The importance of the NABUCCO project in terms of security of energy supply in the EU countries is obvious (Yazar, 2010, p.17).
- With its energy policies, which have been followed in recent years, and the credible partnership with the existing projects it has been involved so far, Turkey has proven that it can play a particular and important role in providing and securing energy for Europe (Yazar, 2010, p.18).
- Turkey is an important player for Europe's energy policy, a central energy corridor and an important negotiating partner (Bareiss, 2010, p.25).
- Lectures and conferences were held on developments in Russia and to develop their own energy supply security, that is to say, to ensure that energy is

delivered on time and adequately as net importer. And as a result, the "Energy Charter Treaty" appeared at the end of 1994, and the "Energy Charter Treaty" entered into force after the signing by the relevant parliaments after 1998. Turkey is part of the "Energy Charter Treaty" (Demirbilek, 2010, p.28). The second point is the issue of South East Europe energy union. This agreement has not yet been signed by Turkey (ibid, p.30).

- When it comes to energy security, there are two specific components. One of these is resource diversity and the other is resource reliability (Suphi, 2010, p.37). East of Turkey, there are sources of gas in Turkmenistan, Iran, Iraq and Egypt as well. Turkey's energy strategy foresees that at some point the gas coming from these sources will pass through Turkey to Europe. For this reason, it is another opportunity for diversity to provide the security of work (Suphi, 2010, p.39).
- As an energy transit country for many member states as it is for the EU's energy foreign policy, Turkey plays an important role as an energy transit country because the pipelines from the Black Sea region, the Caspian Sea and the Middle East pass through the country and reach the Europe. It is still important to link these roots to national energy networks in order to use in energy supply of the EU (Schuss, 2010, p.42).
- Now, looking at the various areas of the energy heading, it is noticed that there is a target harmonization between the EU and Turkey, although not in all the questions. These harmonies are already enabling the regulation and infrastructure to integrate more and more. Therefore, it is difficult to understand the postponement of the opening of the energy chapter (Schuss, 2010, p.45).
- It should be considered that Turkey is a political mediator in the region as an emerging economic power, and that the EU has many opportunities for energy transfer as a revolving platform and that can be considered as important point on negotiations (Schuss, 2010, p.46).
- As the title of today's event shows, Turkey has the potential to play an important role in the EU's energy security (Tvarozkova, 2010, p.49).
- The Energy Community Settlement Agreement, signed in October 2005, extends the EU's internal energy market model to include Southeast Europe and beyond. This treaty aims to provide a stable investment environment based on the rule of law beyond EU borders. Currently, the Parties to the Treaty are Albania, Bosnia and Herzegovina, Croatia, former Yugoslav Republic of Macedonia, Montenegro, Serbia and Moldova, which signed the contract with the European Union. Four countries, namely Georgia, Norway, Turkey and Ukraine, have observer statuses (Tvarozkova, 2010, p.51).

- The role of Turkey in European energy security and its integration into the European energy market is extremely high. The process certainly involves many areas of struggle, but there is no doubt that both Turkey and the EU have unique opportunities to increase energy security (Tvarozkova, 2010, p.52).
- 70% of the world's crude oil and natural gas reserves are located on the right side of an axis drawn through Ankara in the North-South direction, to the left of the axis, are countries that consume more than 50% of the world's oil and gas, especially in Europe. For this reason, Turkey is the center of Eurasia geography (Çapanoğlu, 2010, p.63).
- As hydrocarbon resources are located in the East of Turkey and energy consumers are located in the west, it has been revealed that the shortest and most convenient route is from Turkey (Çapanoğlu, 2010, p.63).
- When Turkey's existing and planned projects are examined, it can be seen that Turkey's energy-related goal is not only to guarantee its own energy demand, but also to get great responsibility to transfer energy to world markets (Çapanoğlu, 2010, p.67).
- Turkey has become a strategic energy partner of the Federal Republic of Germany and the Europe in the energy sector (Çapanoğlu, 2010, p.70).
- Turkey has become an important energy junction and the fact that it can send energy to the European Union and the Federal Republic of Germany thanks to Azerbaijanis, which is the sister nation of the Turks (Çapanoğlu, 2010, p.71).
- A successful international energy policy is linked to a successful national energy policy. Turkey may become one of the top 10 economies in the world in the coming years, but without energy it will be very difficult (Bemberg, 2010, p.78).
- From 2013 onwards, the European Union is planning to adopt a method of procurement from the carbon quota (Carbon Allowances) to tender (Auctioning). What will be the long-term objectives of our country is especially important for the investor who will invest in the domestic power plant based on lignite. Because, depending on the applications, the marginal production costs may be very serious for the electricity generation plants based on carbon source (Aydın, 2010, p.83).
- If the energy junction wants to play an active participatory role in the new international energy business, it must be able to unite and distribute the energy supply. In particular, it should invest in infrastructure and transport lines. Since it is not possible for an individual country to realize these investments alone, close cooperation with buyer and supplier countries should be made. Another point raised is that Turkey does not only want to play the role of a transit

> country or energy junction, but also an energy producer. This is important for countries like Europe and Germany because it creates investment and cooperation opportunities. The subject of renewable energy is also one of the topics discussed today. In this context, it is important that Turkey is included in the European Union's energy network (Senkyr, 2010, p.86).

The recent reconciliation with Turkey is also a case for concern in Europe. A deal between Turkey and Israel for the East Med gas might boost Turkey's role as an energy transit super state and impose pressure on Europe if the Turkish question is not to be resolved soon, and if rapprochement with Russia pulls Turkey away from the EU and exploits the (many) cleavages in EU-Turkey relationship (Popkostova, 2016).

CONCLUSION

The European Union is one of the big player in the world. The EU is the 4th biggest energy consumer in the world. The EU's total energy production is decreasing. Import dependency is increasing. The EU receives supplies of energy from a variety of countries around the world. It works actively with these countries to get the best deal possible, to increasingly diversify its energy sources, and to prevent disruptions to supply. Many countries in Central and South East Europe are dependent on a single supplier for most or all of their natural gas. To help these countries diversify their supplies, the Southern Gas Corridor aims to expand infrastructure that can bring gas to the EU from the Caspian Basin, Central Asia, the Middle East, and the Eastern Mediterranean Basin. EU actions for expanding the Southern Gas Corridor include cooperating closely with transit countries including Azerbaijan, Georgia and Turkey.

Expectations are that the EU's dependence on imports will continue to grow. The European Commission has repeatedly warned since 2000 (Green Paper) that the EU's net energy import dependency will rise from 50.5% in 2005 to 70% of the EU's total energy requirements by 2030. The EU's expected natural gas demand has been growing and a similar trend is expected to continue in the future. In 2030, natural gas is expected to account for 30% of the EU's primary energy consumption. Russia, the EU's most important natural gas supplier, possesses the largest natural gas reserves in the world. However, the country's capability to export natural gas in the future is in doubt due to several internal and external factors. Thus the EU may be forced to diversify its natural gas imports away from Russia in the future and find alternative natural gas sources, suppliers and routes. Turkey's importance is more understood at this point.

In the beginning, energy policy was at the heart of the institutions out of which the EU eventually evolved. Indeed, it could be said that the EU began with a common energy policy. The issue of energy, for the first time at the end of 2009, which entered into force with the Treaty of Lisbon entered into both the TEU (Treaty on European Union) and TFEU (Treaty on Functioning of the European Union) determination of energy policy to the first as the put into a new title. Energy security is one of the main targets of energy policy.

Energy security can be addressed directly through energy-specific policies, but ultimately energy security is inextricable from broader economic and foreign policy challenges.

The main priorities of Turkey's energy policy are; reduce dependence on outside sources, increase regional and global impact of Turkey in the energy field, ensure the diversity of routes, technology and resources. The completion of gas transmission and interconnection projects will help integrate both Turkey and the EU and contribute to the diversification of energy sources. Turkey's geographic position as a bridge between the energy consuming countries to its West and natural gas producing countries to its East enhances Turkey's importance in terms of security of supply.

New gas supplies for Europe will come primarily from North Africa, Russia, and from the Caspian region. The periphery countries from which Europe needs to draw additional volumes of oil and gas, and upon which Europe will become increasingly energy dependent over the next two decades, have significantly lower incomes per capita than do European countries. Large amounts of investment will be required to raise productivity and reduce unemployment in these countries. EU policies designed to make these and other energy periphery countries more attractive to investors will be an important element in ensuring energy security over the long run.

The Western Balkans are a key point for Europe to achieve supply diversity. The Western Balkan states could also provide a route for an extension of the TAP pipeline into their own states and on into Central Europe. Turkey has good relations with these countries.

Considering all this, it is understood that Turkey is an indispensable partner for the EU, and it is on a crucial route to the provision of energy to EU and that it has an important role in closing this deficit. This situation is shaped more day by day with the policies and the projects that Turkey has implemented. Turkey will be a complementary partner for EU, not a competitive one. This feature will improve Turkey's compatibility to EU on energy.

REFERENCES

Amineh, M. P., & Crijns-Graus, W. H. J. (2014). Rethinking EU energy security considering past trends and future prospects. *Perspectives on Global Development and Technology, 13*(5-6), 757–825. doi:10.1163/15691497-12341326

Amineh, M. P., & Guang, Y. (2012). *The Geopolitics of Energy Paths of China and the European Union. Secure Oil and Alternative Energy*. Leiden, Boston: Brill.

Aydın, M. Z. (2010). *Yatırımcıların İhtiyaçları ve Liberizasyon Sürecinde Arz Güvenliği. In Enerji Platformu 2010: Avrupa'nın Enerji Güvenliğinde Türkiye'nin Rolü* (pp. 79–84). Ankara: Konrad-Adenauer-Stiftung.

Bareiss, T. (2010). *Enerji Güvenliği-İşbirliği ve Çeşitlendirme. In Enerji Platformu 2010: Avrupa'nın Enerji Güvenliğinde Türkiye'nin Rolü* (pp. 19–25). Ankara: Konrad-Adenauer-Stiftung.

Belkin, P. (2008). The Europen Union's energy security challenges. *Connections: Quarterly Journal, 7*(1), 76–103. doi:10.11610/Connections.07.1.05

Bemberg, N. (2010). Bir Enerji Konjonktürü Analizi: Türkiye'de Yabancı Enerji Yatırımlarının Geleceği. In Enerji Platformu 2010: Avrupa'nın Enerji Güvenliğinde Türkiye'nin Rolü (pp. 73-78). Academic Press.

Birchfield, V. L., & Duffield, J. S. (2011). *Toward a Common European Union Energy Policy*. Springer. doi:10.1057/9780230119819

Buşe, M. (2017). Energy Security-Priority of the European Union. *International Scientific Conference "Strategies XXI: Bucharest,"* 170-175.

Çapanoğlu, İ. K. (2010). Avrasya Enerji Arenasının Merkezi Türkiye. In *Enerji Platformu 2010: Avrupa'nın Enerji Güvenliğinde Türkiye'nin Rolü* (pp. 61-67). Ankara: Konrad-Adenauer-Stiftung.

Çelikpala, M. (2013). *Enerji Politikası. In Türkiye-Avrupa Birliği İlişkileri* (pp. 385–406). Ankara: Turhan.

Chester, L. (2010). Conceptualizing energy security and making explicit its polysemic nature. *Energy Policy, 38*(2), 887–896. doi:10.1016/j.enpol.2009.10.039

Demirbilek, S. (2010). *Avrupa'nın Enerji Güvenliği Bağlamında Türkiye'nin Konumu. In Enerji Platformu 2010: Avrupa'nın Enerji Güvenliğinde Türkiye'nin Rolü* (pp. 27–35). Ankara: Konrad-Adenauer-Stiftung.

Eriş, Ö. Ü. (2016). *Enerji Politikası. In Avrupa Birliği* (pp. 617–637). Ankara: Seçkin.

European Commission. (2011, November). *Report From The Commission To The European Parliament, The Council And The European Economic And Social Committee, Implementation of the Communication on Security of Energy Supply and International Cooperation and of the Energy Council Conclusions of November 2011*. Author.

European Commission. (2016). EU Reference Senario 2016. Retrieved from https://ec.europa.eu/energy/sites/ener/files/documents/ref2016_report_final-web.pdf

European Commission Progress Report Turkey. (2016). Retrieved from http://www.ab.gov.tr/files/5%20Ekim/20161109_report_turkey.pdf

European Union, Energy Supplier Countries. (2017). Retrieved from http://ec.europa.eu/energy/en/topics/imports-and-secure-supplies/supplier-countries

European Union, Gas and Oil Supply Routes. (2017). Retrieved from http://ec.europa.eu/energy/en/topics/imports-and-secure-supplies/gas-and-oil-supply-routes

European Union, International Cooperation on Energy. (2017). Retrieved from http://ec.europa.eu/energy/en/topics/international-cooperation

European Union, Overview of the secondary EU legislation that falls under the legislative competence of DG ENER and that is currently in force. (2016). Retrieved from https://ec.europa.eu/energy/sites/ener/files/documents/Overview%20of%20ENER-related%20legislation%20%28by%20policy%20areas%29%20update%2014.03.2016.pdf

Gault, J. (2002). *The European Union: Energy Security and the Periphery*. Occasional Paper Series. No.40.

Georgetown Security Studies Review (GSSR). (2017). *Opportunities and Challenges for Europe's Energy Union*. Feb.26, 2017. Retrieved from http://georgetownsecuritystudiesreview.org/2017/02/26/opportunities-and-challenges-for-europes-energy-union/

Green Growth Platform (GGP). (2014). *Advisory Council I: Report to Ministers Energy Security & the 2030 Climate & Energy Package, The Big Picture: A Low Carbon Energy Union Targets Security, Climate, Jobs & Growth*. University of Cambridge.

Guardian. (2014). *Shale and non-Russian gas imports at heart of new EU energy strategy*. Retrieved from https://www.theguardian.com/environment/2014/may/28/shale-gas-russia-eu-renewables-ukraine-crisis

Haghighi, S. S. (2007). *Energy Security: The External Legal Relations of the EU with Major Oil and Gas Supplying Countries*. Bloomsbury Publishing.

International Energy Agency. (2016). *World Energy Outlook*. Paris, France: IEA.

Kruyt, B., Vuuren, D. P., Vries, H. J. M., & Groenenberg, H. (2009). Indicators for energy security. *Energy Policy*, *37*(6), 2166–2181. doi:10.1016/j.enpol.2009.02.006

Lenz, N. V., & Grgurev, I. (2017). Assessment of Energy Poverty in New European Union Member States: The Case of Bulgaria, Croatia and Romania. *International Journal of Energy Economics and Policy*, *7*(2), 1–8.

Makinen, H. (2010). The future of natural gas as the European Union's energy source – risks and possibilities. *Pan-European Institute*. Retrieved from www.tse.fi/pei

Maltby, T. (2013). European Union energy integration: A case of European Commission policy enterpreneurship and increasing supranationalism. *Energy Policy*, *55*, 435–444. doi:10.1016/j.enpol.2012.12.031 PMID:24926115

Metais, R. (2013). *Ensuring Energy Security in Europe: The EU between a Market-based and a Geopolitical Approach. EU Diplomacy Paper (03 / 2013)*. College of Europe.

Ministry of EU (MEU). (2017). Retrieved from http://www.ab.gov.tr/_80.html

Molina-Garcia, A., Hansen, A. D., Muljadi, E., Gevorgian, V., Fortmann, J., & Gomez-Lazaro, E. (2017). *International Requirements for Large Integration of Renewable Energy Sources*. London, UK: The Institution of Engineering and Technology. Retrieved from http://digital-library.theiet.org/content/books/10.1049/pbpo098e_ch2

Popkostova, Y. (2016). *Disruptive change: Europe's energy security and the US LNG*. Retrieved from https://guests.blogactiv.eu/2016/09/08/disruptive-change-europes-energy-security-and-the-us-lng/

Prontera, A. (2017). *The New Politics of Energy Security in the European Union and Beyond: States, Markets, Institutions*. Routledge.

Riley, A. (2014). *The Western Balkans and EU energy security: protecting Europe's Flank*. The Institute for Statecraft. Retrieved from http://www.statecraft.org.uk/research/western-balkans-and-eu-energy-security

Schuss, H. (2010). *Türkiye – Avrupa Ortak Enerji Politikası: Hedeflerde Uyum, Çatışma veya Siyasi Blokaj. In Enerji Platformu 2010: Avrupa'nın Enerji Güvenliğinde Türkiye'nin Rolü* (pp. 41–47). Ankara: Konrad-Adenauer-Stiftung.

Sencar, M., Pozeb, V., & Krope, T. (2014). Development of EU energy market agenda and security of supply. *Energy, 77*, 117–124. doi:10.1016/j.energy.2014.05.031

Senkyr, J. (2010). *Sonuç. In Enerji Platformu 2010: Avrupa'nın Enerji Güvenliğinde Türkiye'nin Rolü* (pp. 85–87). Ankara: Konrad-Adenauer-Stiftung.

Sigot, M. (2013). *Energy Security and the EU: between independency priorities, strategic vulnerabilities and sustainability needs* (Master Thesis). Stockholm University.

Suphi, D. (2010). *Avrupa'nın Enerji Güvenliğinde Hazar'daki Doğalgaz Projelerinin Rolü. In Enerji Platformu 2010: Avrupa'nın Enerji Güvenliğinde Türkiye'nin Rolü* (pp. 37–40). Ankara: Konrad-Adenauer-Stiftung.

{SWD(2013) 334 final}. (2013). Retrieved from http://eur-lex.europa.eu/legal-content/EN/TXT/PDF/?uri=CELEX:52013DC0638&from=EN(COM(2013) 638 final: report from the Commission to the European Parliament, the Council and the European Economic and Social Committee, Implementation of the Communication on Security of Energy Supply and International Cooperation and of the Energy Council Conclusions of November 2011)

Tekin, A., & Williams, P. A. (2011). *Geo-Politics of the Euro-Asia Energy Nexus: The European Union*. Palgrave Macmillan. doi:10.1057/9780230294943

Thieffry, P. (2017). *Environmental Protection and European Union Energy Policy: Energy Transition After the Paris Agreement*. Springer.

Tvarozkova, E. (2010). *Türkiye ve Avrupa Enerji Güvenliği – İmkanlar ve Güçlükler. In Enerji Platformu 2010: Avrupa'nın Enerji Güvenliğinde Türkiye'nin Rolü* (pp. 49–52). Ankara: Konrad-Adenauer-Stiftung.

Umbach, F. (2010). Global energy security and the implications for the EU. *Energy Policy, 38*(3), 1229–1240. doi:10.1016/j.enpol.2009.01.010

Van der Linde, C. (2007). *External energy policy: Oil fears and new dilemmas in a larger Union*. INIS.

Winzer, C. (2012). Conceptualizing energy security. *Energy Policy, 46*, 36–48. doi:10.1016/j.enpol.2012.02.067

Yazar, Y. (2010). *Açılış ve Selamlama. In Enerji Platformu 2010: Avrupa'nın Enerji Güvenliğinde Türkiye'nin Rolü* (pp. 13–18). Ankara: Konrad-Adenauer-Stiftung.

Yolcular, S. (2009). Hydrogen Production for Energy Use in European Union Countries. In Recovery, Utilization and Environmental Effects (pp. 1329-1337). Taylor & Francis.

ADDITIONAL READING

Anger, N., & Zannier, L. (2017). *A new era of EU energy policy? Delivering on the Energy Union by National Plans.* Commentary. Brussels: CEPS. Available at: https://www.ceps.eu/publications/new-era-eu-energy-policy-delivering-energy-unionnational-plans

Bassi, S., Averchenkova, A., & Carvalho, M. (2017). *The credibility of the European Union's efforts to decarbonise the power sector.* Policy report. London: Grantham Research Institute on Climate Change and the Environment and Centre for Climate Change Economics and Policy, London School of Economics and Political Science. Available at: http://www.lse.ac.uk/GranthamInstitute/publications. 19.06.2017 http://www.lse.ac.uk/GranthamInstitute/publications/

Doda, B., & Fankhauser, S. (2017). *Energy policy and the power sector in the long run. Working paper.* London: Grantham Research Institute on Climate Change and the Environment and Centre for Climate Change Economics and Policy, London School of Economics and Political Science; Available at http://www.lse.ac.uk/GranthamInstitute/publication/energy-policy-and-thepower-sector-in-the-long-run/

Dudin, M. N., Frolova, E. E., Sidorenko, V. N., Pogrebinskaya, E. A., & Nikishina, I. V. (2017). Energy Policy of the European Union:Challenges and Possible Development Paths. *International Journal of Energy Economics and Policy*, *7*(3), 294–299. Available at http://www.econjournals.com

Samuela, B., Maria, C., Baran, D., & Sam, F. (2017*). Credible, effective and publicly acceptable policies to decarbonise the European Union,* Final report. Grantham Research Institute on Climate Change and the Environment. Available at: http://www.lse.ac.uk/GranthamInstitute/wp-content/uploads/2017/12/Credible-effective-and-publicly-acceptable-policies-to-decarbonise-the-European-Union-Final-report.pdf

Tagliapietra, S. & Zachmann, G. (2017). *A new strategy for European Union-Turkey energy cooperation,* Bruegel Policy Contribution Issue No 27, October 2017.

Chapter 7
Turkey's Path Towards an Energy Market

Haşmet Gökirmak
Istanbul Sabahattin Zaim University, Turkey

ABSTRACT

This chapter discusses the possibility of developing an energy market in Turkey. Turkey currently serves as an energy transit corridor, with the Baku-Tbilisi-Ceyhan (BTC) and Kirkuk-Ceyhan pipelines and with its seaborne oil trade, where large volumes are carried by tankers through its straits. Turkey also has the potential to become an energy market with new projects connecting producers in Central Asia, the Middle East, and the Caucasus, and major consumers of oil and natural gas in Europe and other regions of the world. Two recent megaprojects, The Trans Anatolian Natural Gas Pipeline Project (TANAP) and Turkish Stream will move Turkey closer to fulfilling this dream. Turkey, however, needs to meet some requirements to be considered a mature energy market. These are related, among others, to factors such as its infrastructure, storage capacity, market reforms, and easy market access for private firms to actively participate in the energy market.

INTRODUCTION

Energy has become an important factor affecting international relations and the economic, political and social developments of many countries. Energy trade is being used as an economic means for countries to achieve their geo-political goals as well. The main goal for many countries with insufficient energy resources, including Turkey, is to access energy resources at a low cost in an uninterrupted and reliable way to ensure sustainable economic development.

DOI: 10.4018/978-1-5225-4203-2.ch007

Economic conditions, energy prices and weather conditions impact energy demand and fuel type selection in the short run. In estimating the long run demand, however, population growth, and economic development projections are taken into consideration.

The world's primary energy demand was 13,559 Million tons of petroleum equivalent (Mtpe) in 2013; It is expected to increase approximately 11% to reach 15,041 Mtpe in 2020, and increase approximately 45% to reach 19,453 Mtpe in 2040 (BOTAŞ, 2016). The oil and gas fields in the region around Turkey are expected to meet this increased demand.

Venezuela and Saudi Arabia have the largest proven oil reserves. Russia, Iran, and Iraq are among the top ten countries with the largest proven oil reserves. Turkey and some of her neighboring countries, Russia, Iraq and Iran, hold the key for the energy future of Europe and the World as a whole. These countries have the richest oil and gas reserves and are also the major producers in the world. The Caspian Basin, the Middle East, and the Russian Federation account for about 65% of world oil reserves and 71% of natural gas reserves (T.C. ETKB, 2016).

Turkey is located between Europe and Asia and connects the World to Europe, Asia, the Middle East, Northern Africa and Continental Africa. Turkey aims for the full integration of the Turkish economy into the global economy by constructing the infrastructure, achieving the development goals, increasing social welfare, and creating a competitive manufacturing industry. Turkey's economy has been one of the fastest growing economies in the world. The country experienced steady growth between 2002 and 2015, averaging a growth rate of 4.9% per year. Turkey's Gross Domestic Product (GDP) was 717 billion USD and per capita income was 9,221 USD in 2015. Turkey had a total population of 77.7 million in 2015 (Focus Economics, 2016). The consumer market and energy demand grow as the population of the country grows and its industrial base expands.

Turkey may play an important role in global energy geopolitics. Turkey is a natural geographical bridge between the world's leading energy suppliers and large consumers. Turkey has been serving as a passage for seaborne oil trade through its straits (the Bosporus and the Dardanelles), originating from countries along the Black Sea, mainly from Russia and Caspian region (Caucasus) to Western markets. In 2013, about 2.9 million b/d of petroleum liquids flowed through the Turkish Straits. About 70% of this volume was crude oil, and the remaining 30% was petroleum products (U.S. Energy Information Agency, 2017).

Oil pipelines from Northern Iraq and Azerbaijan, and natural gas pipelines from Azerbaijan merge at the south eastern port of Ceyhan. Tankers are then used to ship the crude oil and natural gas to its end consumers. There are several pipeline project proposals that might give Turkey an important role as a natural gas transit route.

The next step for Turkey, after serving as an energy corridor for many years, is to become an energy market to serve as an economically and environmentally compatible and reliable bridge between the vast energy resources of the Caspian region, the Middle East and Central Asia and largest consumers of these energy sources in Europe and other regions of the world.

BACKGROUND

Rapid economic development has increased the energy demand in Turkey in recent years. The demand for energy, however, increased at a higher rate than GDP growth, averaging 5.7% per year for the 2002 to 2015 period. Turkey spent $37.6 billion in 2015 on energy imports, which was about 60% of the balance of payments deficit for the same year (Karar, 2016).

Turkey has a limited amount of proven oil reserves and very little natural gas. Turkey would like to keep external dependence at acceptable levels because of energy security reasons. Turkey aims to reduce energy dependency by meeting 35% of its energy needs with its domestic resources of coal and hydraulic energy and by making more investments in renewable energy and by increasing energy efficiency.

Turkey has proven oil reserves of 296 million barrels and natural gas reserves of 241 Bcf as of January 1st, 2015. Domestic natural gas production meets only about 1.6% of the demand and oil production meets about 9% of the domestic demand (Karar, 2016). Turkey imports oil and natural gas (mostly) from its neighbors at increasing quantities. Most of the natural gas comes from Russia, Azerbaijan and Iran via pipelines and liquefied natural gas (LNG) from Algeria, Nigeria, and Qatar by tankers (BOTAŞ, 2016).

In 2015, about 9% of crude oil was produced within the country and the remaining demand for crude oil met by Iraq (41%), Iran (20%), Russia (11%), Saudi Arabia (9%), Kazakhstan (3%), Colombia (3%), and other countries (4%). The average consumption of liquid fuels was 860.000 barrels per day (bbl/d) in 2014 (BOTAŞ, 2016). The International Energy Agency predicts that crude oil imports will continue to increase in the coming years.

The state-owned lignite reserves were about 12.5 billion tons and anthracite reserves were about 1.3 billion tons as of March 2015. The annual lignite production is about 63,3 million tons and anthracite production about 2.8 million tons (T.C. ETKB, 2016). Domestically mined coal, however, is of very low quality (low calorific value) and has high levels of sulphur both in hard coal and lignite.

Consumption of natural gas in Turkey has steadily increased during the last few years and reached 1.7 trillion cubic feet in 2015. Almost all of this consumption is satisfied by imports. Gas imports are contracted with long term agreements with

Figure1. Crude Oil imports
Source: U.S. Energy Information Agency. 2017

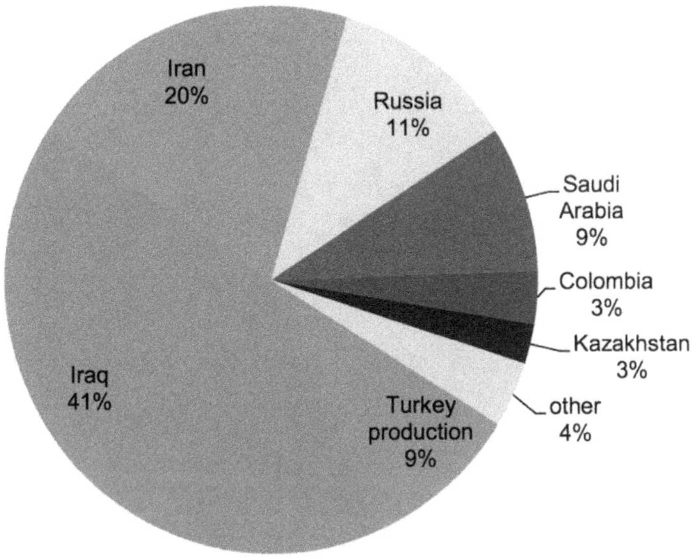

Russia, Iran, Azerbaijan, Algeria, and Nigeria. Russia has been the largest supplier of natural gas to Turkey (56% in 2015). Iran (16%) are Azerbaijan (11%) are the other significant sources for natural gas while Algeria and Nigeria were the main suppliers of LNG.

Azerbaijan became an important supplier in recent years. Turkey will import an additional 6 Bcm of natural gas from Azerbaijan once the Trans Anatolian Pipeline (TANAP) is completed. TANAP will bring natural gas from the second stage of Shah Deniz Field. Iran is ready to supply a lot more oil and natural gas to Turkey once economic sanctions are removed. The National Iranian Gas Transmission Company (NIGTC) reported that Iranian natural gas exports to Turkey have been on the rise and Iran is committed to exporting 30 million cubic meters (Mcb) per day of natural gas to Turkey until 2026. More than 9 Bcm of Iranian natural gas were exported to Turkey in 2014 (Mehr News Agency, 2015).

In 2013, Turkey renewed its LNG contract with Algeria for 10 more years, accounting for about 9% of natural gas imports. Turkey has also recently signed a gas contract with Qatar, where the liquefied natural gas (LNG) purchases have changed from spot purchases to long-term contracts, similar to the LNG contracts with Algeria and Nigeria.

Turkey's energy demand has been growing. Rapid economic growth increased electricity consumption in the last 11 years, averaging about 5.5% per year. The increase in the demand for electricity was about 70% in this period (T.C. ETKB,

Figure 2. Natural gas imports, 2015
Source: U.S. Energy Information Agency. 2017

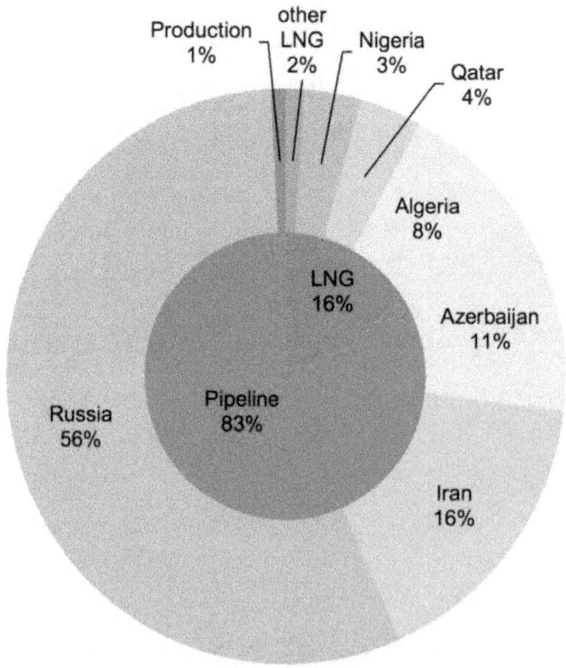

2016). The fast pace of economic growth, however, required quick solutions for energy generation. One way of meeting the ever-increasing energy demand was to construct natural gas turbines. Natural gas turbines can be constructed relatively quickly and inexpensively. These turbines are very flexible; the level of electricity generation can be easily adjusted depending on the demand, without causing problems with the electricity distribution grid. Natural gas is also environmentally cleaner than some of the other fossil fuels used in electricity generation. Demand for natural gas increased steadily while the consumption of petroleum products experienced a decline over the 2004-2014 period.

The installed electricity generation capacity increases as the demand for energy increases. In 2015, the electricity generation capacity was about 259.6 TWh and consumption was about 263.8 TWh. The installed capacity was about 36.8 MW in 2004 but it has almost doubled to reach 73.148 MW in 2015 (Bloomberg, 2016).

In 2004, hydroelectric generation was the number one category in the total installed electricity generation capacity with a 34.3% share. Natural gas had a 30.8% share and coal had a 22.5% share. Wind and geothermal energy had a 0.1% share in the total installed capacity. In 2014, although the rankings for the sources of electricity generation remained unchanged, their shares in the total installed capacity were somewhat different. Hydroelectric energy had a share of 34%, natural gas 31%, coal 21%, and wind, solar and geothermal had a 5.9% share. The installed capacity increased by about 7MW in 2013 and 6.3% in 2014. In the 2004-2014 period the largest growth in installed capacity was experienced in wind, solar and geothermal sources. Starting in 2014, private investments in solar systems to convert sunlight to electricity increased significantly.

By the end of 2014, thermal power plants, natural gas and coal, provided 78% of electricity production. Natural gas had the largest share, 48%, coal generated about 30% and hydroelectric power accounted for 17% of the total electricity generation. While there was a small decline in thermal power generation, geothermal and wind power generated electricity production increased from 3.7% to 4.2% in 2014 (T.C. ETKB, 2016).

TURKEY AS AN ENERGY MARKET

Turkey's energy policy is concerned primarily with meeting the energy demand for its growing economy. The government also wants to take advantage of Turkey's geographic location as an energy transit state and develop an energy market. Accomplishing a fully functioning energy market will generate revenues for the state and help secure energy supplies for the domestic economy.

In an energy market buyers and sellers will interact (directly or through intermediaries) to trade natural gas and oil, or contracts or instruments, for money or barter. There will be mechanisms for price determination and communication, facilitating deals and transactions in the energy market, just like any other market. The energy market will attract existing and potential customers of energy who have the ability and willingness to pay for it (Business Dictionary, 2007).

According to Heather (2012), a gas market needs to have the following attributes to deserve the title of a mature market:

- Good access to the traded market
- A large and growing number of market participants
- A forward curve for financial risk management alongside a very active spot market used primarily for balancing
- A firm and robust gas infrastructure
- Reliable, market based balancing regime
- A single gas quality
- Good access to storage.

Heather (2012) groups hubs into three distinct categories in order to analyze their development: 'trading' hubs, 'transit' hubs, and 'transition' hubs.

- Transit Hubs are actual transit locations, or physical points, at which market participants can choose to trade gas; however, their primary role is to facilitate the transit of large quantities of gas.
- Transition Hubs are based on a virtual trading point but have not yet reached a mature level. They are being used as 'balancing markets' for shippers delivering or taking gas to those grids. They may not develop sufficiently to become more than just national markets.
- Trading hubs (markets) are those which have reached a certain level of maturity and which are already being used for the financial risk management of gas portfolios. They are based on virtual trading points, have open and easy access to a wide number and variety of participants, have good transparency and reporting and have been proven to be reliable markets.

Britain's National Balancing Point (NBP) and the Dutch Title Transfer Facility (TTF) are two of the best examples of trading hubs in Europe that create a competitive market environment.

According to Bilgin (2010), Turkey has become an energy transit country, mainly because it is a natural bridge between Europe and the Caspian region and the Middle East, and it would like to become an energy hub, and even an energy centre. He argues that there are some international and domestic restraints to it becoming an energy hub or an energy centre.

- Turkey as an energy transit corridor implies the existence of pipelines, and other sorts of transportation, carrying oil and natural gas coming from Russia, the Caspian and the Middle East, to Turkey, Europe and other markets. Turkey, as a transit state, receives transit fees but cannot re-export the oil and gas passing through its territories.
- Turkey as an energy hub stresses Turkey's extensive influence on oil and gas pipelines and Liquefied Natural Gas (LNG) trade in terms of its ability to influence transit terms and conditions, and having the right to re-exporting some of the oil, natural gas, and LNG passing through this system.
- Turkey as an energy centre describes a situation in which Turkey's energy hub (market) requirements have been supported by massive investment in other sources of energy, such as nuclear and renewable energy and a well-established infrastructure of refineries, natural gas storage facilities, LNG trains, vessels, marine terminals and ports.

Currently Turkey serves as a transit corridor where oil and gas are transported from east to west. Turkey gets transit fees but does not have the right to re-export natural gas or oil, because of long term agreements with re-export restrictions. Turkey is making progress towards becoming an energy market. The country, however, needs to meet certain requirements to become an energy market. These requirements include a well-built infrastructure, access to storage, oil and gas market liberalization, and a large and growing number of market participants.

Infrastructure

Turkey might have an advantage in its foreign policy dealings with its neighbors and European countries once the necessary infrastructure (pipelines, storage and port facilities) is in place and a stable energy market is formed. This will create a new outlet for oil and natural gas supply and will reduce dependency on Russia for European Union member countries.

The Petroleum Pipeline Corporation (BOTAŞ), a state-owned company, was established in 1974 for the construction and operation of the Kirkuk-Ceyhan Oil Pipeline. BOTAŞ had been solely responsible for the distribution, import, storage, marketing, trade and export of natural gas in addition to crude oil transportation until 2001. Currently BOTAŞ is responsible for the operations of Kirkuk-Ceyhan, Baku-Tbilisi-Ceyhan and the Kurdish Regional Government (KRG) Pipelines (Table 1).

The positive trend in natural gas consumption is expected to continue. Because of this growth, however, Turkey's domestic natural gas consumption is about to reach its annual pipeline and liquefied natural gas (LNG) import capacity limits.

The gas flows from four international entry points into Turkey and there is an exit to international pipelines (U.S. Energy Information Agency, 2017). The national natural gas pipeline network, managed by BOTAŞ, has 7,856 miles of domestic pipeline connecting the metropolitan areas. The network, however, is used almost

Table 1. Turkey's major crude and condensate pipelines

Facility	Capacity (million b/d)	Length (Miles)	Supply Region	Destination	Details
Baku-Tbilisi-Ceyhan	1.2	1,100	Azerbaijan and Kazakhstan	Ceyhan Oil Port	Started in 2006
Kirkuk-Ceyhan	1.5	620	Northern Iraq	Ceyhan Oil Port	Started in 1976
Kurdish Regional Government (KRG) Pipeline	0.7	250	Northern Iraq	Ceyhan Oil Port via Kirkuk Ceyhan Pipeline	Completed in 2013

Source: U.S. Energy Information Administration (2017)

at full capacity during periods of peak demand. The transmission network requires extensive investments to meet the increasing demand within the domestic market and customers in Europe.

Access to Storage

The natural gas and LNG storage capacity is about 5% of the annual consumption in Turkey. These storage facilities are insufficient for the level of consumption in Turkey. This ratio is considerably lower than European countries' averages which range around 20% of their annual consumptions. Turkey's largest natural gas storage facility is located in Silivri, a town close to Istanbul. The storage capacity of Silivri facility is about 90 Bcf. Only a few ports have LNG loading and unloading capabilities and a small storage capacity for LNG.

Table 2. Turkey's major natural gas pipelines

Facility	Capacity (Tcf)	Length (Miles)	Supply Region	Markets	Details
Trans Balkan natural gas pipeline	0.5	600	Russia	South Eastern Europe and Turkey	1987
Tabriz-Dogubayazit	0.5	1,600	Iran	Turkey	2001
Blue stream	0.6	750	Russia	Turkey	2003
South Caucuses Pipeline (SCP)	0.3	430	Azerbaijan	Turkey	2007
Interconnector Turkey-Greece-Italy	0.4	180	Azerbaijan, Iran and Russia	Greece	2007
South Caucasus Pipeline (expansion)	0.6	430	Azerbaijan	South Eastern Europe and Turkey	Expected in 2019
Trans-Anatolian Pipeline (TANAP)	0.6	1,150	Azerbaijan	Turkey and Europe	Expected in 2019
Turkish stream	1.1	500	Russia	Turkey and Southeast via the Black Sea to Europe	Expected in 2019
Iraq-Turkey	0.4-0.7	430	Iraq	Turkey and Southeast Europe	Proposed

Source: U.S. Energy Information Administration (2017.

The construction of new natural gas storage facilities is in progress under Tuz Gölü (the Salt Lake) in Konya and in Tarsus, Mersin. With the completion of Tuz Gölü facility in Central Anatolia at the end of 2019, the total storage capacity will increase to about 10% of Turkey's annual natural gas consumption. The government has a goal for a storage capacity of about 20% of the annual consumption in the long term (Hurriyet Daily News, 2015).

Market Liberalization

Market liberalization is the relaxation of government regulations and restrictions in such areas of social, political and economic policy to encourage a greater participation by private entities. Europe has developed healthy functioning markets for natural gas even though they are dependent on imported gas similar to Turkey. Market depth and transparency should be achieved as Turkey makes progress in market liberalization.

The liberalization of the Turkish natural gas market started in 2001 with the Natural Gas Market Law (No. 4646). This law aimed to secure new supplies, reduce supply costs, and attract private investors to the natural gas market in Turkey. In addition, this law made it easier for private companies to enter the market, ended BOTAŞ' monopoly on natural gas supply, to provide competitive market prices for end-users of natural gas, and initiated the liberalization of the distribution networks. Turkey has achieved a very successful liberalization process, particularly with the privatization of the distribution sector, and created a natural gas market. In 2008, spot LNG imports were liberalized with an amendment to the Natural Gas Market Law.

A more recent law was proposed in 2012 for BOTAŞ to be divided into three independent companies: a company for gas transmission, a company running gas liquefaction plants and a company for the storage of energy resources.

Table 3. Turkey's natural gas storage facilities

Facility	Status	Operator	Working Gas Capacity (Bcf)	Details
Kuzey Marmara (Silivri)	operating	TPAO	90	Capacity to be expand to 150 Bcf.
Marmara Ereğlisi LNG	operating	BOTAŞ	6	LNG terminal storage
Aliaga LNG	operating	Ege Gaz	6	LNG terminal storage
Tuz Gölü	planned	BOTAŞ	35	Salt dome storage
Tarsus	planned	Bendis Energy	180	

Source: U.S. Energy Information Administration (2017)

The energy stock exchange (EPİAŞ) was established in 2014, to participate in foreign energy stock exchanges and to increase liquidity, efficiency and transparency in the energy market (Herdem Attorneys At Law, 2014). The establishment of EPİAŞ was the most important step in the liberalization of the energy market. Turkey is making progress toward a fully functioning gas market.

The liberalization of the petroleum sector has also been in progress. The Turkish Petroleum Company (TPAO) had been in charge of the oil exploration process in Turkey. The Petroleum Market Law, passed in 2013, however, encourages private companies to participate in oil exploration, distribution and retail. With the new law TPAO's involvement was limited to upstream activities by opening room for private initiatives. This law was intended to provide a transparent, non-discriminatory and stable delivery of petroleum to consumers, directly or after processing, within a competitive environment (T.C. ETKB, Petrol İşleri Genel Müdürlüğü, 2016). Preliminary work is in progress for the privatization of TPAO.

Market Participation

Policy changes has made it possible for private companies to participate in the energy sector. State-owned enterprises, such as BOTAŞ and TPAO, however, still dominate the markets for natural gas and oil. The number of private companies active in the energy market is very limited.

Shell Enerji, A.Ş., the first private importer and wholesaler of natural gas, started operations in 2005. In 2012, four private suppliers received import licenses from the EPDK after signing supply contracts with Gazprom. Figure 3 presents some of the major private companies that import natural gas into Turkey including Bosporus Gaz, Enerco Enerji, Akfel, Bati Hatti, Kibar Enerji, Avrasya Gaz, Shell Enerji and Ege Gaz.

Currently, six companies hold petroleum refinery licenses in Turkey, but only four refineries, operated by Türkiye Petrol Rafinerileri A.Ş. (TÜPRAŞ), are in operation (located in İzmir, İzmit, Kırıkkale and Batman). TÜPRAŞ was privatized in 2005. Two other privately owned refineries, Doğu Akdeniz Petrokimya ve Rafineri San. ve Tic. A.Ş. and Star Rafineri A.Ş., are expected to be completed in 2018.

THE SOUTHERN GAS CORRIDOR, TANAP, TURKISH STREAM AND OTHER PROJECTS

There are several pipelines projects, such as TANAP and Turkish Stream, and proposals for new pipelines that may give Turkey an important role in natural gas transit (U.S. Energy Information Agency, 2017). These projects are being evaluated

Figure 3. Natural gas import share of companies in 2014
Source: Energy Market Regulatory Authority (2015)

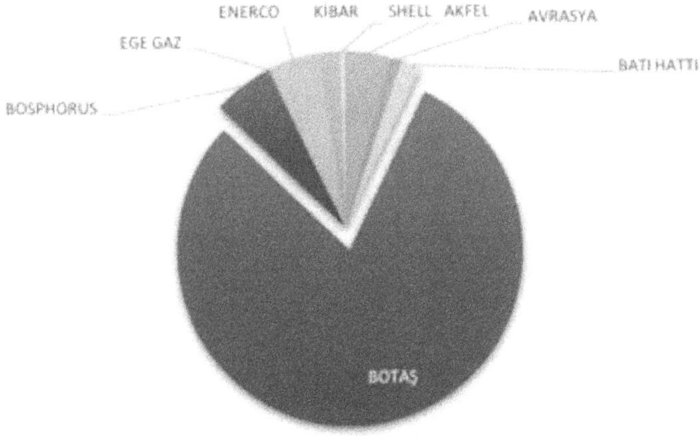

in cooperation with the governments of these countries, which are either producers or consumers of these resources. The projects are in various stages of development and include the installation of pipelines, storage facilities, and oil and natural gas-based plants.

The Southern Gas Corridor

The European Union already has three natural gas supply routes, the Northern Corridor from Norway, the Western Corridor from Northern Africa and the Eastern Corridor from Russia. The fourth route, The Southern Gas Corridor, is an initiative of the European Commission for the supply of gas from the Caspian and Middle Eastern regions to South eastern Europe and the EU (Linke & Vietor, 2010). The initiative was proposed in the European Commission's Communication in 2008. The European Union has identified a number of partner countries for this initiative, including: Azerbaijan, Turkey, Georgia, Turkmenistan, Kazakhstan, Iraq, Egypt, Lebanon, Palestine, Jordan, Syria, Iraq, Uzbekistan and Iran. Turkey will be the key transit country along the southern corridor with multiple pipelines. The southern gas corridor will provide additional security to Europe by diversifying its energy suppliers.

The Southern Corridor will open a new gas supply corridor for Europe and for the transit countries involved and contribute to the security of supply for all partner countries, and Europe as a whole. It will also strengthen the role of the gas pipeline networks of all partners and it will contribute to a well-functioning single gas market, by providing transparency and increasing competitiveness.

Figure 4. International transit pipeline projects
Source: U.S. Energy Information Agency, 2017

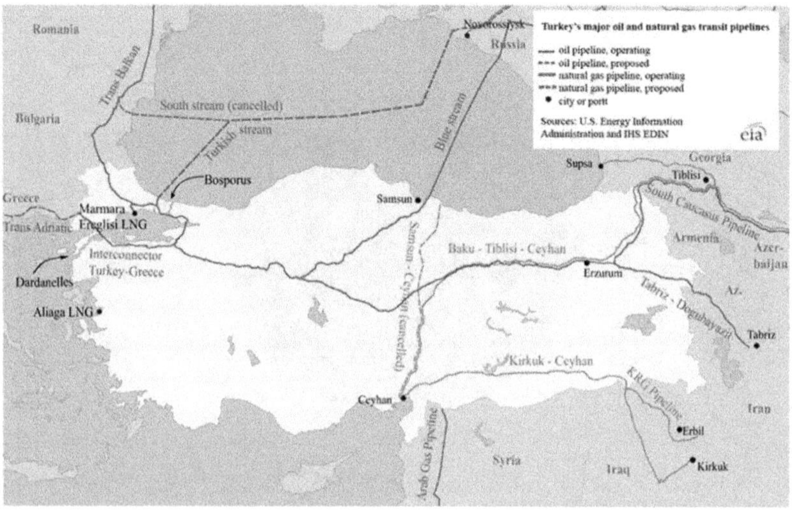

Several projects were developed to establish this corridor from the Caspian and Middle Eastern regions to Europe via pipelines through Turkey. The first mega project was the Nabucco Pipeline Project, from Erzurum in Turkey to Baumgarten an der March in Austria. The original project was backed by several European Union member states and by the United States, and was seen as a rival to the South Stream pipeline project of Russia. The main suppliers would be Iraq, Azerbaijan, Turkmenistan, and Egypt. The intergovernmental agreement between Turkey, Romania, Bulgaria, Hungary and Austria was signed on 13 July 2009 in Ankara. The representatives of European Union and the United States were present at the ceremony.

Shah Deniz gas was to be the main supply source for Nabucco West. Nabucco International signed a memorandum of understanding with the Trans Anatolian Natural Gas Pipeline Project (TANAP) consortium. However, on 28 June 2013 the Shah Deniz consortium announced that it had chosen TANAP over Nabucco for its gas exports, and this concluded the Nabucco Project.

The second failed mega project was The South Stream Pipeline Project (SSPP), which was launched by Russia in 2006 after an intense dispute with the Ukraine. The SSPP aimed to bypass Ukraine and transport natural gas directly from Russia, through the Black Sea to Bulgaria and through Serbia, to Hungary and Slovakia, Austria, and Italy.

According to Türksen and Vihma (2016), the SSPP has been an instrument of a geoeconomic strategy to advance Russia's political ambitions: "to secure Russian influence in Ukraine; to maintain Gazprom's dominant market share in the EU; to maintain Russian influence on the EU member states which rely on Russian energy; and to undermine European unity in energy and foreign policy."

The European Commission, however, did not approve the project on the basis that the same party (Gazprom) may not be both a gas supplier and a network owner in the European Union. EC Directive (2009/73) requires the separation of ownership of transmission assets from supply activities. The European Commission aims to increase competition in national markets and integrate them into regional and, in time, into a single European market for natural gas. President Putin unexpectedly cancelled the SSPP in December 2014 due to the long-lasting disagreement between Russia, the European Commission, and EU members.

The Trans Anatolian Natural Gas Pipeline Project (TANAP)

TANAP is being jointly conducted by Turkey and Azerbaijan. TANAP will bring natural gas produced from Azerbaijan's Shah Deniz-II gas field, and other areas of the Caspian Sea, to Turkey and Europe. TANAP is scheduled for completion by 2018 and in the first phase the capacity will be 16 Bcm per year (10 bcm/y for Europe and 6 bcm/y for Turkey). The capacity will be gradually increased first to 24 bcm/y and later to 31 bcm/y (www.tanap.com).

Figure 5. The Trans Anatolian Natural Gas Pipeline Project (TANAP)
Source: TANAP (www.tanap.com)

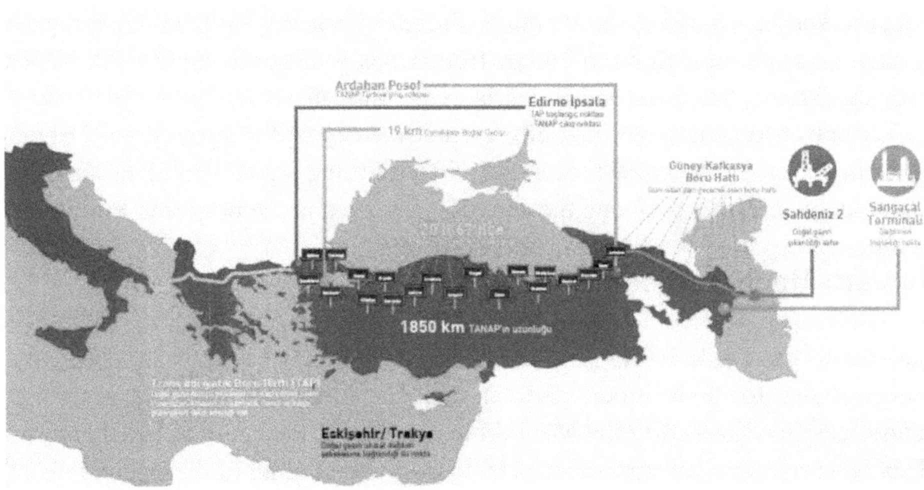

Turkish Stream Gas Pipeline Project

The Turkish Stream pipeline project was announced in February 2015. Gazprom of Russia and BOTAŞ of Turkey agreed on the construction of a natural gas pipeline under the Black Sea in exchange for a discount on gas for Turkey. Turkish Stream will directly connect the large natural gas reserves in Russia to the western end of the Turkey natural gas network, and increase the gas supplies to both Turkey and south-east Europe. Russia aims to avoid one of its existing transit routes through Ukraine, Moldova and Romania. The new agreement, signed on October 10, 2016, reduced the proposed capacity to 32 billion cubic meters of natural gas per year. The estimated cost of the project is 11.4 billion Euros (Sabah.com.tr, 2012).

The planned route includes 569 miles of offshore pipeline under the Black Sea. The offshore section will start from Anapa on the Russian coast, and will extend to Kıyıköy, close to Istanbul, on the Turkish coast. Turkish Stream will, then, be connected from Kıyıköy to the existing natural gas network in Lüleburgaz. The route, then, will be extended from there to the Turkish town of İpsala, close to the Greek border (Turkish Stream, n.d.). The portion of the pipeline on Russian side has been completed. The construction of the Black Sea portion of the pipeline will start on the second half of 2017. Gazprom is responsible for the construction of the offshore component of the system.

One branch of the gas pipeline will provide natural gas to the western part of Turkey and the other will provide natural gas to European countries. The branch to the Turkish market is intended to replace the natural gas flows currently coming from Bulgaria via the Trans-Balkan Pipeline. Currently, about 15 billion cubic meters of natural gas per year flows to Turkey from Trans-Balkan pipeline, which passes through Ukraine, Romania, and Bulgaria to the Turkish border.

The branch to Turkey, with a 16 billion cubic meter per year capacity, will be built in the first stage of the project. The second stage of the project, however, requires a new set of agreements among the countries involved in the project in Europe.

Eastern Mediterranean Reserves

New found gas fields in the Eastern Mediterranean may serve as an alternative energy source for both Turkey and Europe. The Leviathan gas field is a large natural gas field located in the Mediterranean Sea off the coast of Israel and the Gaza Strip. Production expected to start in 2017. According to the estimations of the Israel Ministry of Energy, the Leviathan gas field has about 470 Bcm of natural gas (Globes, 2015). The Aphrodite gas field in the south of Cyprus is estimated to have about 200 Bcm of natural gas (Dünya Gazetesi, 2016). Transporting East

Mediterranean natural gas discoveries to Europe through Turkey is promising in terms of the diversification of sources for both Turkey and European countries.

Northern Iraq Natural Gas Reserves

The Kurdish Regional Government (KRG) is another potential supplier of natural gas. The KRG will start building its first natural gas pipeline in February 2016 to export the fuel to Turkey. The Khor Mor and Chamchamal fields in Northern Iraq, will first be connected to Turkey and later extended to Europe. Turkey will help finance the 831-kilometer network. The new pipeline (831 Km) will run parallel to the existing oil pipeline. The part of the project that runs in the Kurdish territory (181 Km) will cost about $750 million (Financial Tribune, 2016). The initial capacity will be about 10 Bcm/year and will increase to 20 Bcm/year by 2020 (Bloomberg, 2016).

Natural gas from the Eastern Mediterranean and Northern Iraq might reduce dependence on Russia for both Turkey and Europe and contribute to supply security. Transportation of natural gas from new suppliers and increased volumes from existing resources in other countries to Europe may also contribute to the Turkish economy.

CONCLUSION

Turkey has been experiencing steady economic growth in the past 15 years. Economic growth due to government deregulations and increased international trade stimulates manufacturing and service industries. As the electricity and natural gas grids reach out to remote parts of the country, these locations become more integrated into the larger national economy. Economic growth, increased energy consumption for homes and businesses, hence, increases the demand for energy in Turkey.

Turkey, with its large and relatively young population, is an attractive energy market. The country has a very limited amount of proven oil and natural gas reserves and produces a small proportion of its total energy requirements from its domestic resources. Turkey, therefore, is a net importer of these resources.

In addition to meeting the growing domestic energy demand, Turkey has also been serving as an energy transit corridor. The Baku-Tbilisi-Ceyhan (BTC) and Kirkuk-Ceyhan pipelines carry large volumes of oil to the South-Eastern shore of Turkey to be refined and transported to various markets. The Bosphorus and Dardanelle straits facilitate seaborne oil trade from Black Sea ports to Europe and elsewhere in the World.

However, there have been some recent efforts by Turkey to become a prominent energy transportation corridor and energy center to connect energy resources in

Central Asia, the Middle East and Caucasus to major consumption regions in Eastern and Western Europe. TANAP and Turkish Stream are mega projects to transit large volumes of natural gas via Turkey to Europe.

TANAP is a joint project between Turkey and Azerbaijan. TANAP will bring natural gas produced from Azerbaijan's Shah Deniz-II gas field, and other areas of the Caspian Sea, to Turkey and Europe. TANAP will run from Azerbaijan through Georgia to Turkey. The pipeline enters Turkey in Türkgözü in Ardahan, will run through 20 provinces to reach İpsala in Edirne, near the Greek border. From this point, the TAP Pipeline will connect to the existing European natural gas grid to transport natural gas to Southern and Western European countries. TANAP is expected to be constructed by 2018 and the capacity will be 16 Bcm per year in the first phase of the project.

Turkish Stream is a joint project of Gazprom of Russia and BOTAŞ of Turkey. Parties agreed on the construction of a natural gas pipeline under the Black Sea. Turkish Stream will directly connect the large natural gas reserves in Russia to the western end of Turkish natural gas network, and increase gas supplies to both Turkey and South-Eastern Europe. The pipeline will start from Anapa, a small town in Krasnodar Krai, Russia, located on the northern coast of the Black Sea near the Sea of Azov. It will run approximately 570 miles offshore to reach Kıyıköy, in Kırklareli Province at north western Turkey. Turkish Stream will, then, be connected from Kiyikoy to the existing natural gas network in Lüleburgaz. The route, then, will be extended from there to the Turkish town of Ipsala, close to the Greek border. The pipeline will transport 32 billion cubic meters of natural gas per year.

Eastern Mediterranean and Northern Iraq natural gas fields are also promising for both Turkey and Eastern Europe as sources of natural gas. The Leviathan gas field is a large natural gas field located in the Mediterranean Sea off the coast of Israel and the Gaza Strip. The Aphrodite gas field is located in the south of Cyprus. The Khor Mor and Chamchamal fields in Northern Iraq are other potential suppliers of natural gas. The initial shipment from the fields of Northern Iraq will be about 10 Bcm/year and will increase to 20 Bcm/year by 2020.

Turkey, however, needs to meet some requirements to be considered a mature energy market. Turkey has to improve its existing infrastructure, significantly increase its storage capacity, and pursue further market reforms and ease access for private firms to actively participate in the energy market.

The existing gas distribution network is being used at close to its maximum capacity during periods of peak demand. The network needs to be expanded to meet the growing demand in the coming years and also allow transportation of natural gas beyond its borders to markets in Europe.

Currently the storage capacity is only about 5% of Turkey's annual natural gas consumption. The storage capacity for European countries ranges around 20% of their annual natural gas consumption. There are two major projects for natural gas storage that will increase the capacity to first 10% and eventually to 20% of the annual gas consumption.

Turkey has been pushing for important policy changes and market liberalization in the energy sector starting in 2001 with the Natural Gas Market Law. EPİAŞ was established to participate in foreign energy stock exchanges and to increase liquidity, efficiency and transparency in the energy market. The Petroleum Market Law also encourages private companies to become involved in oil exploration, distribution and retail.

These laws intend to reach out to new supply sources, to reduce supply costs, and to attract private investors to the energy market in Turkey. In addition, these laws made it easier for private companies to enter the market, eliminated the monopoly of BOTAŞ on natural gas, provided competitive market prices for end-users of natural gas, and initiated the liberalization of distribution networks. The privatization of the distribution sector was accomplished and spot LNG imports were liberalized. The Petroleum Market law was intended to provide a transparent, non-discriminatory and stable delivery of petroleum to consumers, directly or after processing, within a competitive environment. Market participation by the private firms is low but on the rise. Acceleration of the implementation of these policy changes will help increase the number of participants and the involvement of private sector participation.

The TANAP and Turkish Stream projects along with possible shipments from Eastern Mediterranean and Northern Iraq natural gas fields may help Turkey become an important player in the international energy market. Improving the country's energy distribution infrastructure, increasing its storage capacity and making the necessary market reforms to increase private sector participation will increase the likelihood of Turkey becoming an energy market in the near future.

REFERENCES

Bilgin, M. (2010). Energy and Turkey's Foreign Policy: State Strategy, Regional Cooperation and Private Sector Involvement. *Turkish Policy Quarterly*, *9*(2), 81–92.

Bloomberg, H. T. (2016). *2015 tüm zamanların en çok elektrik tüketilen yılı oldu, 2015–2016.* Retrieved from http://www.bloomberght.com/haberler/haber/1846242-2015-tum-zamanlarin-en-cok-elektrik-tuketilen-yili-oldu

Bloomberg. (2016). *Cash-Strapped Iraqi Kurds to Start Gas Exports to Turkey in 2016.* Retrieved from http://www.bloomberg.com/news/articles/2016-01-13/cash-strapped-iraqi-kurds-to-start-gas-exports-to-turkey-in-2016

FocusEconomics. (2016). Turkey. *Economic Outlook, 15*(June). Retrieved from http://www.focus-economics.com/countries/turkey

Gazetesi. (2016). Türk Akımı'nın ertelenmesi ile enerjide kaynak çeşitliliği hedefleniyor. *Focus*, (8).

Gazetesi, K. (2016). *Türkiye'nin enerji faturası 10.2 milyar dolar azaldı.* Retrieved from http://www.karar.com/ekonomi-haberleri/turkiyenin-enerji-faturasi-102-milyar-dolar-azaldi-295487#

Globes, Israel's Business Area. (2015). *New report sees 20% less gas reserves in Leviathan.* Retrieved from http://www.globes.co.il/en/article-new-report-sees-20-less-gas-reserves-in-leviathan-1001041847

Heather, P. (2012). *Continental European gas hubs: are they fit for purpose?* Oxford Institute for Energy Studies. Working paper. Retrieved from http://www.oxfordenergy.org/wpcms/wp-content/uploads/2012/06/NG-63.pdf

Herdem Attorneys At Law. (2014). *2013 Turkey Energy Report.* Retrieved from http://herdem.av.tr/energy-sector-turkey-central-importance-epias

Hurriyet Daily News. (2015). *Turkey to raise gas storage capacity by 4 bcm.* Retrieved from http://www.hurriyetdailynews.com/turkey-to-raise-gas-storage-capacity-by-4-bcm-minister.aspx?pageID=238&nID=89963&NewsCatID=348

Hürriyet Daily News. (2016). *Turkey, Russia strike strategic Turkish Stream gas pipeline deal.* Retrieved from http://www.hurriyetdailynews.com/turkey-russia-strike-strategic-turkish-stream-gas-pipeline-deal.aspx?pageID=238&nID=104822&NewsCatID=348

Linke, K., & Vietor, V. (2010). *Beyond Turkey. The EU's Energy Policy and the Southern Corridor.* Bonn: Friedrich Ebert Foundation. Retrieved from http://www.fes.kiev.ua/Dokument/beyond_turkey.pdf

Mehr News Agency. (2015). *Export of Iran's gas to Turkey soars.* Retrieved from http://en.mehrnews.com/news/112622/Export-of-Iran-s-gas-to-Turkey-soars

Online Business Dictionary. (2007). *Markets.* Retrieved from http://www.businessdictionary.com/definition/market.html#ixzz3zBQnuZCK

Sabah.com.tr. (2016). *Gazprom'dan flaş Türk Akımı açıklaması*. Retrieved from http://www.sabah.com.tr/ekonomi/2016/12/30/gazpromdan-flas-turk-akimi-aciklamasi

TANAP. (2016). *Major Shareholder in the Project of The Century*. Retrieved from http://www.tanap.com/media/press-releases/major-shareholder-in-the-project-of-the-century/

T.C. Enerji ve Tabii Kaynaklar Bakanlığı (ETKB). (2016). Dünya ve Ülkemiz Enerji ve Tabii Kaynaklar Görünümü. Strateji Geliştirme Başkanlığı, 12.

T.C. Enerji ve Tabii Kaynaklar Bakanlığı, Petrol İşleri Genel Müdürlüğü. (2016). *Turkish Petroleum Law*. T.C. Enerji ve Tabii Kaynaklar Bakanlığı. Retrieved from http://www.pigm.gov.tr/index.php/english/turkish-petroleum-law

The Petroleum Pipeline Corporation (BOTAŞ). (2015). *2015 Sektör raporu*. Retrieved from http://www.botas.gov.tr/docs/raporlar/tur/sektorap_2015.pdf

The Petroleum Pipeline Corporation (BOTAŞ). (2016). *2015 Annual Report*. Retrieved from http://botas.gov.tr/docs/raporlar/faalrap_2015.pdf

Tribune, F. (2016). *KRG Gas Export to Turkey in 2016*. Retrieved from http://financialtribune.com/articles/energy/34341/krg-gas-export-turkey-2016

Turkish Petroleum Corporation (TPAO). (2013). *2012 Annual Report*. Retrieved from http://www.tpao.gov.tr/eng/docs/report/tp-2012-annual-report.pdf

Turkish Stream. (n.d.). *Turk Stream*. Retrieved from http://turkstream.info/project/

Turksen, U., & Vihma, A. (2015). The geoeconomics of the south stream pipeline project. *Journal of International Affairs*, *69*(1), 34.

U.S. Energy Information Agency. (2017). *Country Analysis Brief: Turkey*. Retrieved from https://www.eia.gov/beta/international/analysis.cfm?iso=TUR

ADDITIONAL READING

Bilgin, M. (2010). Energy and Turkey's Foreign Policy: State Strategy, Regional Cooperation and Private Sector Involvement. *Turkish Policy Quarterly*, *9*(2), 81–92.

Chapter 8
The Asymmetry of the Global Changes:
Asymmetric States of the Developing Countries

Guliyeva Aida
Azerbaijan State University of Economics, Azerbaijan

Rzayeva Ulviyya
Azerbaijan State University of Economics, Azerbaijan

ABSTRACT

The object of the study is the asymmetry of interests in geopolitics between developed and developing countries. In the context of the global crisis, the issue of de-dollarization is relevant from the political and economic points of view. What will be the behavior of small oil countries in this situation is a big problem. Also, for them, the question remains how to get off the oil needle in a painless way? The ways of solving the above-mentioned problem by other states are examined in detail, and comparative analysis is conducted in the case of applying these methods to Azerbaijan.

DOI: 10.4018/978-1-5225-4203-2.ch008

INTRODUCTION

The object of our consideration is the meaning of the concept of asymmetry and polarization of hydrocarbon resources in a small group of developing countries and countries with economies in transition.

Historically, commodity markets have used the US currency as means of payment. However, an active anti-dollar campaign is gradually unfolding in the world. These efforts often lead to serious and not always positive developments in global politics. The purpose of the work is to find answers to the following questions: what should be done to ensure that the rejection of the dollar will be real? What can a small developing country's economy expect in a global crisis, with falling prices for raw materials, and after two currency devaluations?

The novelty of this work is studying the structure of the Azerbaijani economy, aimed at getting rid of the "resource curse" - a theory that points to the relationship between large revenues from the export of natural resources and weak economic development.

The aim of research is to investigate this global asymmetry, which is caused by the dominance of the dollar and has the greatest impact on countries whose economies depend on energy exports. Azerbaijan is in the list of such countries.

To achieve this aim it is necessary:

1. Suggest ways to escape the global asymmetry in the light of the diminishing returns from oil sales.
2. Conduct a critical analysis of the structure of the internal markets and find out why raw materials and semi-finished products are mainly dominant in the exports of developing countries.
3. Examine steps that the Azerbaijan government should take towards the development of the non-oil sector (we do not take into consideration such fields of economy as agriculture, tourism, information technologies, etc.)
4. Present the authors' vision for solving the de-dollarization problem from the perspective of a small developing country.

BACKGROUND

The concept of asymmetry i.e. disproportion, discrepancy was included into an arsenal of mathematics and physics long ago. Asymmetry belongs to usual, widespread natural phenomena (geological, biological and other). In the 1970s the head of the Pentagon's obscure Office of Net Assessment, Andrew W. Marshall, was one of the first to apply the term "asymmetry" to the military-political sphere: he paid attention to possible asymmetric warfare with such, apparently, symmetric opponent as the USSR (Lemann, 2011). Actually, at first the term "asymmetric relations" had a military strategic importance and only later became widely used in the political and economic speech.

Scholvin (2016) reveals in detail how asymmetry turned out to be an important aspect mostly in each known geopolitical relations. The asymmetric strategy is always present in relationships (Meinhardt, 2002). All modern asymmetric conflicts can be understood when they are considered through the prism of the strategy and concludes that asymmetry is a strategy and strategy is an asymmetry.

As Bourguignon and Morrison (2002) describes national economies in the framework of the world economy and international economic relations in late XX - early XXI century, these relationships are also characterized by the asymmetry of the states' developments. This was reflected primarily in the greater inequality between them. The system transformation crisis of the 1990s changed the place of the post-Soviet countries quantitatively and qualitatively, throwing them back. With the successful development of China's economy against the background of post-Soviet countries' national economies also increases asymmetry of the world economy (King, 2009; Lau, Qian & Roland, 2000). Investigating relations with Asia Schaffer (2002) determines that even within certain, especially large multinational countries (India, etc.), there is practically no homogeneity of economies and asymmetry in their development is also observed. This fact also applies to such a stable regional community as the European Union.

The greatest asymmetry of interdependence is inherent in the relations of developed countries with the most backward and poor countries of the world. The latter do not have the resources of strategic raw materials, they are not competitive producers of industrial and consumer goods, and are not promising markets for the products of developed countries. Depending on the conjuncture of commodity markets, these countries can influence industry, but a significant dependence on world industrial centers dominates in the system of their economic relations.

Including Azerbaijan, the economic dependence of developing countries on developed countries is manifested primarily in the fact that developed countries are "pure" exporters of capital to developing countries, whereas developing countries are in fact the debtors of developed ones. In foreign trade, economic dependence is manifested not only in the overall asymmetry of exports and imports, but also primarily in their market's commodity structure. Raw materials and semi-finished products are mainly dominated in the exports of developing countries.

Today the USA as the largest importer of oil and the only superstate to have monopolized the right – in the sphere of geopolitics of fuel – to convert the geopolitics of spaces into the geopolitics of streams, geostrategy to geoeconomy and vice versa. In response, there is a tightening of access policy to hydrocarbon resources by oil-producing countries, called "resource nationalism". The geopolitics of asymmetry, unexpected forming the fanciest combinations for reaching the real horizons of multipolarity, today is not restrained by any ideological, confessional, political or social criteria.

According to the title of the study, this paper considers the global asymmetry, which has the greatest impact on countries whose economies depend on energy exports. In the light of the diminishing returns from the sale of oil, this paper requests ways for escaping this asymmetry. The authors consider the geopolitical asymmetry associated with the dominance of the dollar in this market, analyze the existing political situation and offer their vision of the republic's economy, for which oil has become a brake on development.

The Azerbaijan government has taken a firm course towards the development of its non-oil sector. Currently, sectors of the economy such as agriculture, tourism, information technologies and the textile industry are being invested in as an alternative to the "oil needle". The authors focus on the development of entrepreneurship, which occupies an important place in social and economic policy, and is also an active participant in investment processes. In turn, investment processes define scientific and technical progress, modernization of economy, growth rates of branch and regional economies in significant degree. The paper considers the current state of this economic area, presents the critical indicators for Azerbaijan in this sector. As a continuation of the theme of entrepreneurship development, the authors examine in detail the stock market of the republic, considering that the economy is based on the functioning of the corresponding financial system, and the most important place is occupied by the financial market in which there is the movement of money (capital). As a logical conclusion of the study, the authors in the final part again return to the theme of oil's de-dollarization, and, assessing all the shortcomings of the resource asymmetry from the perspective of a weak state, present their vision for solving this difficult problem

MAIN FOCUS OF THE CHAPTER

Issues, Controversies, Problems

Modern research notes a relatively high degree of asymmetry in the geopolitics of fuel. This is because the processes of the global economy are often characterized by both uniqueness and a high degree of uncertainty. As a consequence, the situation with geostrategic instability is difficult to forecast and estimate, and therefore, there are difficulties associated with obtaining reliable information about it. The authors' research aims to assess the degree of information asymmetry in the context of this phenomenon's significance for developing countries, taking into account their absolute dependence on the dollar.

Geopolitical Asymmetry Associated With Energy: US Dollar Oil Policy

The concept of "oil geopolitics" has taken root in global political usage, at least since the late 1970s, especially after the US Senate Committee on Energy and Natural Resources held hearings on this topic in 1980 (Geopolitics of Oil, 1980). The subject of oil geopolitics is the distribution of oil reserves on the Earth and the routes of oil delivery to different points of the planet; these subjects are treated by geopolitics from the point of view of the interested powers' security, in conjunction with world civilization divisions and with the antagonisms which are developed in the civilization (International Energy Agency, 2014). Rumley and Chaturvedi (2015) explore how in the modern world the positions of an exporter and a consumer of fuel are characterized by a fundamental geopolitical asymmetry. First, it is obvious that money is scarcer than fuel (the consequence of this is the danger of the withdrawal of money from oil production into spheres with a more rapid return). The attempt of oil exporters in the 1970s to turn around this situation ended in failure. Secondly, our world is designed in such a way that geopolitics is reduced to geo-economics for exporters – investors and buyers' searches, agreements on volumes of production with other exporters, pipelines' laying (without including geostrategy at all). The producer and exporter of oil does not control the regions to which it flows, and, as a rule, even those areas through which it flows. Meanwhile, the largest oil consumers since the era of the world colonial division have consistently used force to control oil production areas to strengthen and optimize their geoeconomic positions: for them the geopolitics of oil has always been not only geoeconomics but also geostrategy.

Analyzing the structure of the distribution of world oil reserves, Gutiérrez (2007) concludes that the asymmetry and polarization of hydrocarbon resources (along with some regional differentiation) increases in a small group of developing countries with transitional economies, which is one of the prerequisites for the revival of "resource nationalism". Davis (2006) sees the specifics of the current situation on the world oil market in the formation of the new imbalance between supply and demand, and there are two key points: a sharp increase in dependence of energy resources' countries-consumers on countries-producers (growth of market's asymmetry); the achievement of peak growth of world production (oil peak), which leads to additional investments for the development of less profitable and remote fields of alternative energy sources. This explains the instability of the oil market, which manifests itself in an extreme aggravation of the overall geopolitical situation and high price volatility. The consequence was the US's firm orientation toward a unipolar world based solely on forceful methods of solving problems.

Traditionally, oil quotations are exposed in US dollars. You can pay for the goods in any currency, but the estimation is always made in dollars.

As a result, companies and governments of states that buy oil and oil products keep large dollar reserves for oil supply payments, which in turn facilitate the position of the US economy, allowing the US to cover a large trade deficit.

Plummer (2013) determines that the rejection of the dollar in mutual settlements in relations between Cuba, Bolivia, Venezuela, Honduras, Nicaragua and Ecuador occurred in 2009. The states that were members of The Bolivarian Alliance for the Peoples' of Our America (ALBA) had accepted a new currency for mutual settlements, rejecting the US dollar. The first deal with the use of Ecuadorian Sucre was signed between Havana and Caracas.

It was also announced that China and Brazil would make payments in yuan. Moreover, Brazil aimed at a gradual departure from dollar settlements, not only with China, but also with the BRIC countries (Brazil, Russia, India, China). Argentina's experience has confirmed the success of such calculations.

An active anti-dollar campaign is developing in the world. The Chinese monetary authorities proposed their plan for reforming the world financial system, which involved the creation of an international reserve currency. According to the Central Bank of China, special drawing rights (SDR) - IMF's artificial payment facility, tied to a basket of currencies could become an international reserve currency (Xiaochuan, 2009).

At the summit of the Shanghai Cooperation Organization in 2008, the idea of rejecting the dollar in the computations between Russia and China was sounded (Qiang & Xiaokun, 2010). It was proposed to replace the dollar with the Russian ruble or the Chinese yuan. Two countries were able to conduct trade operations among themselves in their national currencies, although it concerned only the border areas.

And yet at the moment there is no tangible threat to the dollar as the only currency of oil contracts. 60-70% of export operations in the world are carried out in dollars. About 60% of all foreign exchange reserves of countries are also made in dollars, 80% of transactions in the world currency market are carried out in dollars and 70% of bank loans are provided in dollars.

The issue of de-dollarization goes far beyond the economy. These attempts sometimes lead to serious and far from positive developments in world politics. For example, as western media presents, the US invasion of Iraq began after S. Hussein had demanded that within the "Oil in exchange for Food" program calculations would be carried out in euro, but not in the American dollars (Proliferation Issues, 1992). "It was the choice that decided the fate of Iraq, – commented William Clark, a security expert and author of labor on the oil market economy. – It was the political step, which brings the income in connection with steady falling of dollar exchange rate. And this was the last argument in deciding whether to seize Iraq. When the American marines entered Baghdad, Iraqi oil again began to be sold for dollars" (Tomberg, 2017).

In 2009, Muammar Gaddafi, the head of Libya and the president of the African Union, invited African countries to switch to a new currency, the "gold dinar", independent from the US dollar and the African franc. Gaddafi proposed the creation of a monetary union in which the main means of paying for oil and other resources would be the "gold dinar". This idea, according to economist Frederick William Engdal, was approved by the leaders of Tunisia and Egypt – Zin el-Abidine Ben Ali and Hosni Mubarak. According to Engdahl (2016), not human rights violations in Libya, but the attempt to de-dollarization was the main reason why Mubarak and Gaddafi were killed.

It is possible that this is not the last war for the stability of the dollar. In 2006, Iran announced plans to open an oil exchange where the price of oil would be quoted in euros. The news aroused serious concern in oil circles. The geographical location of the country in proximity to such important oil and gas importers as China, India and Europe determines the strategic importance of Iran as a world oil supplier. Iran has 158,400 millions of barrels of proven hydrocarbon reserves, or 10% of the world's reserves (ENI, World Oil and Gas Review, 2016). And yet at the beginning of 2016, the Iranian authorities chose the euro in settlements with oil suppliers- in new contracts, the Islamic Republic required payment in Euro.

From this perspective, the greatest danger is political: any trade in a currency other than the US dollar is a blow to the dollar, and the US usually does not leave strikes without an answer.

Now put the policy aside, talk about the economy.

The sale of oil for the national currency is a controversial issue. Let us imagine that a foreign buyer wants to pay for oil in the national monetary standard. He comes with dollars or euro, submits the application for their exchange into the national currency, at the same time our buyer undertakes exchange rate risk. After receiving national currency, the money is deposited into the account of the sellers. However, there is also another party. Oil industry workers buy the equipment abroad for dollars or euro. In case of switching to the national currency, they also have to go to the exchange, sell it, buy dollars again, and take risks for changing the exchange rate. In this situation, as Baumeister and Lutz (2016) describes both sellers and buyers prefer to conduct calculations in dollars in order to avoid unnecessary transaction costs.

What other external factors prevent oil sales from being denominated in a national currency? First, the main factor is already settled system of trade generally for the American dollar. Actually, America remains the main consumer of hydrocarbon raw materials. In addition, the second external factor is the reluctance of European consumers to buy goods for the national currency; in this case it also applies to hydrocarbon feedstock.

Analysis of the Structure of the Economy of Azerbaijan: "A Resource Curse"

It is difficult to overestimate the role of oil in the economy of Azerbaijan. In Azerbaijan, the share of employed in the oil industry accounts for most of GDP, reflecting its predominantly mineral-raw material orientation.

Recently there was a favorable situation on the world market for raw materials. Thanks to the global economic crisis, the "resource blessing" has turned into a "resource curse". This theory points to the interrelation between large revenues from the natural resources' export and the weak economic development of the country, and reflects the suppression of market development and the decline of other economic sectors' competitiveness, the increase in the value of the national currency, inflation and unemployment.

Based on a survey of 20 countries, it is clear that abnormal revenues from mineral resources' export, as well as an excess of this resource within the country distorts the alteration of_economy in all cases, regardless of the political system and the applied economic policy (Karl, 2007). Azerbaijan is not an exception to this rule. The correlation of the rate of Azerbaijan GDP's growth with the oil prices' growth rate on world markets represents a parallel process. It is important to emphasize that this is not even a correlation with the rate of oil prices, namely with the rate of prices' growth, i.e. for the growth of the Azerbaijani economy. How to solve the resource curse problem?

Quickly coping with the Dutch disease, the symptoms of which are evident in Azerbaijan is impossible. There are no simple recipes here. Nevertheless, the problem of price volatility and export revenue is solved quite simply. If we know that oil prices sharply change, we can save them for a rainy day, for those periods when there is a drop in income. Countries that export resources - oil, gas, and so on - can create so-called national welfare funds that have two goals. The first is a savings purpose designed to reserve a part of the income from oil for pensions and tax exemptions in future. The second goal of such funds is stabilization funds, which are directly related to volatility. When oil prices are falling: these funds can be used to meet current needs. When the prices are high, then, on the contrary, we keep our large income in the fund. Accordingly, we smooth out state revenues and solve budget problems.

However, there are problems with stabilization funds. Oil prices are volatile, but, as economists often say, they are sustained, and if they fall, they can stay at a very low level for a very long time. And vice versa.

Not only oil producing countries possess national welfare funds, there are metal funds, copper funds, there are funds completely unrelated to raw materials, as in China and Singapore.

The largest fund is the Government Pension Fund of Norway. The Norwegian government does not spend money coming from oil and gas, but lives only on interest.

The State Oil Fund of the Azerbaijan Republic was established by the decree of President Heydar Aliyev in 1999 (actually began to operate in 2001). The revenues of the State Oil Fund are formed, first, by selling oil remaining at the disposal of Azerbaijan in accordance with production sharing agreements. In addition, the fund receives income from the concession of land in oil-bearing areas, the leasing of fixed assets, oil transit, the difference between the contractual and sales price of exported oil, as well as bonuses and income from the profitable use of assets.

The oil exporting countries are trying to change the development model based on the export of this raw material. However, the strategies of economic modernization need to be reoriented to new sources of national income. A purposeful economic policy can ensure the effective use of foreign exchange earnings from oil exports, and therefore the role of financing advanced sectors of the economy based on ICT raises. Developing countries seek to shift economic growth from the oil and gas production sector to sectors not related to the extraction and sale of crude oil and natural gas in order to reduce dependence on exhaustible oil reserves, the volatility of the world oil market, to change the economic structures based on foreign labor, to solve the problem of the social burden on the budget.

Investment Policy of Azerbaijan: Transition to a Non-Oil Economy

Development of Entrepreneurship as One of the Main Priorities of Economic Development

At the present stage of development of the Azerbaijani economy, foreign investments are directed mainly to the oil industry. However, proportional economic development requires redirection of revenues from the sale of oil and oil products to the development of the non-oil sector. The increase in investments is necessary for the purposes of economic growth, limiting inflation and unemployment, creating new jobs.

Registration of companies in Baku is carried out on the basis of unified principles, which are prescribed at the legislative level. The process of creating a limited liability company is very simple: it requires only 7 procedures and the process is normally completed within 11 days. This is much faster than the corresponding average for the countries of Eastern Europe and Central Asia. The whole process can be done online and the registration usually takes only 2-3 days. Currently, there are no restrictions on the minimum authorized capital. Article 117 of the Civil Code of the Republic of Azerbaijan states says that the amount of the authorized capital cannot be less than the amount established by the relevant executive authority (Civil Code of the Republic of Azerbaijan).

Land is granted only for rent to persons who are not citizens of the Azerbaijan Republic, individuals and legal entities of foreign countries. Some lands constitute the exclusive property of the state; rules for their usage and leasing are carried out in the order prescribed by law. The lease cannot be longer than 99 years. As a rule, contracts are usually concluded for 30 years and give the right to divide the land into sublease or leased land. There are no restrictions on the maximum quantity of land, which can be concentrated in one hand. In Azerbaijan a geographic information system that provides for the collection, storage, processing, access, display and dissemination of spatially coordinated data (spatial data) and, accordingly, centralization of information in one access point is absent.

Foreign direct investments in Azerbaijan in the fourth quarter of 2016 was 1895 million US dollars, reaching a record level of 2231 million US dollars in the fourth quarter of 2014 and a record low 586 US million dollars in the third quarter of 2009 (Figure 1).

Compared to the beginning of the XXI century, the overall investment climate in Azerbaijan is gradually improving, although serious problems remain.

The paradox is that everyone knows about these problems, but they remain unresolved, giving rise, in turn, to new problems. Over the past few years, the Azerbaijani government has worked to integrate the country more fully into the world market, to attract foreign investment, diversify economy and support growth.

Figure 1. Azerbaijan Foreign Direct Investment
Source: Trading Economics, 2018a

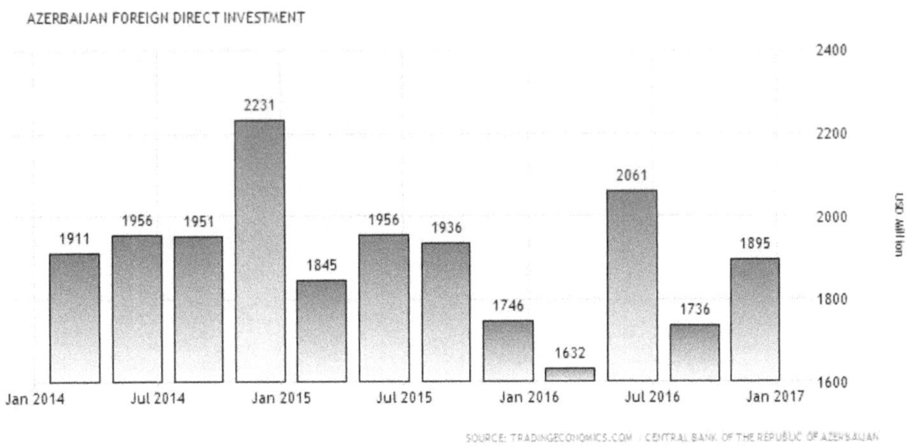

GDP growth of 2.8% in 2014 was due to non-oil economy and long-term public investment (Figure 2).

About 95% of the export income of Azerbaijan is the share of oil and gas products. Economic diversification and large foreign investment for the further development of energy sector are declared by President Aliev as government's goals, but ineffective state bureaucracy, weak legal institutions, requests for illegal payments for cross-border operations and predatory behavior of related monopolistic interests prevent investments in the non-oil sector and present problems for foreign companies. Over the past few years Azerbaijan has been working to improve its regulatory system, but limited transparency and allegations of corruption remain key problems in this area. According to the legislation of Azerbaijan, foreign investors can engage in investment activities not prohibited by law.

Private entities can freely create, acquire and dispose of interests in enterprises. The Law on the Protection of Foreign Investments protects foreign investors from nationalization and requisition, with the exception of certain circumstances.

In February 2015, Azerbaijan devalued its local currency, the manat (AZN), by 35% in order to adapt to the decline in oil prices, putting pressure on the currency reserves of the Central Bank. Then in December 2015 there was a second devaluation of 48%. In general, however, there are no restrictions for converting or transferring of the means related with investments. The sphere of financial services is weakly competitive both on the external and internal markets.

Corporate social responsibility is a relatively new concept in Azerbaijan, and is associated only with assistance programs.

Figure 2. Azerbaijan GDP per capita
Source: Trading Economics, 2018b

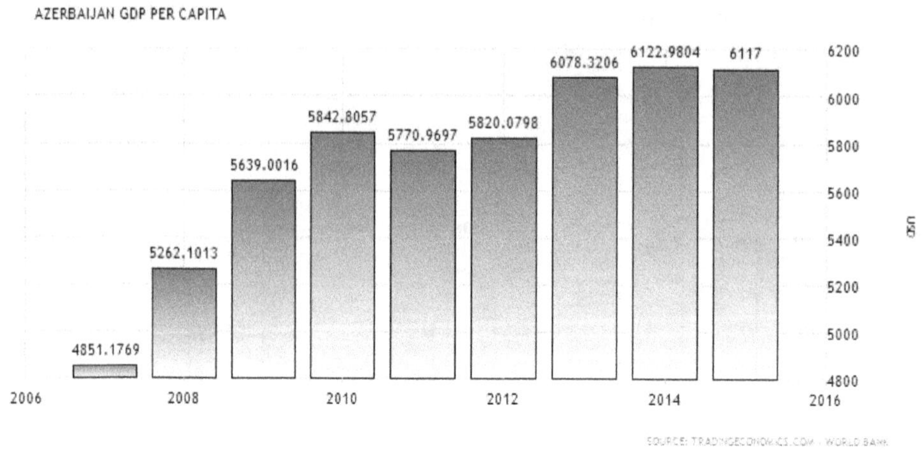

Development of the Securities Market as one of the Priority Areas of the State Investment Policy

In the economy, there is a constant turnover of financial resources, which subjects are households, enterprises, the state and various financial institutions acting as intermediaries in the system of relations. In the process of management, some subjects require means to expand their activities, while others accumulate savings. The distribution of these means occurs through the formation of funds, which can be centralized by the state or owned by the producers.

Until recently, the Azerbaijani government has practically ignored the possibility of attracting the population's financial resources for infusion into the national economy. Practically bank deposits are the only financial instrument to fill up the people's budget. Nevertheless, the two waves of currency devaluation, and the closing of some banks (for today from 45 banks remained acting 33 credit institutions) considerably lowered the population's trust in the banking system. In the developed countries of the world, income from securities is an important additional source of money for the population.

The population provides the bulk of investment in the stock market, if it has incomes that exceed consumption costs. If the consumer has a larger disposable income, then more goods will be produced and/or imported. There are special indicators characterizing the willingness of consumers to spend money on the acquisition of various goods. Some of these indicators are important for foreign exchange markets; however a high consumer demand stimulates the recovery of production in many sectors and can serve as a basis for economic growth. Conversely, the weakness of consumer demand or its decline are signals and may result in a slowdown in the economy. Focusing on these indicators, central banks can change interest rates or use other levers of financial policy, which directly affect the exchange rates. Let's consider some of them for Azerbaijan.

In view of the fact that many economic indicators show economic cycles, but each of them does it in its own way, it is natural to attempt to construct one from several indicators which, thanks to the generalization (averaging), would better predict the cycles than each one does individually. The composite leading indicator – *Leading Economic Indicator (LEI)* combines 11 indicators for this purpose:

1. Average duration of the working week in the manufacturing sector;
2. Average weekly number of applications for payment of state unemployment insurance;
3. New production orders for consumer goods and materials;
4. Efficiency of deliveries;

5. Contracts and orders for capital goods and equipment;
6. Obtaining permits for housing construction;
7. Unfulfilled production orders on durable goods;
8. Change in prices for raw materials;
9. Stock Index S & P500;
10. M2 Monetary aggregate;
11. Consumer expectations index (University of Michigan's Consumer Expectations Index).

The LEI index is based on the idea that the main driving force in the economy is the expectation of future profits. In anticipation of profit growth, companies expand the production of goods and services, invest in new plants and equipment; accordingly, this activity decreases when a decline of profits is foreseen.

In Azerbaijan the GDP YoY (year over year) shrank 0.9 percent year-on-year in January-May of 2017, compared to a 4.2 percent contraction in 2016. LEI in Azerbaijan averaged 9.87 percent from 2006 until 2017, reaching an all time high of 116.10 percent in September of 2012 and a record low of -4.50 percent in April of 2016 (Figure 3).

The Index of Consumer Confidence is an attempt to measure consumer optimism; it has a limited impact on the market, as it may not reflect the real state of the economy. However, it is traditionally used to forecast trends in employment and the general state of the economy. The growth of the index value is a positive factor for the development of the national economy.

Figure 3. Azerbaijan GDP YoY
Source: CEIC, 2018

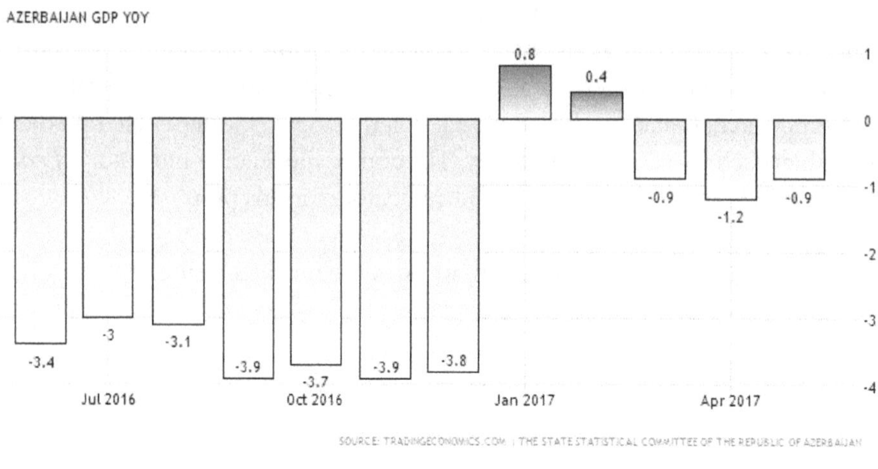

Figure 4. Azerbaijan Consumer Confidence Index
Source: Trading Economics, 2018c

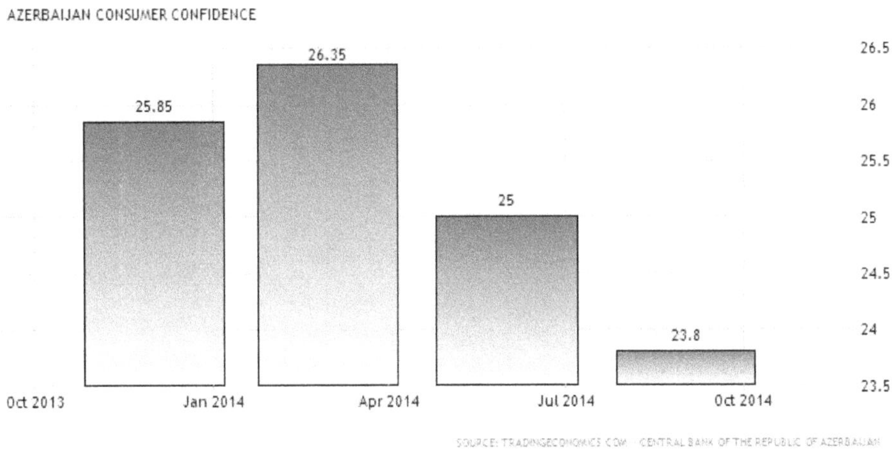

As a reflection of the economic situation, consumer confidence in Azerbaijan declined in the third quarter of 2014 to 23.80 points, being 25 points in the second quarter of 2014.

Consumer confidence in Azerbaijan averaged 25.25 from 2013 to 2014, reaching a maximum of 26.35 points in the first quarter of 2014 and a record low of 23.80 points in the third quarter of 2014 (Figure 4).

The index *Personal Income* includes the salaries of workers and employees, income from rent, dividends, income from bank interest, payments for social insurance, etc. Personal income in Azerbaijan increased from 35.924.80 million manats in 2014 in to 37,948.30 million manats in 2015. This indicator averaged 24101.00 million manats from 2000 to 2015, reaching a record level of 37.948 million manats in 2015 and a record low of 3748.00 million manats in 2000 (Figure 5).

The index *Personal Spending (Consumption)* reflects the change in spending of funds to meet personal needs. Has a limited impact on the market. Consumer spending in Azerbaijan rose_from 9.94 bn manats in the third quarter of 2016 to 9.125 bn manats in the fourth quarter of 2016. This indicator in the period from 2001 to 2016 averaged 3.767.86 million manats, reaching a maximum of 9,190.40 million manats in the fourth quarter of 2015 and a record low of 700 million manats in the first quarter of 2001 (Figure 6).

The *Consumer Price Index (CPI)* is the main indicator of inflation, measures the changes in the goods and services' prices included in a fixed consumer basket (Tsibulnikova, 2013). This indicator increased from 141.40 points in December 2016 to 144.70 points in January 2017, an average of 115.69 points from 2011 to

Figure 5. Azerbaijan Personal Income Index
Source: Trading Economics, 2018d

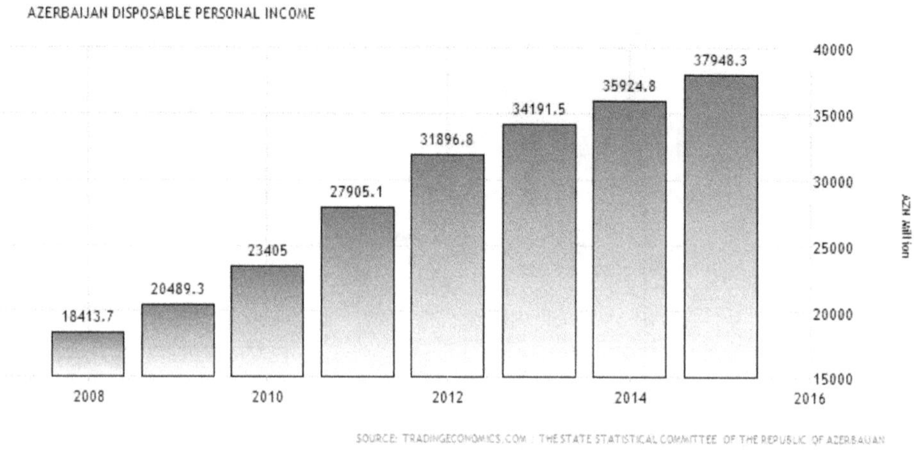

Figure 6. Azerbaijan Personal Spending (Consumption) Index
Source: Trading Economics, 2018e

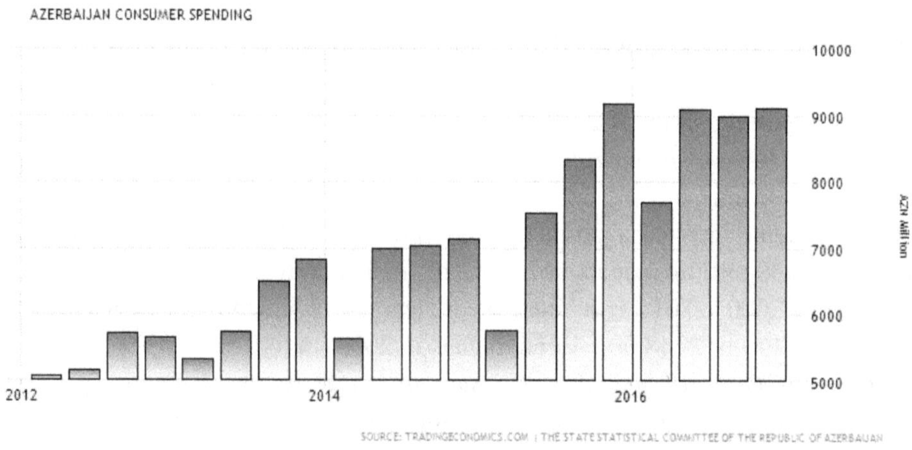

2017 and reaching a record level of 144.70 points in January 2017. The minimum record was fixed at the level of 105.30 points in January 2011 (Figure 7).

As it was said above, the population is the main supplier of resources to the stock market. Credible figures for Azerbaijan, disclosed above, give no chance on successful development of this economic sector.

Let us consider this situation in more detail. The securities market in Azerbaijan was established at the beginning of the 20th century. Back in 1900, promissory notes

Figure 7. Azerbaijan Consumer Price Index
Source: Trading Economics (2018f)

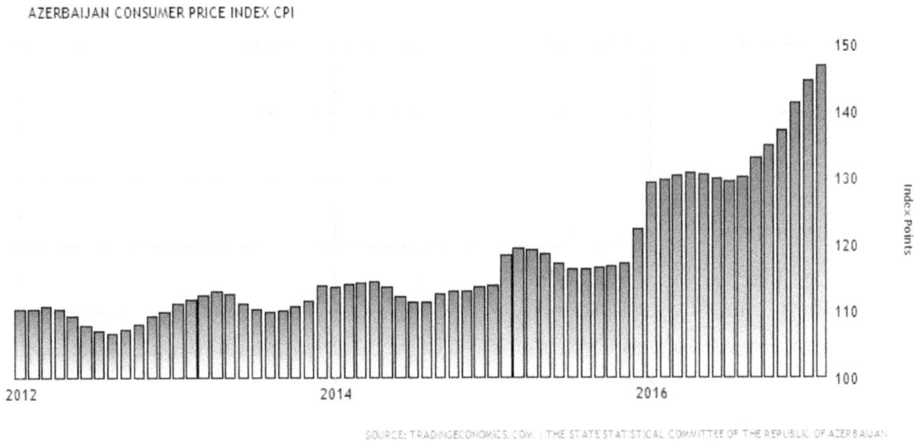

and interest-bearing securities were issued by banks operating in Baku. The first shares were issued in 1913 after the approval of the Baku Merchants Bank's Charter. In the era of the Azerbaijan Democratic Republic, the bill market also developed, but during the Soviet period it was limited.

On December 30, 1998, after the restoration of independence the State Committee for Securities was established to create and develop the securities market.

The Committee is engaged in the adoption of regulations in the securities market, the emission and turnover of shares, the activities of securities market participants and investment funds, the licensing of related activities, as well as the market for debt securities.

As shown in Musayev (2012), the economic reforms carried out in Azerbaijan in recent years have ensured the creation of an economic system operating on the basis of free market relations. The fundamental institutions of the market economy have been formed, including the legislative framework that consolidates new economic relations.

Securities which will be interesting to society have to be issued for work of stock exchange. In Azerbaijan, such large joint-stock companies as the State Oil Company of the Azerbaijan Republic (SOCAR), the organization for providing electricity to the country AzerEnerji, etc. belong to the state. As for public companies, they are open only formally, but in reality, too, belong to the state. Potential buyers do not believe that over time competitive monopoly organizations will give an opportunity to develop joint stock companies.

No one wants to give up part of the own enterprise. However, the shares are essentially related to the transfer of part of the ownership rights to shareholders. In addition, another reason for the underdevelopment of the stock market in Azerbaijan is the lack of qualified and experienced personnel, for example independent brokers.

Thus, the model of the undeveloped securities market in Azerbaijan is characterized by the fact that there are practically no national investment resources, national production cannot effectively consume foreign investment.

Yet the situation in the world economy suits developing countries. They feel much better developed, their markets grow, which makes them attractive to investors. However "overly good" is also not good: capital flows from developed countries can lead to overheating and a crisis.

Azerbaijan does not have such risks - the economy has fallen into a lethargic dream: industry, investment, exports and GDP growth are generally slowing down, and this is the best medicine for overheating.

Oil, Gold and the Stock Market

Let us return to the subject of abandoning the dollar in the computations for oil supply. In our opinion, "rejection of the dollar" is nothing but a desecration. Such "refusals of the dollar" and the emerging systems of local mutual settlements are a distortion because the instrument of mutual settlements is not important. *An instrument for quoting these mutual settlements is important.*

No matter how these countries name their currency, accepted between them for mutual settlements, they will continue to quote the rate of this currency (determine the value of the currency) and to determine the value of the goods (determine the value of the goods) in the same petrodollars. Moreover, this means that, through the dollar exchange rate (the dollar index) the US will retain its economic power over these countries and can still influence the economy and therefore the policies of these countries. What should be done to ensure that the rejection of the dollar will be real and not virtual? To do this, countries need to sell their oil and gas, either for gold or for a national currency tied to gold, and not to the US dollar. Only the termination of quoting (assessment) in dollars of the cost (value) of oil and cost (value) of the used settlement monetary unit is capable to destroy world hegemony of the petrodollar; moreover, with it the entire world hegemony of the United States. Here a reasonable question arises: how to determine the value of oil relative to gold, without the participation of the US dollar, in which today both oil and gold are quoted (estimated) in world markets? The answer is "no way". It's just not necessary. The market itself will determine the relative value of oil against money (gold). It is only necessary to start selling oil for gold, according to the artificial proportions of values that we have today and which are expressed in US dollars. Moreover, in order to do this, there is no need even to announce the rejection of the dollar. It is enough simply to immediately exchange all dollars received from the sale of oil and gas for gold.

As it was said above, oil is a trump card of all geopolitics, and the countries that have this card try to use it. The fact that foreign oil companies with billions of dollars in debts, will pay for Iranian oil in euro, creates the idea of the similar practices' introduction in our country. Nevertheless, can Azerbaijan refuse the dollar in foreign trade and switch to the euro after two devaluations? And do we need this? For Azerbaijan, such a step would be extremely dangerous, because the limited inflow_of dollars is the main cause of instability in the financial market at present. The fact is that 90% of our exports are oil and oil products. The country has very few other dollar revenues. The price of oil is falling: the volume of the dollar coming to the republic is dropping, and the Central Bank is beginning to spend its foreign exchange reserves. The value of gold expressed so far in US dollars, immediately starts to grow rapidly, both against the dollar, and against all goods-works and services. After a while in a certain market-defined trading range the oil-gold ratio will be stabilized naturally and the dollar will simply fall out of the formula which determines the price of oil and gold. On the contrary, the value of the dollar will begin to be measured in gold, as it should be and as it was throughout the entire monetary history of humankind. The value of national currencies has always been valued in money, i.e. in gold.

The changes of oil prices always have a strong influence on the deals. But in the last 10 years the nature of this influence has changed dramatically.

Modern oil exploration is financed through a variety of methods, including issuing shares to increase capital, as well as attracting finance through bonds and bank loans.

If the future expectations of the global economy are bad, as it is happening now in Azerbaijan, people are trying to protect the dollar and gold and sell shares in countries where the stock market is developed. The price of gold is rising; the value of the dollar is rising against the national currency. When the price of the dollar rises, oil prices rise for developing economies. This creates an additional burden on the economy, since inflation increases with energy costs. Because of high inflation, people invest more in gold and less in stocks, which leads to a drop in stock markets.

There are many combinations of oil, gold and shares, but as a rule:

- In developing countries gold and oil are in inverse ratio: decreasing oil prices testifies about bad times, then the price of gold increase.
- In developed countries gold and shares are also in inverse ratio. If shares' prices increase, then the price of gold decreases and vice versa.

SOLUTIONS AND RECOMMENDATIONS

Money is not the exclusive and the main goal of entrepreneurial activity, but also the most important means of this activity, since without the availability of money (initial investments) the emergence of entrepreneurial activity is impossible. At the same time, it is necessary to solve questions about the source of capital and its investment. One of the main subjects of business turnover in financial markets is securities.

There are more restrictions on foreign capital in Azerbaijan than in Eastern Europe and Central Asia. Under current legislation, a controlling stake in companies operating in the mining sector or oil and gas sector remains in the country. Among the most important measures to limit foreign investment are:

1. Establishment of special state control over the admission of foreign capital to the development of mineral and natural resources;
2. The prevention of foreign capital in certain sectors that are most important for the national economy;
3. Establishment of an obligatory share of participation of the national or private capital in the enterprises created by foreign firms (in the mixed societies);
4. Activities directed to use of some of foreign enterprises' profits for the domestic needs of developing country (taxation, restrictions when transferring profits abroad, etc.);
5. Definition of concession policy.

FUTURE RESEARCH DIRECTIONS

Research results indicate that the accumulation of huge monetary and financial resources has an ambiguous impact on the economy of oil-producing countries. On the one hand, it creates unprecedented financial conditions for boosting the economic development of these countries. On the other hand, it generates such problems as, for example, inflation growth, deepening of structural disproportions (flawed development of the "non-oil" sector), and difficulties in the social sphere. The development of entrepreneurship is closely linked with the development of small enterprises, the formation of the middle class and is conditioned by both internal and external factors. Investment activity is associated with the processes of international separation of labor, attracting foreign investment and the development of investment projects.

The modern structure of the world in the postindustrial information space becomes very vulnerable. Therefore, in the very near future the development of oil fields with the correct configuration of the innovation process can move from massive technologies, requiring huge economic and industrial resources to point high-tech modules, whose development requires not so much capital investment, as creative flexibility and avant-garde approach.

The developing countries, to which Azerbaijan belongs, should develop this side of strategic asymmetry as a priority.

CONCLUSION

The main problem is rational using of financial resources within the country for its social and economic development. The desire to master other markets in the face of the declining efficiency of investments in the development of new oil resources can be considered as the most important prerequisites for the transition to a non-primary economy. This process of deliverance of raw dependence can be intensified by increasing investments in entrepreneurship and development of the stock market. The strategic goal of economic development of any country is foreign investments' attraction. Foreign direct investments are of significant importance, they characterize the long-term interest of foreign investors in the development of economic activity. The stock market develops when investors appear on it.

In current conditions, Azerbaijan should treat the idea of de-dollarization more restrainedly. First, if the country switches to calculations for oil and gas in manats, then the budget will lose additional income from devaluation. To date, since the beginning of the year, manats has depreciated more than 100% against the dollar and the euro. The Transformation of the manats into an international currency will probably not lead to the stabilization of the national currency, but to its even greater volatility. the Conversion of currency into a freely convertible one means that its circulation is not limited in any way, the state cannot influence it in any way, and its rate is determined only by speculation on the stock exchange - in this case the dependence on the dollar is only strengthened.

Today, especially during the currency crisis, it is economically unprofitable, not to mention the fact that Azerbaijani raw materials exporters need dollars and euros to pay off their foreign debts and make purchases on imports to support the extraction of raw materials. The transition to manats, if it takes place, will sharply worsen the competitive position of Azerbaijan in the energy markets. The costs will significantly increase for importers: since they need to buy manats for dollars or euros, they lose the difference between the rates of purchase and sale. In addition, they would have to spend money on insurance against risks of a depreciating manats, and this would also be worth a lot, because manats is a currency, which rate is subject to very high fluctuations. The acquisition of fuel is often done at the expense of a loan. There are a very high percentage of manats loans. Here is one more extra cost for those who would have to buy Azerbaijani fuel for manats. When we force customers to pay extra, we lose them.

The results of theoretical research identify essential features of overcoming the dollar dependency and oil dependent countries' strategic objectives. For these aims:

1. The authors consider the geopolitical asymmetry associated with the dominance of the dollar in this market, analyze the existing political situation and offer their vision of republic's economy, for which oil has become a brake on development.
2. The manuscript shows in detail that developing countries including Azerbaijan, depend on developed countries. This is manifested primarily in the fact that developed countries are "pure" exporters of capital to developing countries, whereas developing countries are the debtors of developed countries. In foreign trade, economic dependence is manifested not only in the overall asymmetry of exports and imports, but also primarily in the market's commodity structure.
3. The authors draw attention to the development of entrepreneurship, which occupies an important place in socio-economic policy, and to the development of the stock market, which is an active participant in the investment processes. In turn, investment processes define scientific and technical progress, modernization of economy, growth rates of branch and regional economies in essential degree. The paper considers the current state of this economic area, presents the dominant indicators for Azerbaijan in this sector. As a continuation of the theme of entrepreneurship development, the authors examine in detail the Azerbaijani stock market, considering that the economy is based on the functioning of the corresponding financial system, and the most important place is occupied by the financial market on which there is the movement of money (capital).
4. As a logical conclusion of the study, the authors in the final part again return to the theme of oil's de-dollarization, and, assessing all the shortcomings of the resource asymmetry from the perspective of a weak state, present their vision for solving this difficult problem.

REFERENCES

Baumeister, Ch., & Lutz, K. (2016). Forty Years of Oil Price Fluctuations: Why the Price of Oil May Still Surprise Us. *The Journal of Economic Perspectives*, *30*(1), 139–160. doi:10.1257/jep.30.1.139

Bourguignon, F., & Morrison, Ch. (2002). Inequality among World Citizens: 1820–1992. *The American Economic Review*, *92*(4), 727–744. doi:10.1257/00028280260344443

CEIC. (2018). *Azerbaijan*. Retrieved from https://www.ceicdata.com/indicator/azerbaijan/data/real-gdp-yoy

Civil Code of the Republic of Azerbaijan. (n.d.). Retrieved from http://isb.az/wp-content/uploads/2015/03/mulki_mecelle.pdf

Davis, J. (2006). The Changing World of Oil: An Analysis of Corporate Change and Adaptation. *Environmental Politics*, *18*, 159–160.

Engdahl, W. (2016, May 15). Anglo-American Oil Market Is Broken. *New Eastern Outlook*. Retrieved from https://journal-neo.org/2016/05/15/anglo-american-oil-market-is-broken/

ENI. (2015). *World Oil and Gas Review*. Retrieved from https://www.eni.com/docs/en_IT/enicom/media/press-release/2016/11/Eni_presents_15_edition_World_Oil_and_Gas_Review.pdf

Geopolitics of Oil Hearings before the Committee of Energy and Political Resources (1980). *US Senate, 96 Congress, the 2nd Session. Pt. 1–2. Wash.: US gov., print. off.*

Gutiérrez, D. G. (2007). The Politics of the Interstices: Reflections on Citizenship and Non-Citizenship at the Turn of the Twentieth Century. *Race/Ethnicity. Multidisciplinary Global Contexts*, *1*(1), 89–120.

International Energy Agency. (2014). World Energy Outlook 2014. *International Energy Agency*. Retrieved from https://www.iea.org/publications/freepublications/publication/WEO2014.pdf

Karl, T. L. (2007). Oil-Led Development: Social, Political, and Economic Consequences. *CDDRL Working Papers,* No 80. Retrieved from https://cddrl.fsi.stanford.edu/sites/default/files/No_80_Terry_Karl_-_Effects_of_Oil_Development.pdf

King, M. (2009). Whyte Paradoxes of China's Economic Boom. *Annual Review of Sociology*, *35*(1), 371–392. doi:10.1146/annurev-soc-070308-115905

Lau, L., Qian, Y., & Roland, G. (2000). Reform without losers: An interpretation of China's dual-track approach to transition. *Journal of Political Economy*, *108*(1), 120–143. doi:10.1086/262113

Lemann, N. (2001, July 16). Dreaming about War. *The New Yorker*. Retrieved from: http://www.comw.org/qdr/0107lemann.html

Meinhardt, H. I. (2002). Cooperative Decision Making in Common Pool Situation. Lecture Notes in Economics and Mathematical Systems. Springer Berlin Heidelberg.

Musayev, E. (2012). Securities Market In Azerbaijan: State And Tendencies Of Development. *Russian Business*, *10*, 53-65.

Plummer, R. (2013, August 1). Alba alliance ambitions lay bare Latin trade confusion. *BBC News*. Retrieved from http://www.bbc.com/news/business-23515979

Proliferation Issues. (1992). JPRS Report. JPRS-TND-92-006. *Foreign Broadcast Information Service*. Retrieved from http://www.dtic.mil/dtic/tr/fulltext/u2/a333831.pdf

Qiang, S., & Xiaokun, L. (2010, November 24). Russia quit dollar. *China Daily*. Retrieved from http://www.chinadaily.com.cn/china/2010-11/24/content_11599087.htm

Rumley, D., & Chaturvedi, S. (2015). *Geopolitical Orientations, Regionalism and Security in the Indian Ocean*. Taylor & Francis. Retrieved from http://www.tandfebooks.com/doi/view/10.4324/9781315689487

Schaffer, T. C. (2002). Building a New Partnership with India. *The Washington Quarterly*, *25*(2), 31–44. doi:10.1162/016366002528201l7

Scholvin, S. (2016). *Geopolitics. An Overview Of Concepts And Empirical Examples From International Relations.* FIIA Working Paper.

Trading Economics. (2018a). *Azerbaijan Foreign Direct Investment*. Retrieved from https://tradingeconomics.com/azerbaijan/foreign-direct-investment

Trading Economics. (2018b). *Азербайджан - ВВП на душу населения ППС*. Retrieved from. https://ru.tradingeconomics.com/azerbaijan/gdp-per-capita-ppp

Trading Economics. (2018c). *Azerbaijan Consumer Confidence*. Retrieved from https://tradingeconomics.com/azerbaijan/consumer-confidence

Trading Economics. (2018d). *Azerjaiban Disposable Personal Income*. Retrieved from https://tradingeconomics.com/azerbaijan/disposable-personal-income

Trading Economics. (2018e). *Azerbaijan Consumer Spending*. Retrieved from https://tradingeconomics.com/azerbaijan/consumer-spending

Trading Economics. (2018f). *Azerbaijan Consumer Price Index CPI*. Retrieved from https://tradingeconomics.com/azerbaijan/consumer-price-index-cpi

Xiaochuan, Z. (2009). Reform the International Monetary System. *BIS Review*. Retrieved from https://www.bis.org/review/r090402c.pdf

ADDITIONAL READING

Asongu, S., & Jacinta, N. (2017). Information Asymmetry and Conditional Financial Sector Development. *Journal of Financial Economic Policy*, *9*(4), 372–392.

Conning, J., & Udry, Ch. (2005). *Rural Financial Markets in Developing Countries. Economic growth center*. Yale University.

Grossman, S. J., & Oliver, D. H. (1983). An Analysis of the Principal-Agent Problem. *Econometrica*, *51*(1), 7–45. doi:10.2307/1912246

Hayes, B. (1984). Unions and Strikes with Asymmetric Information. *Journal of Labor Economics*, *2*(1), 57–83. doi:10.1086/298023

Kirkley, J. E., Dale, S., Ferdous, M., Haji, A., & Ishak, O. (2003). Excess Capacity and Asymmetric Information in Developing Country Fisheries: The Malaysian Purse Seine Fishery. *American Journal of Agricultural Economics*, *85*(3), 647–662. doi:10.1111/1467-8276.00462

Mankiw, N. G. (1985). Small Menu Costs and Large Business Cycles: A Macro-economic Model of Monopoly. *The Quarterly Journal of Economics*, *100*(2), 529–537. doi:10.2307/1885395

Mayer, C. (1990). *Asymmetric Information, Corporate Finance, and Investment*. Chicago: U. of Chicago Press.

Mishkin, F. S. (1991). *Financial Markets and Financial Crises*. Chicago: University of Chicago Press.

Newbery, D., & Stiglitz, J. E. (1976). Sharecropping: Risk Sharing and the Importance of Imperfect Information, in J.A. Roumasset, et al. (eds.), Risk Uncertainty and Development, SEARCA, A/D/C, 311–341.

Potoski, M., & Prakash, A. (2009). Information asymmetries as trade barriers: ISO 9000 increases international commerce. *Journal of Policy Analysis and Management*, *28*(2), 221–238. doi:10.1002/pam.20424

Chapter 9
Central Asia's Role in China's Energy Security:
Challenges and Opportunities

Paulo Afonso Brardo Duarte
University of Minho, Portugal

ABSTRACT

Central Asia has gained extraordinary importance in recent years in the framework of global energy security. China is the most significant example of a power that looks to its periphery as a viable option for energy supply. In Central Asia, Chinese companies are dynamic players having even broken the long Soviet and Russian monopoly over regional pipelines. This chapter examines the importance of the region within China's energy security, while not overlooking the potential contribution of the China-Pakistan economic corridor in the energy transit. In addition, Central Asia is likely to help China reduce the energy deficit in Xinjiang, through the import of hydroelectricity generated in Tajikistan and Kyrgyzstan. Although Central Asia's contribution to global energy security is low, it matters in a context of energy diversification, in which China's One Belt One Road brought a more promising dynamics to the cooperation between Beijing and Central Asian countries.

BACKGROUND

Central Asia has often been neglected by many political commentators, although the region offers extraordinary opportunities to help China mitigate the impact of the so-called Malacca Dilemma. The latter refers to the Chinese economy's excessive reliance and vulnerability to pressure at the Straits of Malacca. Much of China's oil

DOI: 10.4018/978-1-5225-4203-2.ch009

comes from West Asia and Africa and around 80 percent of this passes through the Straits of Malacca. Should the Straits of Malacca be threatened or come under the influence of hostile states, China's trade could be choked. The resulting energy crisis could paralyze its economy. The fear of a maritime blockade – in vital chokepoints, as Malacca – by rival powers (United States, Japan, India...), has led China to prioritise the energy issue, raising it to the level of threat to the national interest. This explains the *Going Abroad* policy, whereby the Chinese National Oil Companies (NOCs) are encouraged by the Government to acquire the maximum possible participation in consortia, oil exploration rights, equity oil, and foreign oil companies.

In this sense, the main contribution of this chapter for the development of Science is to make known the role of Central Asia within the framework of the initiative to diversify the sources of energy supply, which China has carried out since 1993, when the country stopped being an oil exporter and started importing this resource from various parts of the world. Among the challenges and opportunities that Central Asia is likely to offer to China, attention will be paid not only to the case of oil and natural gas, but also of water. This is, in practice, another fundamental contribution that this chapter aims to provide to a reader who is generally more used to understanding the importance of the region as strictly associated with oil and gas production, neglecting the other major energy added value of Central Asia: water and the resulting ability to generate electricity from it. Apparently insignificant states in terms of oil and gas production, such as Kyrgyzstan and Tajikistan, are, however, upstream countries with regard to the existing water resources in the region, holding a remarkable potential for hydroelectricity production. As such, and because China's energy needs are not confined to oil and gas, Central Asia can be important in the supply of electricity to China's remote regions which are affected by a high energy deficit, such as Xinjiang, where electricity failures sometimes last for several consecutive hours daily. Another goal of this chapter is to explain the relevance of the so-called China-Pakistan Economic Corridor, susceptible to drain more quickly Central Asian energy production to world markets, and, at the same time, to allow the oil from abroad to arrive to China without having to cross long and sensitive sea routes.

The goals listed above get a greater relevance in a context in which China is committed to make Central Asia a crucial component of its *One Belt One Road* (OBOR). Beijing sees the OBOR as a way to find new markets, to reduce the imbalance of development between its coastal provinces and the poor interior, and to preserve national stability (Yang, 2016). Under the aegis of the OBOR, Chinese political leaders have promoted the creation of important hubs in remote provinces in China, such as Chongqing, Kunming, Xi'an, Xining and Chengdu, which have modern logistics infrastructures that allow a quick connection to the major coastal cities of the country (Lim *et al.,* 2016). Another important goal of the OBOR is the issue of stability. And, in this sense, the OBOR is largely led by Xinjiang's development and stabilization strategy, which Beijing wants to protect from any terrorist or separatist aspiration. Considering Xinjiang's proximity to the Central Asian Republics and Xinjiang's access to the Indian Ocean, Chinese political leaders have adopted the '*Go West*' policy complemented by what Rolland (2015) calls *China's infrastructure diplomacy*. The latter is based on the construction of a *corridor of interconnectivity*, which aims to turn China's remote provinces into economic and logistical hubs in the connection between Asia and Europe (Wang, 2016). Besides, such an achievement is intended to integrate China's neighbors more closely with the Chinese economy, enmeshing them in a network of trade ties, transportation links and multilateral institutions that will have China at their center. Esteban and Otero-Iglesias identify four economic areas in which the OBOR can be decisive: "driving the internationalization of its construction industry, encouraging exports, reducing risks in the supply chain and attracting investments towards the interior of the country" (2015: 6). The OBOR is, metaphorically, a two-way route, either by means of the incitement to expansion of Chinese companies around the world, or in the invitation to foreign investors to bet on the Chinese market. Furthermore, the OBOR can be an important lever regarding the slowdown in construction in China, providing Chinese construction companies with promising profitable opportunities abroad (Xu & Chung, 2016). All these reasons are valid to understand why the OBOR makes sense, although there is another reason, equally or more important: energy. Energy cooperation and the construction of infrastructure will be new engines for cooperation between China and the Central Asian countries.

This analysis is based on a qualitative methodology, supported by the conceptual lenses of the Copenhagen School, in particular the word *securitizes*. According to the Copenhagen School, *securitization* is a process whereby a *securitizing* agent tries to establish, socially, the existence of a *threat* to the survival of a unit. When a subject is *securitized*, it comes out of the scope of *normal policy* and moves into the realm of *emergency policy*, which usually legitimates the use of force (Buzan *et al.*, 1998). In practice, this chapter aims at analyzing how Central Asia can help China to *securitize* the energy issue, mainly oil, natural gas and electricity. Complementing the qualitative (hermeneutic analysis), this study has as empirical support the doctoral research conducted by its author, in 2011 and 2012, both in Portugal, and in Kazakhstan, Kyrgyzstan and Tajikistan, where semi-structured interviews were undertaken with several local experts.

To achieve the aforementioned objectives, this chapter begins by highlighting the contours of China's regionalism, as well as the dimensions of China's energy (in) security. In the second part, it examines how Central Asian energy potential may be appropriate to China's needs and interests. In anticipation of the main conclusions, it is argued that despite all the opportunities that Central Asia may offer in the framework of both Chinese and global energy security, one must not fall into the tendency to over-evaluate the importance of the region, since, for all purposes, in the next decades the Caspian will not replace the Middle East as the main reservoir of the world's oil.

CHINA VIS-À-VIS THE REGIONAL DYNAMICS

China is a latecomer that has dramatically changed the geopolitical and economic state of affairs in Central Asia, having rapidly become a key actor on the regional scene. Although China's relations with Central Asia are far from being recent (more than 3,000 years), China has only been an active participant in Asian regional institutions since the 1990s. A very illustrative example in this regard is the creation of the Shanghai Cooperation Organization (SCO) in 2001, a Chinese initiative. The SCO is used by Beijing to institutionalize its relations with Central Asian countries.

China is seeking to present itself as a responsible stakeholder, which aims to strengthen relations with the states of the region. In this sense, China's good neighborhood policy is based on massive involvement in the construction and upgrading of extraction infrastructure, as well as on transport facilities and communications. With Xi Jinping's rise to power, Chinese regionalism has entered a new stage, aiming at the construction of a "community of common destiny", which is based on the idea that "[China's] dream of revival will come true only in the process of sharing development fruits and making common progress with regional countries" (Ling, 2013: para.18).

For Beijing, the SCO is the instrument that allows China to establish a single economic space in Central Asia, to resolve border disputes and to provide security for Xinjiang. One of the main objectives of the SCO is to ensure energy security in the region, to develop and implement effective strategies in this area, as well as signing of the contracts between member countries. But unlike the economic and security emphasis that China puts on the SCO, Moscow tends to channel, especially, in the Eurasian Economic Community, the dealing with regional economic issues, using, in turn, the Collective Security Treaty Organization for bilateral talks on security with each Central Asian republic. According to the Russians, the SCO is just an interesting and useful forum to manage their relationship with China in Central Asia, so as to institutionalize Moscow and Beijing's regional activities within a multilateral framework. Russia's interest in the SCO aims to undermine momentum behind any Chinese-led economic block in Central Asia that might result from, for example, a China-Central Asian Free Trade Agreement (Reeves, 2014). This instrumentalization and divergent view on the SCO's role, according to the Russian or Chinese concepts, is, basically, a reflection of the tensions underlying the Sino-Russian relations in the region. Indeed, although partners, Russia and China are two direct competitors in the pursuit of power, influence and natural resources in Central Asia.

The next years will be particularly important to test the SCO's resistance in the face of a possible reconfiguration of regional power. Experts like Nopens, consider that "the withdrawal from Afghanistan is accelerating the shift of the balance of power in Central Asia in favor of China" (2014: 1). China's advances are at the origin of this observation – driven by economic, security, geopolitical and geostrategic factors – in Central Asia, where until recently Russia has exercised absolute control on the oil and natural gas flow routes in the region, a monopoly that has been broken, meanwhile, by Western initiatives, as well as by the Chinese, with the construction of regional pipelines.

The Dimensions of the Chinese Energy (in)Security

In April 2015, China surpassed the United States as the world's largest consumer of oil (Reuters, 2015). China's rise marks a new era in the history of energy. In fact, for most of the past decade, rising oil consumption in China has boosted oil prices. The geopolitical importance and potential influence of China in the world have been expanding as a result of its vast oil investments, long-term supply agreements, outstanding bank loans, and a vigorous energy diplomacy (Herberg, 2011). China has long been able to meet its energy needs through the use of domestic reserves. Actually, although China was once a major oil exporter to the East Asian countries, since 1993 its domestic reserves have no longer been able to meet the demand. Since then, China has become a net oil importer. In the case of China, it is unquestionable that there has been both a tremendous growth in terms of population, as well as in terms of the standard of living of its citizens over the past decades, which translates therefore in a greater need for oil.

In order to meet China's energy needs, Beijing's diplomacy has been working actively since the mid-90s, in order to improve its relations with the major oil exporting countries, to ensure safe access to energy sources abroad, and investing in infrastructure projects. The Middle East remains the largest oil supplier to China, although African countries, particularly Angola, have come to occupy an increasingly important place in the context of China's oil imports in recent years. In 2014, the Middle East provided 3.2 million barrels per day (52 percent). Among the other regions that export to China, Africa contributes with 1.4 million barrels per day (22 percent), the Americas with 667,000 barrels per day (11 percent), Russia and the former Soviet Union with 778,000 barrels per day (13 percent), the Asia-Pacific region with 127,000 barrels per day (2 percent), and 27,300 million barrels per day (<1 percent) are coming from other countries (International Energy Agency, 2015).

Like many other countries which import energy resources, China does not have an overall strategy for energy security, being instead guided by a centralized policy that sometimes leads to competing interest groups, exaggerated rhetoric and initiatives at the provincial level. Moreover, China pursues a strategy of petroleum assets, which besides not providing energy security even aggravates instability. Oil is considered a fundamental strategic asset (due to its importance) for China's economic development, but at the moment is proving to be the element par excellence, of China´s present structural energy insecurity (Wang, 2015). China's apprehension regarding the fact that domestic oil production has stagnated in recent years (which corresponds, according to the Copenhagen School, to the perception of a *threat*), reinforces the need to *securitize* the access to new oil sources. After all, although China is supposed to have 32 billion barrels of technically recoverable shale oil reserves, so far shale oil production has not gotten off the ground as it did in the US (Peak Oil Review, 2015). It follows that the first factor contributing to Chinese oil insecurity is, therefore, China's dependence – more than 50 percent – on oil from abroad. China's reliance on imported oil soared to over 60 percent in 2015, which is jeopardizing the country's energy security, to the extent that importing more than 50 percent is a globally recognized energy security alert level (China Daily, 2016).

Another factor linked to China's oil insecurity, and that indicates that the country will continue to depend increasingly on foreign oil, concerns the fact that China has entered a period of rapid development of urbanization and heavy industrialization, which require substantial energy consumption. It is estimated that in the next 20 years, China's urban population will reach 75 percent of the total population (Facts & Details, 2015). However, this data is important with regard to China's energy (in) security, since, as indicated by *The Outlook for Energy: A View to 2040*, an urban resident consumes more energy than a rural inhabitant and, on the other hand, urban dwellers tend to have easier access to more modern and efficient fuel.

Another trend related to Chinese oil insecurity, and that points to a greater dependence on foreign oil concerns is the emergence of a middle class in China, which starts opting for air transport, while simultaneously is fascinated by automobiles (South China Morning Post, 2015). While oil is susceptible of potentially replacing coal, nuclear and hydroelectric power, no other primary source can, however, replace oil at a reasonable economic cost. Since China became an oil importer in 1993, its leaders have considered the development of relations with oil producers in the Middle East as a diplomatic priority. Although there is enough oil abroad to meet the Chinese energy demand in the coming years, one has, however, to acknowledge that most Chinese oil imports come mainly from politically unstable regions (Huang, 2016). One might question the effectiveness of this option (i.e., to resort to these unstable producers), however, it is essential to have in mind that the fact that China has entered late in the international oil market, had already won a first-mover control over the best existing oilfields, leaving Beijing with little choice beyond the establishment of agreements with the so-called rogue states. This also explains why the Chinese state oil companies are so committed to acquire assets in West Africa, Latin America, Central Asia and Southeast Asia. Let us now focus specifically on Central Asia, by highlighting its potential role on decreasing China's energy insecurity.

Central Asia's Role in China's Energy (In)Security

The Issues at Stake

Several geopolitical trends highlight the prominence of Central Asia within the framework of geostrategy and global economy. Authors such as Mackinder (2004) or Brzezinski (1998), underlined the importance of the *'Heartland'* (Eurasian Balkans). In recent years, mainly due to discoveries of major energy reserves in the region and to developments related to world and regional security, several lines of thought have reinforced Mackinder and Brzezinski's assumptions. The debate, itself, on the concept of security, which emerged in the post-Cold War, has assigned other dimensions – surpassing the state-centric and anarchic vision of the international system (defended by authors like Morgenthau,1978) – to the idea of security. In fact, the realist line of thought, for which security was inseparably linked to the possession and use of military capacities, power and interests of state actors, gradually lost its importance (Keohane, 2000). Studies on security, in the 90s, converted it into a global concept. Barry Buzan was important for the current that claimed a wider dimension to the concept of security. This would be composed, henceforth, by a military, political, economic (including energy security) and environmental feature (Buzan, 1991). The way for the *securitization* of the energy phenomenon was then paved.

Several authors argue that Central Asia's energy wealth has transformed the region into a crossroads of tension between states, of competition between companies and regional actors. Consequently, the great powers and multinational companies want to be present in this race to energy (Contessi, 2016). Access to oil reserves, the routes of pipelines and the debate on who should build them are in the premises of what some experts call the *New Great Game*, or better, the return of the *Great Game*.

In the context of the expansion of the Chinese economy and the search for new international markets, China is a regular power in the Central Asian *New Great Game* (Farchy, 2015). Although this interior region (already) has oil and gas pipelines able to drain its energy resources to world markets, there will be, of course, an increasing tendency to strengthen the existing ones, as well as to build new energy infrastructures, susceptible to provide an alternative before the uncertainty of a possible crisis in the Suez, or a saturation of the Panama Canal. The control of these routes, as well as the attempt to direct access to energy resources in Central Asia has been associated with the idea of a *New Great Game*, by analogy to the *Great Game* of the 19th century between the Russian and the British empires (Stegen & Kusznir, 2015).

Central Asia is one of the pivot regions of the world. It is located in the nucleus of the Eurasian continental space and is a crucial link between several robust and dynamic economies, such as China, European Union, India, Japan and Russia. Central Asia owes its importance to the vast economic potential and geostrategic location, becoming increasingly a world economic hub (Khwaja, 2003). Central Asia is the region where the effects of geopolitics and competition between the great powers has been more felt compared to any other part of the world. Indeed, ethnic and religious conflicts, energy competition, the strategic positioning of the various actors and the political unrest in the region, have proved a recurring feature in the Central Asian regional context (Duarte, 2017).

Since the beginning of the 21st century the competition between the great powers around energy resources has intensified, contributing to a rapid rise in energy prices, and also to new outlines in terms of energy security. In this context, and because of its energy reserves, Central Asia has proved to be an area of competition and rivalry between regional and extra-regional powers, which affects the relationship between these, as well as the balance power, influencing thus the international framework that emerged in the post-Cold War. Geopolitics naturally provides an explanation for that fact, considering that it is in large part determined by the dimensions of a region. In fact, great powers need to acquire a large land mass to exert influence in the international arena.

If one looks at the history of oil, the general ambition, since the 70s, by the various consumer countries, has been to rely less on the Persian Gulf, as it is a highly volatile area (Fonseca, personal interview, Lisbon, 2011). In fact, much of the remaining oil reserves in the world, are located in unstable countries in the Middle East, and far from areas of consumption, which raises concerns about the security of oil supplies. When one recalls recent history, it is clear that the North Sea or West Africa, regions that ultimately serve as a counterweight to the dominance of the Persian Gulf and the Middle East in world´s oil production, had been they, too, object of interest on the part of the consuming powers. Nonetheless, if the energy resources of the North Sea proved an attractive option in the period that followed the oil shocks, nowadays it is essential to find other alternatives capable of replacing a production that has been falling, particularly in the UK and Norway. Given this scenario, Central Asia has, therefore, a very important role in the diversification of energy sources (Fonseca, personal interview, Lisbon, 2011). Therefore, it is important to realize the energy potential of the region, in terms of oil, natural gas, water resources and capacity to generate electricity.

Central Asia's Energy Wealth

A special report, prepared for the Caspian region and Central Asia, estimates that the total oil reserves of the region is more than 60 billion barrels, and some predictions bring this figure up to 200 billion barrels (Global Business Reports, 2012). Experts such as Chenoy (2007) believe that the first estimates of the energy potential of the region, purposely overvalued, mainly served political purposes (such as to ease U.S. penetration in Transcaucasia and Central Asia). However, despite the 'instrumentalization' of figures, there is a consensus on the existence of abundant oil and natural gas reserves in the Caspian region. It is estimated that the Caspian contains two to four percent of the world's hydrocarbon reserves, which is comparable not to the reserves of the Persian Gulf, but to the energy potential of the North Sea. Although the Persian Gulf is the energy source, par excellence, of the modern world, Bulat Sultanov (Director of Kazakhstan Institute of Strategic Studies) believes that the Caspian and Central Asia have the resources of the future, from oil and natural gas, to uranium and other various mineral reserves (personal interview, Almaty, 2011). Therefore, according to this expert, the interest in the region will tend to grow, while, at the same time, there will be a fall in the production in other producing regions.

Energy resources are unevenly concentrated in the region. Kazakhstan (the state with the largest recoverable oil reserves) had, at the beginning of 2015, about 30 billion barrels of confirmed oil reserves. Kazakhstan has the world's 10th largest oil reserves and the second largest after Russia, in the Commonwealth of Independent States (CIA - The World FactBook, 2016a). Oil production and exports from the Caspian will tend to grow substantially over the coming years, mainly due to the Kashagan field (in the northern Caspian Sea), which is believed to be one of the most important energy discoveries of the world in the last 30 years (Oil & Gas Monitor, 2015). The offshore Kashagan oil field is generally rated as the 5th or 6th largest in the world, and has the largest reserves of any oil field outside the Middle East. Its reserves are estimated at 38 billion barrels, of which an estimated 10 billion are recoverable. About 70 percent of the Kazakh oil and gas reserves are concentrated in the western region of the country, around the city of Atyrau.

A significant part of the Kazakh and Russian oil is relatively expensive and technically difficult to extract. Those reserves are primarily concentrated at a large onshore depth, which requires special equipment for the operations. Offshore reserves, in the Caspian Sea, are faced with a set of specific problems, as the most promising fields are located in shallow and fragile environmental waters that sometimes freeze. Moreover, many of the Kazakh oil fields are buffeted by frequent seismic activity, as well as extreme weather conditions. The most important deposits are quite distant from population and logistical centers, which requires the construction of long pipelines, and even generates a variety of difficulties, not only in terms of the delivery of equipment and maintenance, but also in terms of supply of the basic workforce in the oilfield (Marten, 2007). Kashagan, Tengiz and Karachaganak are the main target of investments. In what concerns the external involvement in Kazakh oil industry, North American and Western European companies are leaders in the upstream sector of Kazakhstan – Chevron, for example, owns, individually, a 50 percent stake in Tengiz, which is the world's largest oil field with deeper location – about 3,657 meters (Chevron, 2016). However, China has been a dynamic player in the Kazakh oil sector since 1997, investing in oil fields and pipelines. Chinese companies control about a quarter of Kazakh oil production (Fingar, 2016).

Neither Kyrgyzstan nor Tajikistan has substantial oil reserves. While Tajikistan has proven reserves of 12 million oil barrels, Kyrgyzstan, in turn, may have 40 million barrels of proven oil reserves (CIA – The World Factbook, 2016b, 2016c). Furthermore, Tajikistan and Kyrgyzstan's energy infrastructure is extremely limited. There are few refineries to process the oil, and there are several difficulties associated to its transportation to the Tajik market, fruit of the 1992-1997 Civil War (Farchy, 2015). Moreover, Tajikistan's mountainous landscape hinders the oil extraction (this also applies to the case of Kyrgyzstan). Due to these difficulties, Tajikistan and Kyrgyzstan are potential oil importers, which explains, in large part, that both countries are relatively uninteresting from the point of view of foreign investment to China, as well as to other external powers. In both countries the oil industry is state-owned. However, a brief note concerning Tajikistan. According to Christian Mellis, an OSCE expert on water and energy issues, there are strong indications of the possible existence of vast oil and gas reserves in southern Tajikistan, near the border with Afghanistan (personal interview, Dushanbe, 2012).

In mid-July 2012, Tethys Petroleum, that has been prospecting for oil and gas in Tajikistan, Uzbekistan and Kazakhstan, has updated its estimates for oil and gas reserves under the production sharing contract in Bokhtar, Tajikistan, to 27.5 billion barrels of oil equivalent, of recoverable resources, consisting of approximately 3 trillion cubic metres of gas and 8.5 billion barrels of oil (Mashrab, 2012). If, for Mellis, the discovery of oil is a fact "99 percent sure", it is important to verify "if [its] exploration is technically and economically viable", since at the date of the writing of this chapter, experts are not sure about the depth at which these reserves are located (personal interview, Dushanbe, 2012; Tethys Petroleum, 2016). If the exploitation of the Bokhtar reserves is feasible, Tajikistan – a republic with scarce oil reserves – could become one of the major world countries in the production of oil and gas per capita.

Oil production in Turkmenistan and Uzbekistan is relatively insignificant, with a tendency to continue to meet the domestic needs of these countries. In 2015 Uzbekistan had 594 million barrels of proven oil reserves. In turn, Turkmenistan remains a small oil exporter. In January 2015, the country had proven oil reserves of about 600 million barrels (U.S. Energy Information Administration, 2015a, 2015b). Most Turkmen oil deposits are located in the south of the Caspian Basin and in the onshore area of Garashyzlyk, in the west part of the country. In Turkmenistan, the oil industry faces two major obstacles. First, the physical infrastructure has been the subject of corrosion and the sectors of electricity and transport have a level of service increasingly weakened (Peyrouse, 2015). Secondly, most of the Turkmen oil is concentrated both in the margins and at the bottom of the disputed waters of the Caspian Sea. The ongoing debate about whether this is, in terms of international law, a sea or a lake, makes it, in practice, difficult for Turkmenistan, and for any foreign investor, to explore the existing oil in the Caspian.

The majority of the gas reserves in the Caspian region is owned by Uzbekistan and Turkmenistan. Turkmenistan had proven gas reserves of 265 trillion cubic feet in January 2015, being the sixth country holding the largest natural gas reserves in the world, and among the 15 top producers of dry natural gas in 2014 (U.S. Energy Information Administration, 2015b). Turkmenistan produced more than 2.5 trillion cubic feet (bcm) of dry natural gas in 2014, a small fraction of a vast untapped potential. Turkmenistan has increasingly become the focus of international investors, especially since the confirmation by the energy consultant Gaffney, Cline & Associates, that the Turkmen gas reserves may be ranked among the five most important in the world. The giant South Yolotan-Osman, located in southeastern Turkmenistan, holds, by itself (i.e. without taking here into account the other deposits in the country), between 4-14 trillion cubic meters of natural gas. In addition, there are several deposits in the basins of the Amu Darya, the Murgab and the southern Caspian. The two main gas fields are the Dauletabad and the Shatlyk. The existing energy transportation infrastructure in the country is primarily directed to Russia, due to the legacy of over a century of connections, first to Tsarist Russia and later to the Soviet Union. The death of Niyazov, in December 2006, sustained some hope that Turkmenistan would open its energy sector to the international market. Of course, the big players, i.e. the United States, China, the European Union and Russia want to come forward as there seems to be signs of a certain openness (Maza, 2015).

China's need to *securitize* its energy sources would put an end to the Russian monopoly on the control of Central Asian oil and gas routes. Illustrative in this regard, is the Trans-Asian gas pipeline from Turkmenistan to China, through which Turkmen gas is supplied to China via Uzbekistan and Kazakhstan. The creation of energy synergies between China, Turkmenistan and Kazakhstan would lead to the construction of oil and gas pipelines, like a *zero-sum* game, given those gains on the Chinese energy *securitization* would be produced at the expense of losses to the Russian energy *securitization* in the region (Peyrouse, 2016).

The already difficult energy equation would be aggravated for Moscow given the need to satisfy the growing consumption of Russian gas by Europe. Given that Russian gas production has been relatively low in recent years, Russia sought to increase its gas imports from the Caspian region/Central Asia, where Moscow acquired it at a very low price, to sell in European markets for a considerably higher value (Gorst, 2006). For this purpose, Gazprom used, in large part, gas imported from Turkmenistan. However, from January 2016, Gazprom has chosen to stop the purchase of Turkmen gas, because its prices no longer suit the Russian side, due to the fall in export gas prices in Europe connected to the constantly falling oil prices. To compensate for the elimination of Turkmen gas imports, Gazprom announced an increase in the volume of gas imported from Uzbekistan, another important gas producer in the region, in addition to Turkmenistan and Kazakhstan (Azernews, 2016). What is the relevance of this episode within Russia's energy *securitization* in Central Asia? It is possible to infer that, although Russia no longer holds the monopoly of energy routes in the region, Moscow is trying to adapt to a changing gas trading dynamics in Central Asia. Russia is now, like China and the republics of the region, an actor in the Central Asian *New Great Game*. Moscow is currently obliged to negotiate with the states of the region, not being able to impose anything (unlike what happened in Soviet times), because these republics have significant *functional power* and flexibility.

As for Uzbekistan, the country is one of the largest gas producers (about 2 trillion cubic metres in 2014) in the Commonwealth of Independent States, making it one of the 20 largest in the world. Although its oil reserves are not significant – 594 million barrels of proven oil reserves in 2015 – the country is rich in gas and endowed with a geographical position conducive to energy transit to China or Southeast Asia (U.S. Energy Information Administration, 2015a). Moreover, there are promising energy reserves in the Aral Sea basin and the Ustyurt plateau in western Uzbekistan, which can be easily connected to the broader infrastructure of regional transit. Uzbekistan produces gas from 52 deposits, 12 of them responsible for more than 95 percent of the gas production in the country (APS Review Oil Market Trends, 2014). These deposits are concentrated in the Uzbek border of the Amu Darya basin in southeastern and central highlands of Ustyurt, near the Aral Sea, in the west of the country.

Kazakhstan had 85 trillion cubic feet of proven gas reserves in January 2014 (U.S. Energy Information Administration, 2015c). Kazakhstan is an important transit country with regard to the exports of gas from Uzbekistan and Turkmenistan to Russia and China. Most Kazakh gas reserves are located in the western region of the country, especially in the field of Karachaganak, which is estimated to have proven reserves of about 1.2 trillion cubic meters (Daly, 2015).

Kyrgyzstan has about 5.7 trillion cubic meters of gas reserves, although these are technically difficult to explore (CIA – The World Factbook, 2016b). Kyrgyzstan has neither the infrastructure nor the financial capital to increase the exploitation of its gas reserves. Finally, Tajikistan has, like Kyrgyzstan, about 5.7 trillion cubic meters of proven gas reserves (CIA – The World Factbook, 2016c). In 2000, Tajikistan began operations in the field of Khoja Sartez and intensified the exploitation of the Qizil Tumshuq deposit, in the district of Kolkhozobod (Global Security, 2013).

A word concerning the hydrological and hydroelectric characteristics of Central Asia. The hydroelectricity potential in Central Asia is extraordinary. Tajikistan has an installed capacity of 5,190 MW and an estimated hydroelectric potential of 527 billion kWh/year. Tajikistan's hydroelectric potential is ranked the eighth in the world. The two countries with more water resources of the region are Kyrgyzstan and Tajikistan, both upstream states of the main rivers, with capacity to allow for several possible sites for building hydropower plants (Peyrouse, 2015). Therefore, attention will be paid here only to the case of Tajikistan and Kyrgyzstan, as the existing water resources in the other Central Asian Republics are not (as) significant.

Regarding the potential of its hydroelectric resources, Kyrgyzstan ranks third among the countries of the Commonwealth of Independent States, after Russia and Tajikistan. Kyrgyzstan has over 25,000 rivers and streams. The production of electricity in Kyrgyzstan is 14.97 million kWh (CIA - The World FactBook, 2016b). Kyrgyzstan exports electricity to Kazakhstan, to China, to Tajikistan and Uzbekistan. Nonetheless, currently the energy sector in Kyrgyzstan is in critical condition. The existing hydroelectric plants are the result of the Soviet legacy, needing repair due to the lack of maintenance in recent years.

In turn, Tajikistan is one of the greatest countries in the world with regard to the capacity of hydroelectric engineering. Tajikistan has 4 percent of the world's hydroelectric resources and 53 percent of Central Asian resources. The country produces 17.09 billion kWh of electricity, with a total capacity of 4,476 megawatts at its hydroelectric plants (CIA - The World FactBook, 2016c). However, the Government is seeking funds for the proposed Rogun project, which, if completed, will become the largest Tajikistan hydroelectric project of 3,600 MW, and would turn Tajikistan into a net exporter of electricity.

Having analyzed the energy potential of the region, attention will be paid to a key logistical link – the China-Pakistan Economic Corridor – in the context of the Chinese energy *securitization*. This corridor enables a more effective flow of Central Asian oil and gas to the major consumer world markets, reinforcing, therefore, global energy security (not only China's).

The China-Pakistan Economic Corridor

The China-Pakistan Economic Corridor (CPEC) is another crucial component for the development of Central Asia and Xinjiang. Gwadar is one of the Indian Ocean ports with overland links to western and southern China that can help Beijing to avoid the *Malacca Dilemma* (Xu and Chung, 2016). Figure 1 shows the Strait of Malacca and other major oil transit chokepoints.

In fact, the fears of a disruption in oil supply to China are the major reasons that explain Beijing's interest in the construction of the CPEC. Although China has never been the object of a maritime blockade, and in a context where the power outages and the pollution generated by coal constitute a real internal threat, Beijing seems to believe it is from abroad that the greatest *threat* to its energy security can come (Xu & Chung, 2016). Although 'more psychological' than real, the *threat* of interruption of the oil supply lines is, in any case, a crucial factor in the development of Chinese foreign policy. Thus, Gwadar could be the beginning of a southern route for the OBOR, through which China can more easily exchange goods with Central Asia and the Middle East (Weitz, 2013). Besides the economic *securitization*, the CPEC also assumes a physical/military *securitization*. In practice, by offering Xinjiang and Central Asia an access to the sea – contributing simultaneously to

Figure 1. Malacca and other major oil transit chokepoints
Source: (Business Insider, 2015)

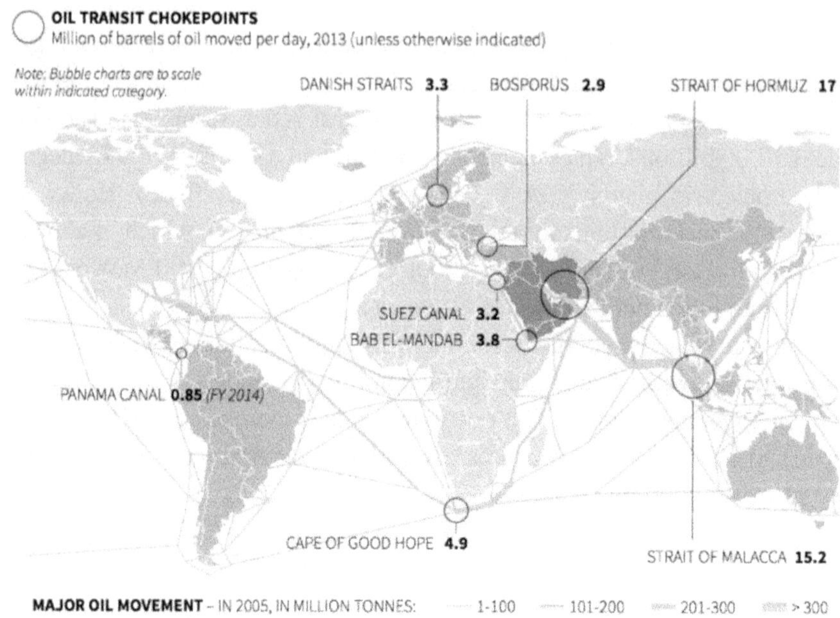

make Kashgar and Gwadar major commercial hubs – Beijing conceives the CPEC as an instrument likely to mitigate the separatist feelings that undermine Xinjiang's stability (Zhiqin & Yang, 2016).

If at the level of the logistical and energy *securitization*, the CPEC will include the construction of highways, railways and pipelines linking China to the Middle East, at the geopolitical level, China's participation in Gwadar will also allow it to expand its influence in the Indian Ocean, a vital route for oil transportation between the Atlantic and the Pacific. Pakistan provides China with a trade and energy corridor, by Gwadar, through which the oil from the Middle East will get to China through pipelines and railroads (Zhiqin & Yang, 2016). This corridor offers a shorter route between Western Asia and China, allowing considerable savings in time and shipping costs. In fact, the current route for transporting oil and other commodities from western Asia to Chinese ports, which is via the Straits of Malacca, is roughly 13,000 km long. It is another 3,500 km of overland travel from Chinese ports to Xinjiang. In comparison, the route from Gwadar Port to Xinjiang is just 2,000 km (Ramachandran, 2015).

By adopting a critical attitude, that also admits the contradictory, let us also present here the opposite point of view, i.e., that Gwadar is not as fundamental as it may, eventually, seem in the context of China's OBOR. Ghulam Ali (2013) points out some reasons which tend to overestimate Gwadar's geo-economic importance. The author begins by explaining that even though China has built several roads in Pakistan, it should need, however, to create infrastructure such as railways, gas pipelines and oil pipelines to link both countries; that is, extraordinarily expensive projects, at a time when China is reluctant to invest in the province of Balochistan, because of the insecurity that affects it. Yousaf (2013) highlights the existence of tensions with locals demanding an equal share of their resources, and royalties for the extraction of natural resources. Therefore, given the insurgency, along with military deployments, sectarian militancy, and frequent infrastructure sabotage, it seems improbable that work on the project will be smooth, in part because for Baloch insurgents, the port is just another attempt at a 'conspiracy' to exploit the region's resources. Several feudal lords in Pakistan are opposed to large-scale foreign investment, fearing it will bring an influx of outsiders. Growing Indian influence in Afghanistan and interference in Balochistan, as alleged by Pakistan, could also be important factors hindering the port's progress. If China wants to use a Pakistani port for military purposes, Ali (2013) believes that Karachi is a viable alternative, despite its proximity to India. In conclusion, for the supporters of this vision – which criticizes the overvaluation of Gwadar's importance to China – the port (and by extension the CPEC) still have a long way to go before being fully operational.

Despite Gwadar and the CPEC undeniable importance within the OBOR, it would be simplistic to limit the analysis of the *Grand Chinese project* regarding Central Asia, to a systematic emphasis on gas and oil. Thus, the analysis below will focus on the water and hydroelectricity issues, which are, in any case, also important energy issues in the context of the synergies between China and the Central Asian republics.

Central Asia's White Gold

The author of this chapter believes that it is worth mentioning that before his doctoral field research in Kazakhstan, Kyrgyzstan and Tajikistan, he had unwittingly underestimated the water energy (in)security, as well as the remarkable contribution that Central Asia can give to China, especially in the electricity supply to Xinjiang. Only in the field and after being in direct contact with great local experts, did this author realize the serious lapse it would be to ignore the dynamics inherent in the production of hydroelectricity, even because these have an impact on the relationship that the region's countries maintain among themselves, as well as with powers such as Russia, China and the US. Countries apparently insignificant in terms of oil and gas reserves, such as Tajikistan and Kyrgyzstan, become centerpieces in the dynamics of regional conflict and cooperation, as a result of what they have as most precious: *white gold*. In fact, the hydroelectric sector is their great asset, since, unlike the other regional states – Uzbekistan, Kazakhstan and Turkmenistan, which are rich in oil, natural gas and several mineral resources – Kyrgyzstan and Tajikistan have vast water reserves, some mineral resources and small oil and natural gas reserves.

That said, how does China *securitize* its interests in Central Asian hydroelectric resources? China is an increasingly important actor in what concerns the management of water resources in Central Asia. But, unlike China's interest in Central Asian oil and gas – which has been widely reported – Beijing's commitment in the sector of Central Asian hydroelectric resources has been almost unnoticed. A possible explanation for this is the priority of energy *securitization*, in terms of the diversification of oil and gas sources – Central Asia being a viable alternative, either by its geographical proximity vis-à-vis China, or by the existence of abundant reserves of these resources – in what concerns Central Asian hydroelectric resources. Indeed, the excerpt below is interesting because it highlights the *non-perception*, by China's *National Energy Administration,* of power shortages as a *threat* as serious as the one concerning the disruption of oil supply. Thus, one can argue that oil *securitization* is more urgent for Chinese policy makers than electricity *securitization*:

Our country does not really have an electricity supply problem. Although we sometimes face some power shortage problems during peak seasons and in some provinces, these problems are short-lived and localized, and [...] can be solved by ourselves. Oil imports are different. If our oil imports are cut off, it affects the whole nation [...], and we no longer maintain self-reliance... (in Leung et al., 2014: 322).

Another explanation, less obvious, but still appropriate, why Chinese participation in large hydroelectric projects in the region is still not noticed, concerns the fact that China is seeking to act with caution in a context where there are several conflicts and tensions around the management of water and hydroelectric resources in the region. A concrete example: Rogun. It is not by chance if China is reluctant to invest in the construction of what would be the highest dam in the world – 335m. Chinese leaders and businessmen know that this can be seen as a hostile enterprise to Uzbekistan (which wants to prevent upstream countries – Kyrgyzstan and Tajikistan – from achieving electrical autonomy) or Russia, whose cultural, security and economic influence is still predominant in the region. To what extent can Central Asia's hydroelectric potential be suitable for China's energy interests? Although China has arrived late to the Central Asian hydroelectric market, Beijing has, in recent years, cooperated with countries such as Kazakhstan, Kyrgyzstan and Tajikistan in terms of *know-how* and investments in the construction of new electric lines. Contrary to China's hydrocarbon policy, the aim of the Chinese here is not to have this hydroelectricity delivered to their large cities in the east (the electrical lines required would need to stretch over at least 6,000 km), but rather to make up for the energy shortfall in Xinjiang. In fact, it is this province of the Chinese *Far West* which seems to be at the heart of the Chinese hydroelectric interests in Central Asia (Peyrouse, 2016). Xinjiang is in dire need for additional water and energy resources as evidenced by its interest in the hydro energy potential of the Tarim River, which flows to mid-western Xinjiang from Kyrgyzstan.

In addition to the interest, already mentioned, in ensuring energy supply to Xinjiang, China is considering selling surplus energy produced in Tajikistan and Kyrgyzstan to the countries located in the southern corridor, i.e. Afghanistan, Pakistan, Iran and India because of the significant transit fees it would accrue (Peyrouse, 2015). To achieve these goals, Beijing has focused on the construction of hydroelectric power plants and on the installation of high-voltage lines, likely to distribute electricity both to the region and outwards. China follows a *securitization* strategy similar to the one used in the exploration of Central Asian oil and gas. What changes here is just the energy source, in this case, electricity. Instead of oil and gas pipelines, China undertakes, in this case, the construction of high-voltage lines likely to provide Central Asian electricity to Xinjiang, and in the future, to the southern corridor described above, at the same time it invests in the modernization of the Central Asian electric lines, in most cases with serious problems or even obsolete (Peyrouse, 2015).

As the reader might also be interested to know about the contours of the water (in)security in other parts of China, we find it useful to make an overall assessment of such a pressing topic below.

Due to over-exploration and inefficient consumption, China's water resources are declining as more rivers disappear and aquifer water levels drop. The situation is so serious that in the last 20 years China has lost more than 28,000 rivers (Zhang, 2014). Besides, 60 percent of China's underground water is polluted, and about 300 million people in China drink contaminated water every day. To the scenario of uneven distribution of water resources, we add the fact that China's agricultural production and industries are shifting from the southern regions to the central, western, and northern regions where water resources are even scarcer. Among the various measures that the Chinese Government has taken to fight the water scarcity, we highlight the controversial *South-to-North Water Diversion Project*, which involves drawing water from southern rivers and supplying it to the dry north, being the largest of its kind ever undertaken. It is estimated to be completed in 2050, at a total cost of approximately $62bn, which is more than twice as much as the country's Three Gorges Dam. Nonetheless, several experts perceive the South-to-North Water Diversion Project more as a panacea to the taste of a generation of leaders with a background in engineering than a long-term commitment to solve the real problem of water shortage, which is mainly a problem of preservation and not of extraordinary or gigantic works (The Guardian. 2014). One reason why China's water crisis is so dire is that the central government hasn't been able to coordinate national efforts to conserve water. Local environmental bureaus are often weak. Companies fined for breaking pollution rules often ignore the fines or renegotiate them with local officials. It follows that governance of water resources in China is often a murky affair, with bureaucratic restructuring, semi-privatization of many former state-owned enterprises, and changes in legal institutions resulting in many ambiguities and areas of overlapping jurisdiction. The water crisis that China is facing is, in essence, a governance crisis (Duarte, 2016).

FUTURE RESEARCH DIRECTIONS

OBOR's accomplishment depends to a large extent on the receptiveness of the Central Asian elites to China's projects, as well as on the process of institutional reforms, the struggle against corruption and the way countries like Uzbekistan (the most populous) and Kazakhstan (the most stable and developed) will prepare political succession. Indeed, no investor feels tempted to engage in large-scale projects if Central Asian leaders do not create the necessary confidence conditions, that range from preparing Nazarbayev's succession and implementing in Uzbekistan an environment (more) conducive to economic and political interaction between neighboring countries, on the one hand, and extra-regional investors, on the other. With this effort, one can speculate that projects, until now stagnated, such as cooperation in the field of water resources or electricity transmission into and out of Central Asia (the future of the controversial Rogun project, is here hostage), will be, in the medium/long term a reality, eventually feasible. Now, to understand how governance evolves in Kazakhstan and Uzbekistan, certainly has relevance in the winds of change that may take place there in the coming decades. This chapter encourages thus researchers to explore the cooperation trends among regional leaders, its impact on the water dynamics in Central Asia, and the remarkable contribution that the region can make to China's energy security.

CONCLUSION

Central Asia has an important role in China's energy security. The *securitization* of energy supply (oil, gas, but also electricity from Central Asia) is a sign of the unquestionable strategic relevance of this region to China, which will complement the *securitization* of the sea lanes. The strategic location of the Central Asian hydrocarbon reserves allows them to be transported through specially constructed pipelines, and the current infrastructure links these resources mainly to Russia, other CIS countries and Eastern European countries. The possibilities of relatively easy transportation by land – in fact, there are several projects of pipelines involving various countries of Central Asia, which even seek to avoid Russia – makes it unnecessary, or limited in good part, the use of maritime lines, and, consequently, the passage through sensitive chokepoints. After all, pipelines on land can be subject to greater control that, in the case of the ocean cannot be protected effectively. Hence China has all the interest in attracting the goodwill of Central Asian peoples, by means of an effective *soft power*, combining economic and diplomatic incentives, as the Beijing consensus offers. By doing so, China can effectively demonstrate that, apart from Russian or European markets, the path of Central Asian gas, oil and electricity can perfectly, as the logic of geography supports, be that from the East. The perspectives for the relationship of Central Asian republics regarding major powers, has taken predominantly commercial contours, in which these rentier states try mainly to see them as customers, in a seller-buyer relationship, in which the situation of Russian monopoly in the control of routes and, eventually, of resources, has, over the past few years, been counterbalanced by American and Chinese interests, among others.

Despite all the challenges and opportunities that Central Asia may offer in the framework of both Chinese and global energy security, one must not fall into the tendency to over-evaluate the importance of the region, since, in the next decades, the Caspian will not replace the Middle East as the main reservoir of the world's oil. Thus, there is a need of relativising the importance of Central Asia for Beijing, because the resources that the region can offer to China, considering its needs, are very limited. China will continue to depend, unquestionably, on the Middle East. Nevertheless, any central decision-maker in China, who thinks strategically, will consider that to have a balance in energy supplies made by land, in addition to the supplies arriving by sea, is certainly positive. Although Central Asia cannot replace the Middle East as the main world oil producer, it is, for all purposes, another piece in the strategy of diversification of energy resources that China has been carrying out over the last few years. Besides, it must be acknowledged that the production resulting from the Caspian will bring more oil to international markets, thus contributing to the strengthening of global energy security. Moreover, the contribution of the energy potential of Central Asia to the world oil market is likely to weaken the pricing policy of the Organization of Petroleum Exporting Countries, as well as political manipulation. In addition, the Caspian oil production may be able to help to mitigate the impact of a temporary reduction of the flow of Gulf oil in the world oil markets.

REFERENCES

APS Review Oil Market Trends. (2014, October 20). *Uzbekistan - the Gas Production System*. Author.

Azernews. (2016, January 6). *Ashgabat can face challenges in gas transportation*. Author.

Brzezinski, Z. (1998). *The Grand Chessboard: American Primacy and its Geostrategic Imperatives*. New York: Basic Books.

Buzan, B. (1991). *People, States and Fear: an Agenda for International Security Studies in the Post-Cold War*. London: Hasvester Wheatsheaf.

Buzan, B. (1998). *Security – A New Framework for Analysis*. Boulder, CO: Lynne Rinner Publishers, Inc.

Chenoy, A. (2007, September). India's Foreign Policy Shifts and the Calculus of Power. *Economic and Political Weekly*.

Chevron. (2016). Retrieved from http://www.chevron.com/

China Daily. (2016, January 27). *Slump in oil prices fuels increase in China's crude imports*. Author.

CIA - The World FactBook. (2016a). *Kazakhstan, Last Updated on May 5*. Author.

CIA - The World FactBook. (2016b). *Kyrgyzstan, Last Updated on May 5*. Author.

CIA - The World FactBook. (2016c). *Tajikistan, Last Updated on May 5*. Author.

Contessi, N. (2016). Central Asia in Asia: Charting growing trans-regional linkages. *Journal of Eurasian Studies, 7*(1), 3–13. doi:10.1016/j.euras.2015.11.001

Daly, J. (2015, July 10). Kashagan Oil Field Comes Back to Life. *Silk Road Reporters*.

Duarte, P. (2016). Ahuyentando sus Espantapájaros: Una evaluación de la (in) seguridad alimentaria y hídrica en China. *Revista Geográfica Venezolana, 57*(2), 298–317.

Duarte, P. (2017). *Pax Sinica*. Lisbon: Chiado Editora.

Esteban, M., & Otero-Iglesias, M. (2015). What are the prospects for the new Chinese-led Silk Road and Asian Infrastructure Investment Bank? *Elcano Royal Institute., 23/2015*, 1–10.

Facts & Details. (2015). *China's City Dwellers Now Outnumber the Rural Population*. Retrieved from http://factsanddetails.com/china/cat11/sub72/item151.html

Farchy, J. (2015, October 14). China's Great Game: In Russia's backyard. *Financial Times*.

Fingar, T. (2016). *The New Great Game: China and South and Central Asia in the Era of Reform*. Stanford, CA: Stanford University Press.

Global Business Reports (2012). *Special Report: Caspian Region*. Author.

Global Security. (2013). Retrieved from http://www.globalsecurity.org/military/world/centralasia/tajik- energy.htm

Gorst, I. (2006). Struggle for Gas Intensifies. Petroleum Economist, 73(5).

Herberg, M. (2011). *China's Energy Rise and the Future of U.S.- China Energy Relations*. New America Foundation.

Huang, C. (2016, February 2). China's Silk Road ventures carry financial risks as well as political rewards. *South China Morning Post*.

International Energy Agency. (2015). *China. International energy data and analysis*. IEA.

Keohane, R. (2000). *Realism, Neorealism and the Study of World Politics*. Basic Books.

Khwaja, A. (2003). Pakistan and The New Great Game. *IPRI Paper*, *5*(April), 5.

Leung, G., Cherp, A., Jewell, J., & Wei, Y.-M. (2014). Securitization of energy supply chains in China. *Journal of Applied Energy.*, *123*, 316–326. doi:10.1016/j.apenergy.2013.12.016

Lim, T. (2016). *China's One Belt One Road Initiative*. Imperial College Press. doi:10.1142/p1076

Ling, W. (2013, October 9). A community of common destiny. *China Daily*.

Mackinder, H. (2004). The geographical pivot of history. *The Geographical Journal*, *170*(4), 298–321. doi:10.1111/j.0016-7398.2004.00132.x

Marten, K. (2007). Russia Efforts to Control Kazakhstan's Oil: The Kumkol Case. *Post-Soviet Affairs*, *23*(1), 18–37. doi:10.2747/1060-586X.23.1.18

Mashrab, F. (2012, August 9). Tajikistan holds back on cheering huge oil, gas find. *Asia Times*.

Maza, C. (2015, September 24). Central Asia's pipeline politics: a quest for energy independence. *The Christian Science Monitor*.

Morgenthau, H. (1978). *Politics Among Nations: The Struggle for Power and Peace* (5th ed.). New York: Alfred A. Knopf.

Nopens, P. (2014). The Impact of the Withdrawal from Afghanistan on Russia's Security. Security Policy Brief, n.º 54, March, Ergmont Royal Institute for International Relations, 1-10.

Oil & Gas Monitor. (2015, September 30). *The World's Largest Oil and Gas Projects – What You Should Know*. Author.

Peak Oil Review. (2015). Retrieved from http://peak-oil.org/peak-oil-review-13-jul-2015/

Peyrouse, S. (2015, March 18). China as an Economic Actor in Central Asia: Between Development And Concerns. Testimony before the U.S.-China Economic and Security Review Commission.

Peyrouse, S. (2016). Discussing China: Sinophilia and sinophobia in Central Asia. *Journal of Eurasian Studies*, *7*(1), 14–23. doi:10.1016/j.euras.2015.10.003

Ramachandran, S. (2015). *China-Pakistan Economic Corridor: Road to Riches?* China Brief.

Reeves, J. (2014). *SCO: A Tenable Provider of Security in Post-2014 Central Asia.* Asia-Pacific Center for Security Studies.

Reuters. (2015, May 11). *China becomes world's top crude buyer despite economy stuttering*. Author.

Rolland, N. (2015). *China's New Silk Road.* The National Bureau of Asian Research.

South China Morning Post. (2015, December 21). *China's growing upper middle class to drive consumption by 2020, research finds*. Author.

Stegen, K., & Kusznir, J. (2015). Outcomes and strategies in the 'New Great Game': China and the Caspian states emerge as winners. *Journal of Eurasian Studies*, *6*(2), 91–106. doi:10.1016/j.euras.2015.03.002

Tethys Petroleum. (2016). *Bokhtar Production Sharing Contract Area*. Retrieved from http://www.tethyspetroleum.com/operations/bokhtar-psc-area

The Guardian. (2014, December 12). *China's water diversion project starts to flow to Beijing*. Author.

U.S. Energy Information Administration. (2015a). *International energy data and analysis. Uzbekistan*. Author.

U.S. Energy Information Administration. (2015b). *International energy data and analysis. Turkmenistan*. Author.

U.S. Energy Information Administration (2015c). *Kazakhstan. International energy data and analysis*. Author.

Wang, H. (2016, January 15). China's Triple Wins: The New Silk Roads. *Forbes*.

Wang, Z. (2015). *Securing Energy Flows from Central Asia to China and the Relevance of the Energy Charter Treaty to China*. Energy Charter Secretariat.

Weitz, R. (2013). Beijing Builds its Eurasian Transportation Network. *China Brief*, *13*(8), 6–9.

Xu, Q., & Chung, W. (2016). *China's Energy Policy from National and International Perspectives: The Energy Revolution and One Belt One Road*. City University of HK Press.

Yang, M. (2016). Understanding the One Belt One Road Initiative: China's Perspective. In China's One Belt One Road: Initiative, Challenges and Prospects. Vij Books India Pvt Ltd. Delhi.

Zhang, H. (2014). Confronting China's Water Insecurity. *RSIS Commentaries, S. Rajaratnam School of International Studies*, (57), 1-2.

Zhiqin, S., & Yang, L. (2016, December 21). *The Benefits and Risks of the China-Pakistan Economic Corridor*. Carnegie-Tsinghua. Center for Global Policy.

ADDITIONAL READING

Aminjonov, F. (2016). New Russian Gas Politics or Adapting to a Changing Gas Trading Dynamics in Central Asia?. Eurasian Research Institute, January 22, n.º 420.

Foreign Affairs (2015). A Strategic Seaport: Is Pakistan Key to China's Energy Supremacy?.

Lan, L. (2014). Experts say energy and infrastructure will drive Silk Road policy. China Daily, June 17. Retrieved from http://usa.chinadaily.com.cn/epaper/2014-06/17/content_17594605.htm

Musuraliev, M. (2016). Marat Musuraliev: Problems in the Electro-energy Sector of Kyrgyzstan: Is There a Way Out? *Central Asian Bureau for Analitical Reporting*, April 22.

Pigato, M., & Tang, W. (2015). China and Africa: Expanding Economic Ties in an Evolving Global Context. *Investing in Africa Forum*, March.

United Nations Development Programme. (2016). *Setting ground rules for investment in hydropower*. Kyrgyzstan.

Zhao, H. (2009). Central Asian Geopolitics and China's Security. *Strategic Analysis, Routledge*, *33*(4), 475–477. doi:10.1080/09700160902930979

KEY TERMS AND DEFINITIONS

CIS Countries: A group of countries that form the Commonwealth of Independent States (CIS). Created in December 1991, after the collapse of the Soviet Union, CIS currently unites Azerbaijan, Armenia, Belarus, Georgia, Kazakhstan, Kyrgyzstan, Moldova, Russia, Tajikistan, Turkmenistan, Uzbekistan, and Ukraine.

Energy Security: A state's need to ensure a stable energy supply at reasonable prices.

Malacca Dilemma: This refers to China's fear of a maritime blockade at the Straits of Malacca. Since most of China's oil imports pass through the Straits of Malacca, a maritime blockade here could paralyze China's economy.

New Great Game: The attempt to control Central Asia's oil and gas, as well as the regional pipelines, has been associated to the idea of a 'New Great Game', by analogy to the "Great Game" of the 19th century between the Russian and the British empires.

One Belt One Road: A multifaceted plan that combines soft power, hard power, political, social, and economic aspects. The OBOR aims to relaunch the Chinese economy, preserve the continuity of the Communist Party and to create a favorable environment at the international level to foster the reemergence of China.

Rogun: One of the planned hydroelectric power plants to be built in southern Tajikistan. The dam is controversial as it has drawn complaints from Uzbekistan, which fears it will give Tajikistan an almost absolute control over water resources in the region.

Securitization: According to the Copenhagen School, the *securitization* is a process whereby a *securitizing* agent tries to establish, socially, the existence of a *threat* to the survival of a unit.

Chapter 10
Maximizing the Potential of a Japan–Turkey Strategic Relationship:
The Security of Middle Eastern Energy Supplies

Devrim Şahin
Eastern Mediterranean University, Cyprus

Ahmet Sözen
Eastern Mediterranean University, Cyprus

ABSTRACT

The international system has experienced a shift from a Western-centric world dominated by the United States to a decentralized world. This accompanies a shift in the distribution of energy wherein China overtakes the US as the biggest oil importer. Energy is vital for the survival of countries' economic facilities. Japan is a country that needs to import the energy resources to run its industry. Thus, the security of its economic activities is dependent on the security of energy supplies. The fact that Turkey is a country with ambitions to become a regional energy center and has strong ties to the Middle East makes it vital to Japanese interests. Also, Japan's advanced technology and economy are vital to Turkey's ambitions of becoming an energy hub. These reciprocal interests make it possible for both countries to attain a high level of cooperation. Nevertheless, this depends on their self-awareness and political will in order to exhibit a more independent behavior which transcends their previously US-dominated foreign policy.

DOI: 10.4018/978-1-5225-4203-2.ch010

INTRODUCTION

What is the potential of a Turkey-Japan strategic partnership and how can they reach a high level of cooperation? Rapidly changing international circumstances bring to the fore a Turkey-Japan strategic partnership which is vital for the security of energy supply, particularly in their engagement with the Middle East. A strategic partnership can be accomplished as a result of a partnership which includes Japan's advanced technology and economy, and Turkey's strong ties in the region. Nevertheless, this depends on the self-awareness and political will of both Turkey and Japan to exhibit a more independent behavior which transcends their previously US dominated foreign policy. The ambition to control valuable energy resources fuels protracted historic tensions as experienced in Cyprus, Iraq, Syria, South Sudan, the Crimea, Ukraine, and the South China Sea. This proves how timely it is to reconsider how Japan and Turkey can reach a high level of cooperation on the security of the Middle Eastern energy supply.

Energy security is a complicated issue that offers opportunities as well as perils for contemporary politics. While classical economists define energy as a basic good that influences every process of manufacturing and services, scholars of politics define it as a strategic good with a comparatively high utility which is difficult to replace. Nowadays more than ever, uninterrupted access to petroleum and other liquids restructures more intensely contemporary politics. The supply of energy is related primarily to pipeline security. This raises questions regarding the stability of the supplying and transit countries. It has been a high priority the way in which this stability and security of supplying and transit countries will be provided locally, regionally and globally. Turkey as a transit country has ambitions of becoming a regional energy center. It can use its hard power as complementary to its soft power in the Middle East. Turkey has strong bonds and influence within the supplying countries such as Qatar, where it has a military base and is ready to play a mediating role. Thus, this role of Turkey is vital to Japan's interests in order to ensure the latter's uninterrupted access to energy supplies. Turkey, nevertheless, needs Japan's energy technology and the money of Japanese investors to carry out its projects, such as a greater investment in Turkey's atomic sector, as soon as possible. Turkey can model its best experience from partnering with Japan in projects in various fields such as energy, technology industry or university education.

Japan as the world's third largest economy has experienced advances in industrial, information and energy technologies. Japan has achieved the most efficient energy utilization in the world by a very efficient use of imported raw materials and saving in resource consumption. Japan has highly efficient energy technologies including nuclear power, natural gas and coal fired power generation, fuel cells, energy storage systems and conversion materials, heat pumps, and renewable energy systems.

Japanese high technologies when coupled with its experience and investment power provide the country with the capabilities to deliver solutions to common problems in the energy field such as maximizing the efficient use of resources. Japanese technology can be used to construct highly efficient energy plants in any place of the world, particularly in the Middle East region, which is rich in fossil fuel energy sources, and pipelines to transfer the energy produced in these plants. The country has the investment power to finance these constructions. Herein, the problem that Japan faces is to securely transfer energy through long distances from production sites to where it will be used. Japan's dependency on secured energy sources, as well as Turkey's dependency on the transfer of advanced technology skills generates a mutual dependency. They can complement each other's power in the Middle East. Japan's technology and investment can be used to build pipelines to transfer Middle-Eastern energy resources, while Turkey can provide security of these energy routes. This mutual inter-dependency stems from the search for stability in the Middle East, which can be fulfilled through a Turkey-Japan strategic partnership based on shared economic benefits.

A Turkey-Japan strategic partnership is essential for the stability of the Middle East, a region that faces increased challenges stemming from systemic changes. The international system has shifted from a bipolar system to a unipolar one in the aftermath of the dissolution of the USSR. What soon followed that was a rapid transition into a multi-polar system through the rise of China in the aftermath of the 2003 Iraq war. The US has been gradually abandoning the Middle East through its withdrawal of its military presence in Iraq. This has deepened the regional 'power vacuum', which is defined by the Cambridge dictionary as "a condition that exists when someone has lost control of something and no one has replaced them". Turkey, along with other regional powers such as Iran, Israel and Russia as well as the petro-monarchies of the Gulf Saudi Arabia and Qatar are competing to promote order and control the energy resources in the Middle East. The growing sense of insecurity among these rivalries has a direct effect on the choices of a Turkey-Japan strategic partnership. The decline of US hegemonic influence, coupled with the rise of China, provokes a potential US-China rivalry over global oil supplies. This generates a pressure on Japan to increase its engagement over control of the Middle East's energy supplies in order to defuse the changing patterns of China's demands. A strategic partnership throughout the Middle East would allow Turkey to benefit from the advanced technology and economy of Japan, while the latter can benefit from the former's ties in order to reach rich regional oil and natural gas, energy sources which are vital for Japan's economy. Turkey has geographical, economical, historical, cultural and religious ties with the region. Domestic changes within both Turkey and Japan through a more independent foreign policy designed independently from the US, offer a potential for mutual cooperation in the Middle East.

To explore the limits and potential success of a Turkey-Japan strategic partnership, the article firstly scrutinizes the historical relationship between the two countries. Then, the article investigates the changing relations of Turkey and Japan due to systemic shifts and due to domestic shifts. Furthermore, the article analyzes the possibility in the two countries of reaching a high level of cooperation on energy issues in the Middle East. The article will conclude by highlighting future directions in a Turkey-Japan strategic partnership.

BACKGROUND OF THE RELATIONS BETWEEN TURKEY AND JAPAN

Turkey's relations with Japan date back to its predecessor the Ottoman Empire, a multi-ethnic and multi-religious state which existed between the years 1299 and 1922. In 1887, Prince Komatsu Akihito, the uncle of the Japanese Emperor Mutsuhito Meiji who reigned between 1867 and 1912, arrived in Istanbul, which was the then capital of the Ottoman Empire, to seek support for the potential war between Japan and Russia. When the ongoing war between Japan and Russia ended with the victory of Japan in 1904, the Ottoman Empire demonstrated a very sincere commitment by declaring on the said day, an official holiday. Just as there is recently a competition between Russia and Turkey in locations like Georgia, Nagorno-Karabakh, Syria and even Cyprus, where they have a strong influence, the Ottoman Empire saw Russia as a rival for its zone of influence and perceived it as a threat back then.

Emperor Meiji initiated diplomatic relations with Abdul Hamid II, who was the Ottoman Sultan between 1842 and 1918. An important event widely accepted as a symbol of the solidarity between Turkey and Japan is the sinking of Ertuğrul Frigate during its return voyage to İstanbul – which was sent to Japan by Abdul Hamid II in 1889 as a gesture of goodwill. The efforts of the Japanese villagers in the area aimed at rescuing Turkish sailors, were always recited with sympathy by Turkey. Japan subsequently transported the survivors to Istanbul with its ships. Also, there is a monument and a graveyard kept in Ertuğrul's memory in Kushimoto, the Japanese town where the ship sank together with 580 of its crew, totaling 650 sailors (Kalın, 2011).

Another important event is the evacuation of 215 Japanese nationals out of Iran in 1985. Turkey played an important role in their transportation back to Japan with a Turkish Airlines airplane. They were stuck in Tehran, which was shaken with bombs when the Iran-Iraq war was about to break out. The one-hour program called 'Project X' broadcasted on the first channel of the semi-official Japan Broadcasting Corporation (NHK) and which drew considerable interest, reenacted the operation by way of interviews and told the story of the heroic Turkish pilots. Sarıboğa and Sarıkaya (2004) points to a news report appearing on February 5, 2004 in Sabah, a Turkish newspaper with a high circulation, the article started with the following statement:

There was little time left before the end of the time period allowed for over 450 people comprising employees of Japanese companies, technicians and families living in Iran to leave the country. Yet, they were unable to leave. Foreign airlines companies were giving priority to their own citizens. Japanese people stuck in Iran were under the threat of air strikes. With the initiatives of Yutaka Nomura, the then Ambassador of Japan to Tehran, Turkish Airlines organized a special flight. This was the cross-border rescue operation of the century.

The Tokyo Bank official Satoru Nori, one of the Japanese rescued by the Turkish Airlines airplane, collected 5 million yens after the Marmara earthquake which took place on August 17, 1999 in Turkey, 14 years after his rescue and thus showed his gratitude. Following the Marmara earthquake of 1999, Japan has been one of the countries which aided Turkey the most. Owing to the above, perceptions of both countries about each other have always been based on positive impressions.

Toshiki Kaifu, the Japanese Prime Minister in office between 1989 and 1991, visited Turkey in 1990 during the 'Gulf War'. In 2000, with a view to strengthen and diversify the economic, political and cultural collaboration, the then Foreign Minister of Turkey İsmail Cem visited Japan as the guest of the Foreign Ministry of Japan, and a 'Joint Action Plan' which would serve as the pillar of the collaborative relations between the two countries, was signed. The governments of both countries also show political will in terms of promoting the relations between Turkey and Japan. The year of 2003 was celebrated as the 'Turkey Year in Japan'. Following the official visit of Abdullah Gül, as Foreign Minister and Deputy Prime Minister, in December 2003, current President Recep Tayyip Erdoğan officially visited Japan in April 2004 when he was the Prime Minister of Turkey. These official visits took place in a cordial atmosphere reflecting the traditional ties of friendship between Turkey and Japan and the possibility of potential cooperation in various areas was reiterated.

Horiguchi Mitsuru, the Deputy Consul General of Japan, attended the TV program called 'Alternatif' in Cem TV and stated that Atatürk was mentioned in all the books on world history as the founder of modern Turkey and that he was known as a genius in military strategy in the Japanese War College where the Gallipoli land battles were studied. This is again an indicator of the genuine attachment of Japan to the perceptions of Turkish people.

On October 1, 2004, Tomoyuku Abe, the Ambassador of Japan to Turkey, made statements during the inauguration ceremony of 'Events for the 80th Anniversary of the Diplomatic Relations between Japan and Turkey', which are significant in terms of manifesting the dimension and perception of mutual relations. Mütercimler (2010) quotes the following statements of Tomoyuku Abe about the three main issues pertinent to the relations between Japan and Turkey:

The first one is the issue of developing a mutual understanding between our countries. "2003 as Turkish Year in Japan" has been a great success in Japan and it can be said that the affection of the Japanese people towards Turkey, the reason of which is not known, played an important role in that regard...

The second one is the issue of convergence of economic relations. The Japanese-Turkish relations go back over a hundred years and while eighty years have passed since the establishment of our interstate relations, it has only been fifteen, sixteen years since the Japanese industry established factories and started production in Turkey. In that sense, the economic partnership activities of Turkey and Japan have just begun. The number of Japanese companies in Turkey have not yet reached (20) twenty. For Japan, the most important thing is the transfer of the experience gained by the Japanese companies operating in Turkey to the economic world of Japan. By this way, Japanese companies seeking production facilities abroad would have an idea about the facilities provided by Turkey. Apart from that, there are several issues which Turkey should be cooperative about. It is especially important that the private and public sectors act conjointly and devise a campaign to invite foreign investment to Turkey...

The third issue, on the other hand, is the economic cooperation of Japan with Turkey. As Turkey is a country with very sound economic foundations, she is not a country to receive grants from Japan, however, loans in yen are provided and solid technological cooperation is ongoing...

The Former Japanese Prime Minister Koizumi who was in office between the years 2001 and 2006 paid a visit to Turkey in 2006 during the religious Feast of the Sacrifice, the main purpose of which was the continuation of the Middle East Peace Process and cooperating with Turkey for the restructuring of Iraq. In his statements before the visit, Prime Minister Erdoğan stated that Turkey could provide Japan with significant opportunities to have a presence in Eurasia, and that Turkey attached great importance to cooperation with Japan since the common efforts of both countries were designed to contribute to the economic and political stability much needed in both sides of Asia.

In 2007, the 'Turkey-Japan Business Forum' was established to develop the economic and commercial cooperation between Turkey and Japan. and to ensure the convergence of the Turkish-Japanese business world. The visit paid by President Abdullah Gül to Japan during 3-8 June 2008 together with 150 Turkish businessmen was a significant step for cooperation in terms of bilateral relations. Katsuya Okada, the Foreign Minister of Japan during 2009-2010, however, visited Turkey in early January in order to attend the inauguration ceremonies of the 'Year of Japan' celebrated in Turkey in 2010. In its statement, the Executive Committee of the '2010 Year of Japan in Turkey' stated that 'the Year of Japan' had three priorities which were bringing closer the attractions of Japan, extending the scope of the Turkish-Japanese friendship and carrying the cooperation between Turkey and Japan into the future. Subsequently, the relations between Turkey and Japan were upgraded to the level of 'Strategic Partnership' in 2013.

The fact that there has been no damaging memory in the deep-rooted relations between Turkey and Japan and that they have positive perceptions about each other, enhance the potential of cooperation in many areas. Japan is a former empire like the Ottoman Empire, which was the predecessor of the Turkish Republic. Despite its devastation in the Second World War, Japan developed both politically and economically owing to a successful modernization process. To the Turkish public, Japan does not appear as a state with imperialist aspirations. The similar values of Turkey and Japan such as strong family ties, respect for the elderly, work ethics, discipline and respect for others arouse sympathy in the Turkish public opinion towards Japanese people. Japan with its common characteristics with Turkey is a fine example of success for the latter, which is seeking a model of modernization that would be an alternative to Westernization.

Despite all these sympathetic approaches and common strategic interests, cooperation between Japan and Turkey has some limitations. It would not be right to assert that the relations between the two countries are progressing in the strategic sense. There is no sustainable strategic partnership in which Japan and Turkey protect each other's interests. Their communication and cooperation is not as frequent as a strategic partnership would demand. In 2009, Nabuoki Tanaka, the Ambassador of Japan in Ankara, stressed in the statement he gave to the press, that Japan was regarded with sympathy in the Turkish public opinion; yet the fact that this sympathy was not based on an information infrastructure was a problem. In Turkey, there exists a public opinion which is not knowledgeable about the world other than the West. Insufficient perceptions (or misperceptions) about Japan is the case not only for the average person, but also for high-level decision-makers.

CHANGING RELATIONS DUE TO SYSTEMIC SHIFTS

The international system has experienced a shift from a Western-centric world that the US dominated to a more decentralized world shaped through the rise of the East. Brzezinski (2016) points out the signals of a "global alignment" as: the decline of US hegemony, the imperial devolution of Russia, the gradual rise of China, the failure of the EU to be a global power and the violent political awakening in the Middle East. The future is uncertain, whether the US and China will struggle over the terms of global order. Ikanberry et al. (2015), who is known for his characterization of US policy during the presidency of George W. Bush as the "neo-imperial grand strategy", points to "the specific ideas and values that each will bring to the table". Realist scholar Mearshmeimer (2015), on the other hand, insists that it is not China that "alone will drive the security competition that lies ahead. The United States is also likely to behave in aggressive ways." Whatever happens in the relationship between the US and China, the restructuring of the global power balance is already one of the most consequential issues. It also accompanies changes to energy fields which is of particular concern to Japan and Turkey in their relationship with each other as well as with the US. The need to address these changes urges Japan and Turkey to provide their strategic alliance with the maximum resources they can muster.

The 9/11 terrorist attacks in 2001 have triggered a security quagmire, where the US foreign policy drifted towards unilateralism, which was a key factor in declaring war against Iraq in 2003. A very dramatic consequence of the 2003 Iraq War depicts a huge budget deficit in the US economy which further worsened the troubling situation of US mortgage firms. A crisis of the American mortgage financing industry emerged within the borders of the US. This, in turn, accompanied the 2008 global financial crisis that has socio-economic and political outcomes. Yet, all analyses are not of the same opinion about the causes of the 2008 financial crisis. Hamilton (2009), who is known for his research in the energy market area, links the 2008 recession to the oil shock of 2007-2008. Hamilton assumes that "the experience of 2007-08 should thus be added to the list of recessions to which oil prices appear to have made a material contribution." There are many scholars such as Kilian (2009) who claim the opposite to Hamilton. Yet, it is a crystal-clear fact that the price of oil and gas are already intertwined with national defense, socio-economic development and political power. "Control oil and you control nations" says former US diplomat and political scientist Henry Kissinger – in 1970 when he was National Security Advisor under the presidential administrations of Richard Nixon (Heinz, 2011).

Many who do not agree with the argument that the rise of the oil prices was the cause of the 2008 recession would point out to the costs of the 2003 Iraq war as well as mismanagement by the Bush administration. However, this raises another question about the main US logic behind the 2003 Iraq war. If Vice President Dick

Cheney's position is valuable enough to give an idea, it would not be irrelevant to remind us about his warnings. Cheney accused the Iraqi leader Saddam Hussein as someone who might "seek domination of the entire Middle East" and "take control of a great portion of the world's energy supplies" (Sanger, 2002). A misunderstanding was to simplify US motivation into an interest for oil that "76% of Russians, 75% of French, 54% of Germans and 44% of British believe the desire to control Iraq's oil lies behind Bush's bellicosity" (McAllister, 2003). Although Iraq's oil reserves are the second largest in the world and the cheapest to extract, the economic burdens of the war were too high. These burdens include: a high amount of investment required to increase production; a high amount of debts of the Iraqi government; the need for investment for the infrastructure and building democracy; and, obviously, the cost of the US military action. Thus, the primary motivation of the US behind the 2003 Iraq war was not simply oil revenue but something more valuable: to secure the energy supplies.

The rise of China is consequential in rebalancing the share of energy imports between the East and the West. Graaf (2013) explains the causes of the high oil prices in 2008 as combination of an "exploding energy demand, rising extraction costs, and shrinking low-cost reserves". When we look at the driving force behind the exploding energy demand, it is the rising demand from China and India – according to the findings of Mehdi Siamak Monadjemi in his 2017 article "Oil Price Rise and the Great Recession of 2008". The US along with its Western European allies had long been the top buyers of energy sources. China nevertheless overtakes the US as the biggest energy importer. While this has an optimistic spin-off in the type of global economic development, it has further perils. This significantly increases the chances of multiple rivalries between energy-importing countries and energy exporting countries, and competition for supply also intensifies. As Johnson (2015) notes in a Foreign Policy article entitled "China Tops U.S. as Biggest Oil Importer" in May 2015: "Beijing's big and growing reliance on energy from the Middle East leaves China vulnerable to regional strife and worries over sea lane security". This also goes for Japan, whose dependency on energy from the Middle East leaves its manufacturing vulnerable to regional rivalries and pipeline security. Herein, Turkey's strong bonds with the region makes it a very important strategic partner for Japan to secure transportation of the Middle Eastern energy resources.

Energy security is very important for global political economy. During Barack Obama's Presidency, the US started to export oil for the first time in its history which "comes as the Organization of Petroleum Exporting Countries trims output in an effort to end a glut that battered the economies of global energy exporters" (Traywick & Tobben, 2017). This strategy of the US also sought to dampen the political influence of Russia and Saudi Arabia, who used their energy supplies as a political tool. As the US oil was more competitive than OPEC's (Organization of

Petroleum Exporting Countries), it cornered a market that was long dominated by Saudi Arabia and other Middle East producers. China becoming the prime importer of US oil has already turned into both a cause and a result of the rebalancing of the share of energy imports between the East and the West. These energy rivalries are not absent in Eurasia wherein the geopolitics is shaped not only by identity, but by the energy policies – about energy fields, pipelines and commercialization of raw materials. The President of Russia Vladimir Putin is the architect of Russia's use of energy as a foreign policy tool of "containing the EU so that it achieves de facto recognition of a Russian 'sphere of influence' within the former Soviet Union" (Graaf, 2013). As the world's largest gas exporter that owns 35% of the world's natural gas reserves, Russia extends its control over the Eurasia strategic energy flow as well as Ukraine, Georgia and Crimea. This power derives from the energy policies that were implemented in line with strategies defined in Putin's PhD dissertation titled "The Strategic Planning of Regional Resources under the Formation of Market Relations".

Besides the shifts in global power balance, attention also should be paid to the changes in the share of the global energy mix which is dominated by fossil fuels such as oil, coal and natural gas. There is a move towards the renewable energy sources due to the obvious consequences of the climate change – such as glacier melting and decrease of water sources. But as Chazan (2013) says in a Financial Times article in 2013, "despite the inroads made by wind, solar, biofuels and nuclear, energy continues to be dominated by fossil fuels". Natural gas has become the primary fuel choice of many end-users. Its portion grows quicker than that of oil and coal in the basic energy combination. Further, the initiation of LNG projects stretches the landscape of the natural gas. In addition, changing technologies carry on transforming the oil and gas industry. The US exports its continued innovations and developments in drilling, extraction and refining technologies that are adapted by other energy exporting countries. This makes the US together with Canada the technological superpowers of oil. The US and Canadian technologies are adopted by other countries such as Egypt that had pushed the employment of innovative methods to discover gas onshore before ENI's discovery of Zohr field in the Egyptian EEZ (exclusive economic zone) 2015. This discovery is expected to turn Egypt into an exporter of gas again. Egypt together with other potential exporters such as Israel and Cyprus can provide an alternative against the oligopoly of natural gas exporters such as Russia, Iran and Qatar. At this point, the need for common pipelines and secured energy routes becomes a prominent issue where Turkey can play a vital role.

To summarize, the restructuring of the global power balance accompanies flexibility to the foreign policy preferences of Japan and Turkey in their relationship as well as with the US. Turkey as a country having strong ties and influence within its region can play a very important role to secure and transfer energy pipelines which is crucial for Japan's production and services sector. And, Japan's technology and finance can be used to build up these pipelines which would strengthen Turkey's economic and political power.

CHANGING RELATIONS DUE TO DOMESTIC CHANGES

During the Cold War, the superpower competition between the US and the Soviet Union intruded into all regions through the bipolar international system. While Turkey feared being a target of the geographical and ideological expansion of the Soviet Union, the US came to conclude that Turkey was very important for an effective defense of Western Europe on the southeastern flank of NATO. Meanwhile the US military bases in Japan have been signs of the US domination in Eastern Asia as Japan's security depended on the American-Japanese coalition. These security interests of Japan and Turkey provided the US with an opportunity to play a crucial role in its alliances with them. This facilitated reasonable respect of Japan and Turkey to US choices in planning the strategy and the wide framework of the foreign policies of these alliances. Firstly, the collapse of the Soviet Union and then the power shift from the US to China dampened the US domination of Japan and Turkey. These systemic power shifts are determinants on domestic changes in Japan and Turkey which, in turn, affects their relationship.

During the Cold War, Turkey was dealing with its problems within the context of the NATO alliance. But, in the aftermath of the 2003 Iraq War, Turkey realized that its security concerns will not be addressed under the NATO umbrella. Turkey sought to find a solution by new diplomatic openings through unilateral means. The Justice and Development Party (AKP) who came into power in 2002 has a direct effect on Turkey's abandoning of its 'isolationist policy'. Turkey had previously turned in on itself and avoided regional and global developments due to its isolationist policy implemented since 1923 when the Turkish Republic was established. In fact, Turkey has strong historic and political ties with the Middle East and played a significant role in the geographic structure of the region through population expansion. Thus, a country like Turkey had no other option but to continue to isolate itself from the growing sense of regional insecurity and increasingly complex conflicts of the Middle East.

Starting with the AKP administration, one of Turkey's key capabilities has been its ability to establish relations with various ethnic and religious groups in the Middle East as exemplified by Hamas in Palestine. It is a fact that the characteristics of the AKP, which had peculiarities in terms of relations with the Middle East, have catalyzed Turkey to play an effective role in this region. All these initiatives of Turkey may also be explained with the multidimensional diplomatic efficiency principle which reflects a more liberal foreign policy rather than the interests imposed by the

US which reflect the logic of the Cold war. Turkey also pursues more proactive, multidimensional and constructive foreign policies to become a regional energy center. Turkey can fulfill its promise to ensure energy supplies, as the Turkish government has come to exert very effective power thanks to the transformation of the republic from a parliamentary to a presidential system. Despite the new presidential system being widely criticized for being less (or not) democratic, Erdoğan can more easily fulfill his promises thanks to his increased powers. To illustrate the case, if Abe agrees with US President Donald Trump on any issue, the legislation of the issue would be challenged as it depends on the approval of both the US Senate and the House of Representatives. But, this is not the case for Turkey due to the expanded powers of the president as well as the strong leadership capabilities of Erdoğan. When Erdoğan makes promises on an issue, the fulfilment of his promise would be less challenged as he exerts extended powers over the other institutions.

While Turkey carries out its energy cooperation with Russia, Azerbaijan and Iran, new projects are pursued by Israel and Egypt. Turkey's plans to facilitate transportation of gas from alternative markets such as Iraq's Kurdish Regional Government, Turkmenistan and Qatar will strengthen the potential of Turkey to become a regional energy center. The Baku-Tbilisi-Ceyhan (BTC) pipeline, Nabucco and Trans-Anatolian pipeline projects are significant initiatives which indicate Turkey's aspiration to play a global role by serving as an energy transit state in an indispensable position. As the first pipeline purposely intended to transfer Caspian oil to the EU without passing through Russia, BTC sought to replace Russia with more dependable suppliers (Said, 2005). Turkey signed energy agreements with Iran in 2007 for Turkey's Petroleum Corporation (TPAO) to drill in Iran and for the transportation of natural gas from Turkmenistan to Turkey and Iran to Europe. There is a strained relationship regarding the energy agreement between the two countries, since according to Turkey Iran seeks to use energy as a political leverage. Yet, Turkey has sought to strengthen its ties with Iran through bilateral meetings. (Hürriyet, 2007). The Trans-Anatolian pipeline project within the so-called Southern Gas Corridor seeks to carry Azerbaijani natural gas through Turkey to European countries. These pipeline projects aim to become an energy corridor for oil and gas fields in the Caucasus and Central Asia. Turkey seeks to expand the security of energy supply and to become an important alternative to the oligopoly of gas exporting countries such as Russia and Saudi Arabia.

Now there is a more independent minded and self-confident Turkey that aims to become a regional energy center. Yet, Turkey lacks advanced technology and economic development. Strategic partners should get used to dealing with an ally which includes sharing problems, being inclusive and trying to solve these problems together. The security of Middle Eastern energy supplies is a shared problem of both Japan and Turkey. Thus, they should try to bring the potential pipeline projects mentioned above to a successful completion.

After its defeat in the Second World War, thanks to the modernization strategies adopted in production, Japan has achieved a very successful modernization outcome. Japan managed to become the second biggest economy in the world and challenged US supremacy in production as well as US economic hegemony in the 1970s and 1980s. Schwartz (1994) explains Japanese modernization strategies as: "at the level of firm organization, the *keiretsu*; at the level of production processes, *kanban* (just-in-time inventory); and at the level of the worker, responsible multi-skilled workers via *kaizan* (continuous improvement) and total quality management." "Kanban refers in English to the three zeros: zero buffering of inventory (sometimes called 'just-in-time inventory'); zero buffering of labor; and zero defects. Kaizen refers to continuous improvement of the three 'P's: production, products, and producers". The Japanese model of modernization in production was not simply an imitation of the US model, but rather a range of unique managerial techniques. Through furthering its relations with Japan, Turkey can also learn from the Japanese modernization process which is unique to the latter as a prime example of success.

Japan struggles to become more independent from the US also in political terms. It successfully accomplished the transition from the single-party to the multi-party system. Turkey has also gone through similar political processes as Japan. But Turkey's experience has been uneasy while Japan successfully accomplished the transition from the single-party to the multi-party system. During 50 years when the Liberal Democratic Party was in power, Japan traditionally assumed a passive role in the international arena. However, opposition parties were accusing the Liberal Democratic Party of complying with each and every request of the US in line with American interests and not advocating Japan's own interests. The Democratic Party of Japan led by Former Prime Minister Yukio Hatoyama was victorious in the general elections held in August 2009 by a wide margin and put an end to the rule of conservatives which lasted for half century (Financial Times, 2012). It was expected to devise policies where Japan would be more active and efficient in the restructuring of the international system. While the Democratic Party of Japan was established by those who left the Liberal Democratic Party, the former was not thinking differently about the highly prioritized issues than the latter. However, during its election campaign, certain changes were promised including a revision of Japan's relations with Washington. Hatoyama once stated that the US-Japanese alliance would continue to be the pillar of the Japanese diplomatic policy, yet they should not forget their own identity as an Asian state.

However, things did not go as planned. When Hatoyama's actions did not match his words, members of the parliament were concerned that their party could be defeated in the parliamentary elections of July 2010. Thus, they put pressure on Hatoyama to quit temporarily. He realized during his tenure as Prime Minister the impossibility of keeping the promises he made against the establishment of a US military base in his campaign. There were thousands of US troops in the base in Okinawa. In the face of US objections and the opposition of the residents of the island thought as an alternative to Okinawa, Prime Minister Hatoyama was forced to take a step back which was regarded as humiliating. In a joint written statement by Japan and the US, it was stated that Futenma Base will be relocated from its then location which was close to residential areas, to a less populous part of the island. While causing the resignation of Hatoyama, this also showed that Japan actually faced challenges about completely leaving the sphere of influence of the US to make and execute policies of its own.

In the following general elections, it was believed that the Democratic Party of Japan did not stand much of a chance. Since December 2012, a coalition between the Liberal Democratic Party and the Komeito Party under the leadership of Prime Minister Shinzo Abe is in power. It has become a current issue to change the pacifist security doctrine of Japan and to adopt the bills allowing the Japanese military to fight overseas in the case of a crisis. There are continued debates to amend the Japanese Constitution, which has never been changed since it came into force. Abe proposed an amendment of Article 9 of the Japanese Constitution which forbids having potential resources for war. He claims that the sole aim of his proposed amendment is to clarify the legal status of the Self-Defense Forces (SDF). An editorial article in the Japan Times on May 28, 2017 reminds the position of successive governments "that the war-renouncing Article 9 does not deny the nation the right to defend itself against enemy attacks". Japan also seeks to obtain the status of a permanent member of the UN Security Council. Even though its current policy looks politically inept, Japan is zealous to become a political actor. It strives to bring about its transformation from an economic giant to being politically influential as well. However, to reach this goal, Japan needs to move from a pacifist approach that relies on the US to a proactive approach that relies on a multilateral network for the defense of the country. Thus, Japan needs support from many countries, including those of the Middle East. This brings Turkey as vital to Japanese interests thanks to the former's strong ties with the Middle East.

In short, the US military bases in Japan and Turkey are signs of the ongoing leadership of the US in Eastern Asia and in the Middle East. Yet, the US' use of these military facilities for operational purposes would depend on operations that clearly serve Japan or Turkey's national interests. US policy makers cannot maintain typical Cold War logic, assuming that the US is the single power that desires something from the allies all the time. During the Cold War Japan and Turkey were feared to be targets of the expansion of the Soviet Union; nowadays they do not fear to be a target of the expansion of China. But, they rather perceive the domination of the US as a bigger threat. The Japanese public still has such expectations that in case US hegemony continues to decline at a time when global dynamic changes are occurring rapidly, a completely different landscape may emerge in the years to come. Thus, Japan might become more independent of the US. These changing internal dynamics when coupled with shifting external dynamics push Japan further towards playing a global role. Therefore, these changes make Japan approach to Turkey to play the role it is expected to play in the Middle East: providing security of energy supplies.

MAXIMIZING THE POTENTIAL IN THE RELATIONS

How will Turkey and Japan maximize the potential of their strategic relationship? And, on what grounds will they base their reciprocal relations to ensure their common interests will be met in a stable and consistent way? This is a question of their strategic preference. They should show political will and determination to act independently of the US when needed. The US policy towards Eastern Asia has been to maintain the US domination in the region. Yet, the irresistible rise of China makes the US dependent on its alliance with Japan to confront China. On the other hand, both the US and Japan expect to maximize their gains from the developed economy and huge market of China. The future of Japan's economic power depends on China. Beyond its huge market, China is transforming to be the production base of Japan. So, if the US behaves in aggressive ways towards China, this would not always be in the interest of Japan. Turkey can create an environment of energy security for Japan, while Japan can supply Turkey with the capital and technological support it needs. Strategic relations can be examined from two perspectives: one of them being economic relations regarding cooperation in the fields of trade and investment; and, the other one being political cooperation. These domains interact with each other and political relations play an important role in the development of economic relations.

The data pertaining to economic relations between Japan and Turkey are way below the level that such a long established relations dating back 120 years should be. Turkey`s imports from Japan amounted to $3,3 billion in 2010, $4,26 billion in 2011, $3,6 billion in 2012, $3,45 billion in 2013, $3,2 billion in 2014, and $3,14 billion in 2015. On the other hand, Turkey's exports to Japan were worth $272,2 million in 2010, $296,4 million in 2011, $332,2 million in 2012, $409,3 million in 2013, $375,7 million in 2014, and $335 million in 2015. According to the data available in the web-page of the Ministry of Foreign Affairs of Turkey (MFA), Turkey's major exports to Japan are "fish, crustaceans, mollusks, aquatic invertebrates;

carpets and other textile floor coverings; vehicles other than railway, tramway; cereal, flour, starch, milk preparations and products". Japan's major exports to Turkey, on the other hand, are "machinery, nuclear reactors, boilers, etc.; vehicles other than railway, tramway; electrical, electronic equipment; optical, photo, technical, medical, etc. apparatus" (MFA, 2017). The figure below reveals the stark trade imbalance between Turkey and Japan in which the latter dominates in the trade relationship between the two countries.

Turkey, which serves as a market for Japan, suffers a constant and unbalanced trade deficit. Hence, Turkey is both trying to develop its trade with Japan in favor of exports and also endeavoring to balance this deficit by attracting more Japanese capital and exporting to third countries from the facilities created by such investments. Also, the number of tourists visiting Turkey in recent years has consistently increased and has reached almost 200 thousand. Nevertheless, foreign capital investments made by Japanese investors in Turkey are much lower than those they make in the EU. While Turkey was initially seen as a bridge to reach EU markets by Japanese investors, Caucasian, Balkan and even Central Asian countries have recently taken over this role. Japan has always been a very reliable trade partner for Turkey. However, trade and investments with Japan is far from balanced and cooperation is inadequate. Furthermore, economic relations are maintained unilaterally through trade and investments of Japan. Japanese foreign direct investment (FDI) in Turkey was a little more than $1.52 billion during the period 2002-2015 (Ministry of Foreign Affairs of Japan (MOFA)). This amount is low compared to the enormous investment power of Japan which derives from the country's gross domestic product (GDP) that

Figure 1. Exports of Japan and Turkey

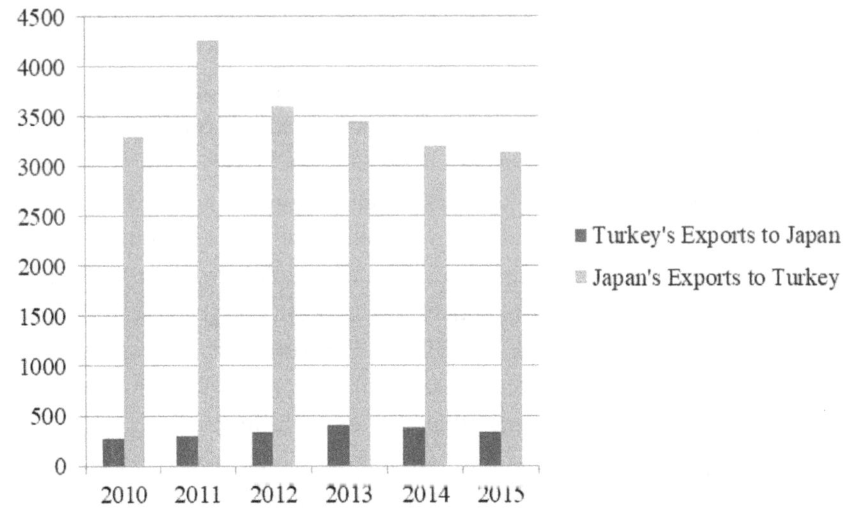

amounted to $4,960 billion in 2016 while the country's GDP per capita was worth $39,089. In the same year, Turkey's GDP was $856 billion while the country has a GDP per capita of $10,731.

Several Japanese multinational companies like Toyota, Isuzu and Honda have a close interest in Turkey and are making investments in Turkey in various sectors. "At the beginning of 2015, 193 Japanese companies operate in Turkey, while 2 Turkish companies operate in Japan" (MOFA, 2016). Furthermore, the Japanese International Cooperation Agency has given considerable support to Turkey by financing important projects. These projects include Marmaray, the Bridge of Haliç, Hasan Uğurlu Dam, Altınkaya Barrage, the Second Bosphorus Bridge, that have been completed in collaboration with Japan. Japan has also been the country with the most advanced and reliable technology in the construction sector, owing to its experience in earthquake-resistant and anti-seismic construction. Japan was relied upon for the construction of earthquake resistant bridges in Turkey, the construction and inspection of tunnel systems within the scope of the Ankara potable water project, the electrical commuter train project and in the construction of the nuclear plant to be built in Sinop to meet the need for energy. Japanese Mitsubishi Heavy Industries has been awarded the 22-billion-dollar project together with Areva, the French nuclear energy giant. Abe and Erdoğan have also signed a joint declaration to enhance cooperation in science and technology in addition to the cooperation regarding Turkey's need for infrastructure and nuclear energy (Miller, 2014). Japan is also a shareholder in strategic investments such as the Baku-Tbilisi-Ceyhan (BTC) Pipeline. These are all examples of Japan's exercising its economic power as an element complementing its rise as a political power.

The political side of relations between Turkey and Japan has been traditionally warm. As it was made explicit from a public survey that the Japanese Foreign Ministry conducted in Turkey, more than 80% of Turkish people view the relationship between the two countries as friendly or almost friendly (MOFA, 2016). Japan hosts a population of 4,444 Turkish people that constitute a significant part of the Turkey-Japan relationship. Turkey and Japan have diplomatic relations since the 1920s, while cultural and other types of exchange activities between the two countries exist since 1873. The Japan-Turkey Society has been organizing seminars as well as cultural and other types of exchange activities to promote friendly relations between the two countries since 1926 when the association was established. Turkey and Japan have some other common interests and values as well. They are both democratic countries with friendly ties with the US. The AKP government has exhibited a more authoritarian approach in the recent decade and suppressed legitimate opposition groups, and this has been further exacerbated after the military coup attempt of July 2016. Yet, Turkey is still governed through democratic elections and referendums (İdiz, 2012). Although Turkey had problems with the US in 2003 during the war in Iraq, policies and principles advocated by Turkey are generally not opposed to those of the latter. Both countries are members of the Organization for Economic Co-operation and Development (OECD) and the World Trade Organization (WTO), while Turkey is a member of the Council of Europe and the Organization for Security and Co-operation in Europe (OSCE), in which Japan is an observer.

Despite the warm relationship between Turkey and Japan, there has not been any noteworthy cooperation in bilateral relations for a long time. The fact that the political relations between Turkey and Japan remain at low-intensity has a negative impact on the flourishing of trade and the increase of Japan's investments in Turkey. A strategic roadmap for the short and medium-term should be devised and a strategic agenda should be set to develop more fruitful relations between the two countries. After the relations between the two countries were upgraded to the level of strategic partnership in 2013, the Turkish-Japan Business Forum was organized during Erdoğan's visit to Japan on 7-8 October 2015. Then Abe joined the Turkish-Japan Business Forum on 16 November 2015 which was organized in Turkey during his visit. Turkey and Japan should increase these kinds of forums as well as other platforms such as symposiums or meetings where politicians, entrepreneurs and business people from both countries can meet, sign agreements and develop strategies to maximize cooperation with each other. Turkey and Japan started to take concrete steps to establish a Turkish Japan Science and Technology University in Turkey, as agreed by the governments of the two countries and signed in Ankara on June 30, 2016. The aim of establishing this university is such that the strategic partnership will produce tangible results in the fields of education, culture, science and technology. The idea for the establishment of this university was suggested in a meeting between Erdoğan and Abe, hence proving the importance of all kinds of dialogue between Turkey and Japan including meetings, forums or symposiums.

Turkey has strived to increase its sphere of influence especially in the Middle East during the rule of the AKP. Turkey aims to expand it in Europe, Africa and Caucasia and needs the strategic cooperation to be maintained with Japan in order to be effective in the Asia-Pacific basin where it has a weak presence. Turkey and Japan, which have the potential to become more effective actors in the system of international relations, need to demonstrate a more enterprising and collaborative foreign policy approach. They do not to have the luxury of keeping Turkey-Japan strategic relations within the limits of bilateral relations. Turkey and Japan should rather focus on a global level to develop mechanisms in order to solve regional and global problems in a way that both countries will benefit, as a function of their strategic relationship. This also requires the proliferation of intellectual platforms and symposiums in which Turkish and Japanese intellectuals will participate to discuss areas of cooperation for the solution of global issues. Another important issue is the urgent need to conclude the ongoing meetings between Turkey and

Japan for a free trade agreement between the two countries. This would strengthen Turkey's economic stability which is vital for its political stability and for Japanese interests in Turkey.

Japan's economic power is weakening as a result of China's rapid rise. If the Chinese economy continues to grow at its present rate, the control of East Asia will be split into two poles between the US and China. And, Japan is far from being a third pole. It might be claimed that the continuing economic transformation in the East Asian region would make Turkey more dependent on industrial inputs from China at the expense of the declining importance of Japan. But, China is so huge not only in economic capabilities but also in hard power that it would easily impose its power on Turkey in case of clashing interests between Turkey and China. It is a potential threat for Turkey that China seeks to deepen its relationship with Cyprus, a country that Turkey has problematic relations with. China's Coscgo group seeks to gain control of Cyprus' main port of Limassol. Then, Chinese companies would seek investment in the EEZ of Cyprus, where Turkey has already made clear its objections by sending its frigates into the region whenever a multinational company has begun natural gas explorations. The fact that China increasingly emphasizes its capabilities related to its projection of power in the seas as well as to its weapons deployed in space, drives Turkey and Japan to devise a joint strategic approach to harness the increasing military influence of China in the Middle East. The military power of China has not reached a point where it can challenge the US-Japan coalition in East Asia. This limits the potential of hot conflicts between Japan and China. What is more important for Japan is the competition with China, the US and the EU in terms of energy resources. Japan's energy needs gradually increase in the context of its competition with China, the US and the EU in terms of energy resources. Japan is among the biggest global consumers of energy as well as natural gas and petroleum importers. It purchases almost all of the petroleum it consumes from other countries. This is one of the factors which increase the significance of Turkey as a strategic country as well as Japan's efforts for rapprochement.

Following the end of the Cold War, particularly after September 11, 2011 and the war in Iraq, Japan reshaped its Middle East policy and maintained a strong diplomacy to secure its energy supply from the region. Japan, which previously avoided the Middle East peace process, is trying to improve its image and assuming a more active role in the region by providing economic aid to Palestine. Japan as the biggest country in Asia and the second country in the world in terms of extending foreign aid, is drawing attention with the support it gives to Afghanistan and Palestine through Foreign Development Aid. Japan has been increasing its aid to the Middle East since 1972 and using the 'Official Development Assistance' as the most important tool to further its interests in the region. Also, the Japan - Arab Dialogue Forum was established in order to deepen the relations between Japan and the Gulf countries, which Japan ascribes strategic importance to. In this context, with an active role in the regional economic life, Japan has a significant trade partnership with Iran. Japan needs to become a more active actor in the global system regarding Iran's nuclear program, bringing stability to the region primarily beginning with Iraq, and thus improving the security situation and helping developing countries in the resolution of humanitarian crises and security.

Owing to its neutrality, Japan, as one of the countries which gives the greatest support to the Middle East peace process, may play an effective role in the Israel-Palestine conflict Through the years, the reason for Japan's adopting a balanced policy vis a vis the US policy towards the Israel-Palestine conflict is that peace and stability are much needed in the region to ensure the uninterrupted flow of Middle Eastern petroleum for long periods. Japan perceives stability in the Middle East as a security problem with political, economic and strategic dimensions both internally and externally. An energy bottleneck in the Japanese economy caused by disruptions in the flow of energy may have a big negative impact due to the slowing down of business activities and a reduction in employment. These developments taking place in the dynamics of the international system in the last decade have caused Turkey's value to increase in the eyes of Japan, which has been more willing to play a global role beyond its territories. For a long time, Turkey failed to adopt a more strategic and integrated approach in the area of energy and has considerable budget deficits due to the increasing need for energy. But, it now seeks to become a regional energy center.

Japan is also striving to achieve the transformation from being merely an economic giant following the Cold War to being influential also on the political front. Yet, Japan cannot rely on the US support in every situation since their interests may diverge. In order to fulfill this aim, Japan needs the support of a number of countries including Middle Eastern countries. At this point, Turkey which is an influential power in the Middle East region can become a key actor as a strategic country for Japan which seeks a stable Middle East. Turkey may also devise a joint strategy with Japan regarding the operation of the Middle East's energy resources. Turkey is open to cooperation with respect to the operation of oil and gas resources of the Middle East. While all these serve Turkey's strategies to become an energy center, the required technology and assistance may be obtained from Japan and energy security may be provided in return.

CONCLUSION

The restructuring of a power balance further complicates the Japan-Turkey relationship owing to the rise of China as a counterbalance against the US hegemony in the Middle East as well as in Eastern Asia. China claims that its rise is not a challenge or threat to any country. Yet, it is also likely that the rise of a rival brings a power transition and increased tensions in both Eastern Asia and the Middle East. Worldwide tensions recently signal that future wars will be based on energy supplies. This is one of the most important consequences of the rise of the East which seeks a rebalancing of the distribution of energy sources. The management of the potential effects stemming from this power shift from West to East depends on the strategic choices of Japan and Turkey, whether to get the US involved or to act independently. Japan struggles to become more independent of the US and acquire a bigger military and political power in order to secure the energy supply its economy depends on. This requires Japan to maximize the potential in its relations with Turkey that seeks to become an energy center in the Middle East where it has strong ties.

The strategic partnership of Turkey and Japan has the potential to offer an alternative to the US dominated world model, or the multipolar world model of China, by which the US or China occasionally imposes its interests. Turkey and Japan are non-Western countries seeking to establish a democratic state of affairs. Turkey and Japan need to extend the international role they have been undertaking and use more flexible and efficient approaches in foreign policy. A harmonization or balance of powers is required to prevent the US, China or another country from establishing their hegemony of the international system. To maximize the potential of their relationship, Turkey and Japan must understand and use well what they mean for strategic and broader interests of each other.

REFERENCES

Brzezinski, Z. (2016, April 17). Toward a Global Realignment. *The American Interest.*

Cambridge Dictionary. (2017). Retrieved from http://dictionary.cambridge.org/dictionary/english-chinese-traditional/power-vacuum

Chazan, G. (2013, June 12). Changing Sources of Energy. *Financial Times.*

Graaf, T. V. (2013). *The Politics and Institutions of Global Energy Governance*. New York: Palgrave. doi:10.1057/9781137320735

Hamilton, J. D. (2009). *Causes and Consequences of the Oil Shock of 2007-08, NBER Working Paper no. 15002*. Washington, DC: Brookings

Heinz, A. (2011, November 27). Henry Kissinger: "If You Can't Hear the Drums of War You Must Be Deaf". *The Daily Squib*.

İdiz, S. (2012, October 30). Erdoğan's 'Ankara Criteria'. *Hürriyet Daily News*.

Ikanberry, J. (2015). *America China and the Struggle for World Order: Ideas, Traditions, Historical Legacies and Global Visions*. New York: Palgrave. doi:10.1057/9781137508317

Johnson, K. (2015, May). China Tops U.S. as Biggest Oil Importer. *Foreign Policy*.

Kalın, İ. (2011, March 7). Türkiye ve Japonya: Değerli Bir Ortaklık [Turkey and Japan: A Precious Partnership]. *Today's Zaman*.

Kilian, L. (2009). Not all Oil Price Shocks Are Alike: Disentangling Demand and Supply Schocks in Crude Oil Market. *The American Economic Review*, *93*(3), 1053–1069. doi:10.1257/aer.99.3.1053

McAllister, F. 0. (2003, January 20). Mad at America. *Time*.

Mearshmeimer, J. (2015). *The Rise of China and the Decline of the US Army*. San Diego, CA: Didactic Press.

Miller, J. B. (2014, March 18). Japan's Strategic Push with Turkey. *The Diplomat*.

Ministry of Foreign Affairs of Japan. (2017). Retrieved from http://www.mofa.go.jp/region/middle_e/turkey/data.htm

Mütercimler, E. (2010). *Ertuğrul Faciası, Ertuğrul Disaster*. İstanbul: Alfa Printing House.

Portrait of Japan's Main Political Parties. (2012, 17 December). *Financial Times*.

Questions that Abe's Article 9 Pitch Will Raise. (2017, May 28). *Japan Times*.

Republic of Turkey: Ministry of Foreign Affairs. (2017). Retrieved from http://www.mfa.gov.tr/turkey_s-commercial-and-economic-relations-with-japan.en.mfa

Said, Y. (2005). *Oil and Activism, Global Civil Society*. Sage. doi:10.4135/9781446211908.n5

Sanger, D. E. (2002, November 1). The debate over attacking Iraq heats up. *The New York Times.*

Schwartz, H. M. (1994). *States versus Markets*. New York: Palgrave. doi:10.1007/978-1-349-13536-3

Traywick, C. & Tobben, S. (2017, April 4). China Surpasses Canada as Top Buyer of U.S. Crude. *Bloomberg*.

Turkey, Iran set to increase energy ties (2007, February 22). *Hurriyet.*

Turkey, Iran vow to boost bilateral, regional cooperation. (2006, December 4). *Hurriyet.*

Veli, S. V., & Sarıkaya, E. (2004, February 5). Yüzyılın kurtarma operasyonu [The rescue operation of the century]. *Sabah Newspaper.*

Chapter 11
Analysis of the Robustness of Norway's Economy and Energy Supply/Demand Fluctuations

Maryam Hamledar
University of Tehran, Iran

Reza Saeidi
University of Tehran, Iran

Alireza Aslani
University of Tehran, Iran

ABSTRACT

The high dependency on fossil fuels, fluctuations in prices and supply have macro/micro-economics effects for both energy exporters and importers. Therefore, understanding economic stability based on energy market changes is an important subject for policymakers and researchers. Norway, as a fossil fuel exporting country, is a good choice for the analysis of the relationships between economic robustness and fossil fuel economic fluctuations. While the country is one of the pioneers in the field of sustainable energy utilization, they have tried to provide a robust economic environment for oil export revenues. In this chapter, the impacts of energy changes on the economy are investigated in Norway. In this regard, first, the impact of oil prices on macro-economic parameters is discussed. Afterwards, the main issues related to energy economics including resilience of the energy sector, energy policies, economics analysis of the energy sector, and the electricity markets are discussed.

DOI: 10.4018/978-1-5225-4203-2.ch011

INTRODUCTION

Norway is one of the Nordic countries with a population of 5.22 million and one of the richest in economic terms and social welfare provision in the world (Fløttum, Dahl, & Rivenes, 2016). While the country is one of the richest in fossil fuel reserves and production, more than 96% of electricity is generated by hydropower ("International Energy Agency," n.d.). Although available hydropower resources in Norway can provide low-cost electricity, the costs of electricity usage are more expensive than in other countries like Sweden. Hence, the government has decided to reduce the dependence of the budget on oil income. Norway as one of the energy exporting countries plays a key role in securing oil supplies to the European Union; thus, its economic stability is crucial, especially for European countries (International Energy Agency, 2014b; Claes, 2010). Therefore, it can be stated that investigating the relationship between the energy and economics in Norway has a great importance (Yousefi, Hamlehdar, Tabasi, & Noorollahi, 2017).

Energy as the main input of products/services has a significant effect in improving living standards. On the one hand, energy price and world energy demand have increased, on the other hand, fossil fuel replacement with renewable energies to meet energy demand has developed gradually. For these reasons, energy security has become a major concern for most societies (Mehrpooya, Mohammadi, & Ahmadi, 2018; Mohammadi et al., 2018; Noorollahi, Itoi, Yousefi, Mohammadi, & Farhadi, 2017; Noorollahi, Saeidi, Mohammadi, Hosseinzadeh, & Amiri, 2017; Yousefi, Roumi, Tabasi, & Hamlehdar, 2017; Yousefi, Tavakkoli-Moghaddam, Oliaei, Mohammadi, & Mozaffari, 2017). Because Norway is a wealthy country in terms of energy resources, thus the policy of using indigenous energy resources has helped the industrialization and economic growth in this country (Tønne & Tonne, 1983). In 2013, Norway was the first country in the world based on sustainability management and economic growth ("Norway - the official site in Canada," n.d.). Given the relationship between energy and economic growth, energy price plays a decisive role in economic activity (Stern, 2004).

BACKGROUND

Many empirical studies have investigated the relationship between oil price fluctuations and macroeconomic activity. Moreover, the importance of taking the two-way causality into account with regard to international shocks is strongly emphasized (Baumeister & Peersman, 2013; Kilian, 2009; Kilian & Murphy, 2012). (Kilian, 2009) and (Kilian & Murphy, 2012) demonstrated that oil demand volatility was the most significant factor in oil price fluctuations throughout the 1974–2009 period. (Esfahani, Mohaddes, & Pesaran, 2014) explained the direct effects of oil price shocks on domestic production for 9 major oil exporters, without considering the variance effects of demand on oil price shocks.

According to research, macroeconomic parameters can be affected by oil price shocks. In this context, similar results were obtained in several Asian countries and indicated that oil price fluctuations have an effect on economic growth (Cunado & Perez de Gracia, 2005; Du, Yanan, & Wei, 2010; Hanabusa, 2009; Jayaraman & Choong, 2009; Rafiq, Salim, & Bloch, 2009). Also, the impact of an oil price shock on exchange rates has been observed in both developed and developing countries (Narayan, Narayan, & Prasad, 2008; Ozturk, Feridun, & Kalyoncu, 2008; Rautava, 2004; Shehu Usman Rano Aliyu, 2009). Some current researches have modeled a system of demand and supply in global oil markets by using DSGE models. They have considered the fluctuations in macroeconomic variables in response to oil shocks while taking into account oil importers mostly the U.S. economy. (Bodenstein & Guerrieri, 2011; Nakov & Pescatori, 2010; Peersman & Stevens, 2013). Nobili (Lippi & Nobili, 2012) analyzed the unstable correlation between oil prices and the US economic activity which revealed that the supply shocks in the global economy have a greater effect on oil price fluctuations than canonical oil supply shocks.

Beckmann and Czudaj, (2012) studied the relationship between oil prices and exchange rates based on the U.S. dollar. They employed the Vector Error Correction method in 10 oil-exporting and oil-importing countries. In the exporting countries, the positive relationship between oil prices and exchange rates was very significant. However, a negative correlation was demonstrated for oil-importers. Kopytin (2014) reported the impact of oil prices on the Norwegian stock market by an autoregressive approach. They evaluated that oil price shocks have not had a negative influence on the stock market, because of the significant role of USD exchange rate and S &P 500 stock index.

Mehrara, (2007) investigated the causal relation between the gross domestic product (GDP) and the energy consumption in 11 oil exporting countries and suggested that the energy protection by improving energy price policies has no damaging effect on economic growth. Yildirim, Sukruoglu, and Aslan, (2014) analyzed the causal relationship between energy consumption and economic growth in the 11 countries. Energy saving policy findings may lead to a drop in the economic growth of Turkey. Another study examined the relationship between economic growth and energy consumption in Pakistan. The results showed the relationship between economic growth, nonrenewable energy consumption and renewable energy consumption (Shahbaz, Zeshan, & Afza, 2012). In addition, the relationship between oil prices and economic growth for different panels of OECD countries was studied in a comprehensive investigation. Furthermore, a robust negative connection between oil price and economic growth was found (Kim & Willett, 2000). In another study, the relationship between government revenues, oil prices, and real GDP was assessed in Russia. It was found that Russia's real GDP and government revenues were negatively influenced by oil price changes (Rautava, 2004).

In this chapter, the effects of energy prices on the economy are investigated in Norway. First of all, the impact of oil prices on macroeconomic parameters is assessed. Second, the main issues related to the energy economics in this country are presented including the resilience of the energy sector, energy policies, economy analysis of the energy sector, and electricity markets.

OIL PRICES AND ITS EFFECT ON MACROECONOMIC IN NORWAY

In this part, based on the relationship between energy and economic growth by taking into account macroeconomic parameters, the impact of oil price shocks on the Norway economy is explained. Oil price drops have a significant negative impact on Norway as an exporter. So, the Norwegian economy has faced a recession caused by falling oil prices. The important symptoms of this economic recession are seen in real wages and unemployment (Hemmings, Koutsogeorgopoulou & Sorsa;, 2016). In contrast, an increase in oil price significantly contributes to raising Norway's consumption, decreasing unemployment and enhancing welfare (Nilsen, 2015). Figure 1 illustrates the impact of oil price on selected macroeconomic parameters. The impact of oil price shocks on GDP from 1975 to 2008 shows that the rising oil prices increased Norway's GDP (OECD, 2016). As for the Figure 1-a, in recent years, with the sharp fall in oil prices, GDP is increased, because GDP depends on other factors in addition to oil prices.

The world and particularly Europe were affected by the economic downturn of America in the middle of the first decade of the twenty-first century. Hence, depression was established in Norway, and its industrial growth declined; however, based on Figure 1-b after 2010, industrial growth has increased in line with a recovery in the Europe economy.

According to Figure 1-c, rising oil prices has led to an increase in the oil industry investment and consequently created new job opportunities in oil companies such as Statoil and reduced the unemployment rate as well. In early 2015, the decline in oil prices resulted in an increase in the rate of unemployment, which reached over 4% of the labor force (Hemmings, Koutsogeorgopoulou, & Sorsa;, 2016). In other words, it can be stated that businesses and sectors that are heavily dependent on the petroleum sector, are most affected by oil price shocks. The oil and gas industry directly affects the Norway economy by modest direct employment. Employment in oil and gas extraction has increased by over 1 percent. The unemployment rate is

Figure 1. Current macroeconomic indicators in Norway

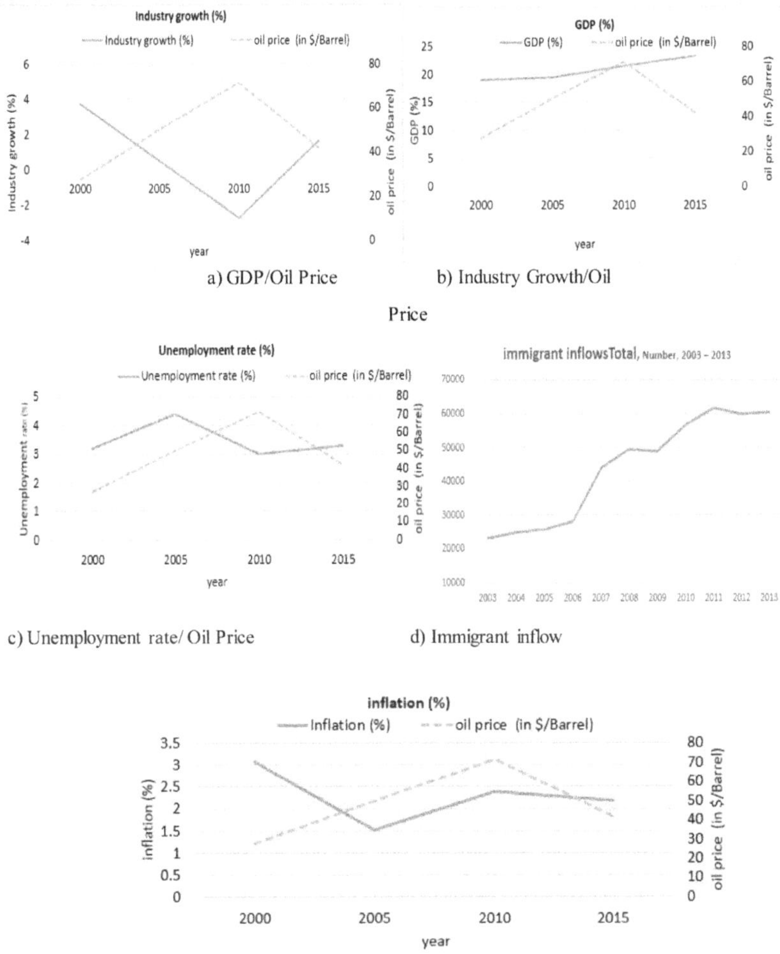

a) GDP/Oil Price b) Industry Growth/Oil Price

c) Unemployment rate/ Oil Price d) Immigrant inflow

e) Inflation/Oil Price

associated with immigration and asylum which based on the Figure 1-d, Immigration increased dramatically from 2000 to 2005 in Norway. For this reason, despite the increase in oil prices, the unemployment rate increased during this period.

Regarding the Figure 1-e, Norway is an oil exporter in which by oil prices rising, the inflation is dropped and decline in oil prices cause an increase in the inflation. In Norway, Government Pension Fund Global (GPFG) plays a significant role in inflation reduction through buffering of financial challenges, including oil price shocks and non-petroleum budget shortages (Nilsen, 2015). In the late 1990s, the government gained considerable revenues from oil and gas, and this issue was synchronous with the establishment of the GPFG and the fiscal rule. In 2000, the income related to petroleum increased to more than 10 percent of GDP and 25 percent of total revenue when it reached its highest level. Norway's fiscal rule efficiently determines its budget from oil price fluctuations in the short term. Fiscal policy is taken from the fiscal rule to stipulate gradual phasing-in of oil income into Norway's economy with the expected returns on the GPFG which is evaluated to be 4 percent. According to the rule, oil and gas revenues should not directly cover government expenses, but instead, 4 percent of GPFG*'s assets* contribute to the budget yearly as the expected income from this blessing. The fluctuations in oil price and production volumes are significant factors that change the value of the GPFG every year. But, in recent years, changes in financial market returns have influenced the value of the GPFG more than the oil price fluctuations. GPFG saves Norway's petroleum revenues to finance in international possessions. The fund has grown in the last 15 years as a share of mainland GDP (see Figure 2) (Bergholt & Larsen, 2016).

Oil price fluctuations, or in the other words, oil shocks have played a significant role in the US economy and the global economy since 1974. In this context, there is an extensive literature studying the economic consequences of oil shocks on the real output and real equity prices. The change in macroeconomic variables in the OECD countries especially oil importers is adequately considered by the researchers (Cashin, Mohaddes, Raissi, & Raissi, 2014; Hamilton, 2009; Kilian, 2009; Mohaddes & Pesaran, 2016; Mohaddes & Raissi, 2016). The Norwegian maritime industry is one of the institutions affiliated with the oil market. Throughout the 2004 - 2014

Figure 2. The Government Pension Fund Global (GPFG) in Norwegian kroner (left) and in US dollars (right)

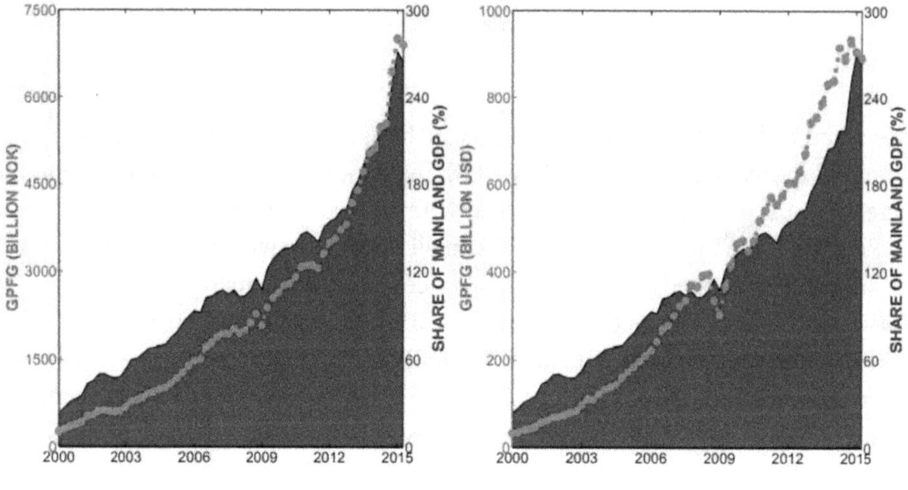

period, because of the strong oil price environment, the maritime industry turnover related to oil improved by more than 200 percent. But in the second half of 2008 and in 2009, the global financial crisis which was followed by the sharp reduction in the oil price (from USD 145 per barrel to USD 35 per barrel for Brent Crude) decreased the number of offshore vessel orders. From 2009 to 2011, bank lending fell by 50 percent while the government lending grew by 10 percent, for this reason, lending to the shipping industry dropped by 57 percent. This decline was mainly due to the downturn and the financial crisis following the fall in oil prices that occurred during this period ("PEER REVIEW OF THE NORWEGIAN SHIPBUILDING INDUSTRY," n.d.). In general, oil price shock affects the Norwegian' economy considerably, Norwegian oil production only constitutes around 2% of global oil production. Therefore, changes in the Norwegian oil production have no significant impact on the global fuel prices, and the resulting income level related to energy in the developing countries. For instance, a drop of 50% in the Norwegian oil production has increased fuel prices by 1% ("statistics Norway," n.d.).

Resilience of the Energy Sector

Total primary energy supply (TPES) in Norway was 32706 ktoe in 2013, in which the share of crude oil, hydropower, and natural gas was41.5%, 33.3%, and 17.3% respectively ("International Energy Agency," n.d.). From these energy resources, most of the crude oil and natural gas production is exported and only a small amount of it is stored. In the recent years, Norway has moved to the use of renewable energy sources so that, yearly the share of RE production accounted for 120- 135 TWh. As a matter of fact, between 95% and 99% of RE resources are utilized to generate electricity (Gullberg, Ohlhorst, & Schreurs, 2014). Norwegian electricity production is mainly based on the hydropower; nonetheless, wind and thermal energy cannot be ignored. For instance, in 2013, Norway produced 134 (TWh) electricity, of that wind power and thermal power had a share of 1.4% and 2.42% respectably, on the other hand, the share of hydropower was 96% (Hamlehdar & Aslani, 2017).

As regards wind resources, the number of wind turbines increases year by year and in 2014, the installed wind power was 811 MW in Norway (Stern, 2004). Also, the government provides a 25% investment grant for wind energy based electricity production ("World nuclear association," n.d.). On the one hand, the expansion of hydropower in Norway causes economic improvement and consequently industrial growth. Nevertheless, investing in hydropower development is faced with many risks, for instance, exchange rate risks, interest rate risks and foreign market risks. Recently, Norway has focused on the technology of small hydropower that the related investment contains risk factors. Moreover, subsidies are provided by the government for the small hydropower plants or in other words taxes regarding this type of plants are considered a political risk. Furthermore, the exchange rate risk is the most important risk for foreign investors, especially European investors in the field of electricity generation. This is clear when a significant oscillation against the Euro_have rather exchange rate risk decreases (Aarrestad & Hatlen, 2014). The resilience of the energy sector in Norway can be discussed from different aspects based on the following.

Energy Security

According to the International Energy Agency, energy security for importers is defined as the continuous availability of energy at an affordable price. Energy security is of vital importance for the energy importers in terms of energy supply and recently has become a major concern ("International Energy Agency," n.d.). On the other hand, energy security for energy exporters is a constant energy export flow at an appropriate price that can exert positive effects on both existing energy investment and economic growth. The demand security should be considered as a vital part of energy security. For example, many oil exporters faced energy insecurity of demand when oil prices declined, especially in the years 1986, 1998, 2009, and 2014. At this time, when oil exporting countries were facing an economic downturn following the falls in oil prices, oil importing countries effected_(Secretariat, 2015).

Norway with a large share of hydropower based electricity supply can be regarded as a successful case in renewable energy development. Research shows that the diversification of energy supply sources (DESS) in total energy supply for Norway was 0.66 in 2009 (Aslani, Antila, & Wong, 2012). This has caused Norway to stand among the secure countries in term of energy supply (US Chamber of Commerce, 2015). In fact, Norway is an energy independent country which means that its domestic production is more than the country's requirement, and thus a huge part of this energy which is oil and gas is exported. Besides, Norway itself helps the global supply disruptions in the world. Norway government had saved some amount of oil for emergency situations up to 2006, but in 2007 after announcing a new strategy, it sold these reserves, that forced industries to save and keep their products in the domestic market after 20 days (International Energy Agency, 2014a).

Norway has 26 main storage facilities, more than 17 localities and also about 50 warehouses. They use the huge amount of water resources spread all around the country in many useful ways. To illustrate this point, Hydro itself includes 96% of production in this country. However, throughout 2002- 2003, in winter, Norway experienced a drought by entering an intensive cold wave that led to the decrease in hydro storage and the cost event. In this year, a commercial onshore gas-fired power plant was built with the capacity of 420 MW, and it was one of the cleanest fossil fuel power plants in Europe. Theoretically, this project can supply 3% of production at a cost of around EUR 253 million (NOK 2 billion). The project can theoretically deliver up to about 3% of Norway's total electricity production (equivalent to around 175 000 households) (International Energy Agency, 2014a).

Review of Selected Energy Policies in Norway

Energy policy comprises of energy efficiency and renewable energy policy so that energy efficiency can be considered as an essential tool to manage energy challenges (Bukarica & Tomsic, 2017). The optimization of the energy system in the production sector can not only improve efficiency but also it can contribute to economic development (Zhang, Rahbari-Asr, Duan, & Chow, 2016). In many countries, a renewable energy policy with the aim to develop the low-carbon economy model has resulted in energy consumption reduction and the tackling of climate change (Chen, 2010).

Taxation: energy tax is used for different purposes, especially in government's income, executing energy policies and environmental issues. Energy taxes on fossil fuels are mostly added to government's income in Norway. In 1991, a CO_2 tax on fossil fuels was introduced. This plan is also executed for oil and gas, which makes the technology look for more ways to decrease CO_2 production and it also promotes more investment (Gavenas, Rosendahl, & Skjerpen, 2015; Larsen & Nesbakken, 1997). The tax also helps to increase efficiency for power supply from offshore installations. In this regard, some industries are exempted from electricity tax in Norway; however, the mineral oil tax is used to prevent the consumer from changing taxation on petroleum. In the petroleum section, a special tax (50%) of the income is received from extracting products. Furthermore, the income tax on petroleum and hydropower generation sector is 28 percent. As a result, the total tax rate in the petroleum sector is 78% and the final tax on hydropower generation which its excess return is 30% becomes 58% (IEA, 2011).

R&D Strategy

Norway has two major strategies for R&D, one of which is OG21, for oil and the other one is energy 21 which is for other energy resources. The government also supplies the bodies that support energy R&D such as Gassonva and Innovation Norway (IEA, 2011).

Strategies

- **OG21:** The "Oil and Gas in the 21st Century" strategy that MPE started executing it in 2001 and updated it in 2010 (Gass, 2016).

- **Energy21:** started in 2007 after the success of OG21 and its goal was R&D in different sections of energy (Lives & Roads, 2014). New technologies have been created to provide a safe platform for growth in economic activities and safety of supply in each part of energy. It was run by RCN's, RENERGI and CLIMIT programmers to demonstrate development, and commercialization (IEA, 2011).

Plan

Petromarks: The PETROMAKS was implemented in 2004 to execute OG21 plans that both by academia and industry and it was supposed to end in 2013. The total budget for 2010 was NOK 260 million (EUR 30 million) (Program, 2013). DEMO 2000: a pilot both to increase the value of exploration and development of hydrocarbon resources on the Norwegian Plan for helping specific projects. The budget for 2010 was NOK 100 million (around EUR 11 million) (Norway, 2013). RENERGI: RENERGI is a research plan in the field of clean energy. It was launched in 2004 and is planned to run until 2013 (*Preliminary work programme for the RENERGIX programme - abbreviated English version*, 2012).CLIMIT: CLIMIT was a project which started in 2005 and it had a particular focus on Carbon capture and storage (ccs). It is a long-term policy to support the development of gas-fired power plants with ccs. The program has had an annual budget of NOK 180 million (around EUR 20.6 million) since 2010 (IEA, 2011).

Supporting Bodies

Innovation Norway and Enova: These provide the financial sources for the projects. Innovation Norway covers all the sections, while Enova was established to increase efficiency and also renewable energy development. The ENOVA budget for 2010 was NOK 137 million (around EUR 11 million) (NAKSTAD, 2013). Gassnova: the purpose of establishing this organization was carbon capture and storage (CCS). Norway government invested in Gassnova, to find a solution and technology for capturing and storing CO_2, technology related to CCS, and the deployment of complete CCS businesses. The State enterprise Gassnova had a budget of NOK 91 million in 2010.

Transnova: Transnova is a trial R&D funding program for limiting CO_2 emissions in the transport sector. The Transnova budget is NOK 50 million (EUR 5.7 million) for 2010. Figure 3 shows the structure of the R&D in Norway (IEA, 2011).

Figure 3. R&D system in Norway (MPE: Ministry of Petroleum and Energy)

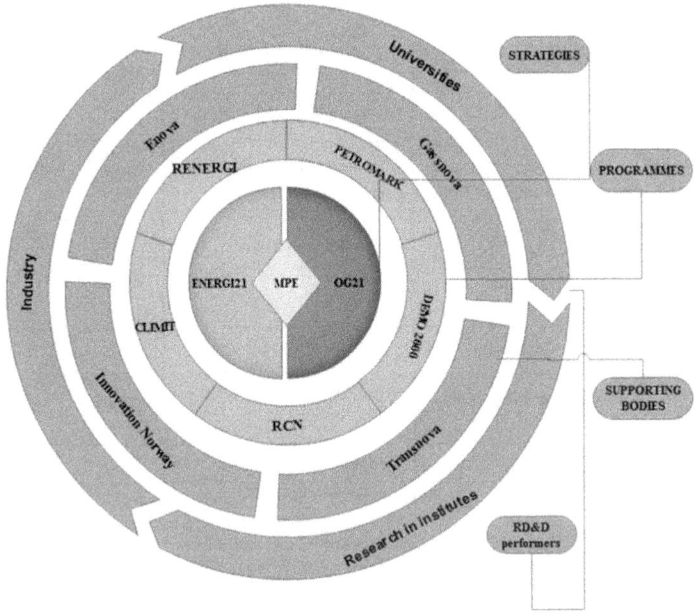

Overview of Energy Regulation in Norway

In order to increase the resilience of the energy sector and its effect on the economy and social welfare in Norway, different regulations have been launched by the government. An overview of the policies and measures to promote the use of energy from the renewable resources is shown in Table 1 (OED, 2013).

Economics Analysis of the Energy Sector

Regarding the economy, elasticity is a measure of how sensitive one variable is to a change in some other, which is represented in Equation (1). Specifically, elasticity is calculated as a one positive/negative percent change in a dependent variable in response to a one percent change (positive/negative) in another independent variable (Subhes, 2011).

$$e_t = \frac{\frac{dQ}{Q_t}}{\frac{dP}{P_t}} \tag{1}$$

Table 1. Energy regulatory

Name and Reference of the Measure	Expected Result	Targeted Group and/or Activity	Start and End Dates
Planning and building Act and technical building regulations (TEK). Requirements for energy consumption and energy supply in new constructions and for extensive rehabilitation	Energy efficient buildings with a greater share of renewable energy consumption	Construction sector	Revised in 2007, changed in 2020.
Energy Act with appurtenant regulations. Energy rating of residences and buildings and energy assessment of technical facilities in buildings.	The assessment will provide more information about the residence's	Building owners	2010-
Royal Decree on requirements for energy-flexible heating systems in State-owned buildings, as well as in private buildings that are rented to the State	More flexible heating systems in buildings used by the State	State building owners	1998-
Eco-design requirements	Promote development and application of energy-efficient and environmentally friendly energy related products	Consumers	2011
The obligation to connect producers	Facilitation of new renewable power production.	Power producers and grid companies	January 2010
Introduction of AMS	Increase flexibility with end-users. Enable increased integration of intermittent power production.	Producers consumers and grid companies	2006-2017
Schedule for receiving and processing grid connection requests	Increased predictability for new power producers that want to connect to the grid	Producers, consumers, and grid companies	January 2012
Reporting of operational measures that limit renewable power production	Prevent undesirable reduction in renewable power production	Systems manager, Regulators and renewable power producers	January 2012
Mandatory sale requirement for biofuel in road traffic	Increased use of biofuel in road traffic	Transport Fuel converters	2009

Q and P are a set of data for energy consumption (EC), GDP, and oil price (P) at time t. Energy consumption-GDP elasticity shows the changes in EC according to the changes in GDP or GDP-oil price elasticity indicating the changes in GDP, according to the changes in energy consumption-GDP. In fact, elasticity is a key indicator in projecting energy needs in economic development. The increase in GDP is usually related to a growth in industrial development and higher capital intensity, both of these factors contributing to higher energy consumption. The percentage change in energy consumption associated with a 1% change in GDP plays a key role in such forecasts (Zilberfarb & Adams, 1981).

The Relationship Between GDP and Oil Price

Elasticity between GDP and oil price is calculated from Equation (2), Suggesting that a 1% change in oil price will increase GDP by 0.72. It is obvious that changes in oil prices do not have a significant influence on GDP.

$$e_t = \frac{\frac{d(GDP)}{GDP_t}}{\frac{d(P)}{(P)_t}} \tag{2}$$

The Relationship Between Energy Consumption and GDP

In the calculation of elasticity from Equation (3), it is determined that with 1-unit growth in GDP, energy consumption will change only by 0.17. An increase in GDP causes a growth in social welfare, transportation, and gasoline consumption. Owing to the fact that Norway is a developed country with a stable economy, GDP does not have a profound influence on energy consumption. Also, some researchers have analyzed energy as a significant element in the economy, taking into account the regression data and the correlation coefficient. The results demonstrated that GDP, population, and industrial growth rate are not correlated with energy consumption.

$$e_t = \frac{\frac{d(EC)}{(EC)_t}}{\frac{d(GDP)}{(GDP)_t}} \tag{3}$$

It is not considered as a surprising matter, because that the value of this factor is under 1 in most of the developed countries. Also, we can say an energy crisis does not make a significant difference in elasticity. Therefore, EC variation with respect to GDP over development is logical.

The Relationship Between GDP and Energy Consumption in Industry (ECI)

A 1% change in GDP changed energy consumption by 0.21% from 2005 to 2014 which proves that GDP does not have an effect on energy consumption in industry.

$$e_t = \frac{\dfrac{d\left(ECI\right)}{\left(ECI\right)_t}}{\dfrac{d\left(GDP\right)}{\left(GDP\right)_t}} \tag{4}$$

If energy causes an increase in GDP, then reducing consumption in the industry could lead to a fall in the income and employment (Chontanawat, Hunt, & Pierse, 2008).

$$e_t = \frac{\dfrac{d\left(GDP\right)}{GDP_t}}{\dfrac{d\left(ECI\right)}{\left(ECI\right)_t}} \tag{5}$$

GDP- EC elasticity (from 2005 to 2015) is 3.42 in Norway, which means that the unemployment rate will be increased by reducing the energy consumption in the industry sector.

Table 2 shows the calculation of elasticity from 2005 to 2014 ("International Energy Agency," n.d.). The geometric average elasticity, for all these years, is shown in the last row. In this table, GDP is expressed in U.S. dollars (Worldbank, 2015). Also, energy consumption is ktoe, thousand tons of oil equivalent, and its data has been extracted from the IEA ("International Energy Agency," n.d.). In the fifth column, the price of petroleum is in the form of Brent Crude oil price in U.S. dollar per barrel average annual. There are also relations between economics and energy parameters which its elasticity is calculated according to equation (1). According to the oil-GDP elasticity section for 2011, it increased to a great extent, and it reached to 6.97. In fact, most of it resulted from the increased income that oil companies gained because of the rise in oil prices. However, in the value of elasticity between GDP and energy consumption, a trend is observed, the least of which occurred in 2005 with a value of 2.02, and the most of it happened in the year 2014 by 0.66. In

Table 2. Indicates calculation of elasticity from 2005 to 2014.

Year	GDP	EC	EC in industry	Oil price	GDP → OIL PRICE elasticity	GDP → EC elasticity	GDP → ECI elasticity	ECI → GDP elasticity
2005	3.09E+11	20453	6732	54.57	0.61	0.02	0.21	4.71
2006	3.45E+11	20406	6562	65.16	1.44	0.14	0.06	16.21
2007	4.01E+11	20872	6497	72.44	0.45	0.03	0.13	7.92
2008	4.62E+11	20963	6622	96.94	0.45	0.27	1.05	0.96
2009	3.86E+11	20031	5490	61.74	0.38	0.60	0.99	1.01
2010	4.29E+11	21351	6085	79.61	0.41	0.27	0.06	15.69
2011	4.98E+11	20424	6022	111.26	6.97	0.11	1.75	0.57
2012	5.10E+11	20475	5778	111.63	0.93	0.41	0.26	3.79
2013	5.23E+11	20691	5817	108.56	0.48	0.66	0.02	49.47
2014	5.01E+11	20114	5812	98.97				
	The geometric mean of elasticity				0.76	0.17	0.21	3.44

the column showing the elasticity between GDP and energy consumption in industry, the highest value occurred in 2011 which raised by increasing in the oil price as a result of GDP, hence led to more investment in the industry. As a result, the energy consumption in this part has risen. In the last column of table 1, unlike the previous column, the GDP changes based on EC in industry, shows a large fluctuation and in 2013 it reached a maximum value, 49. Also, this shows the intensive dependence of GDP on energy consumption in industry.

The Effects of Economic Factors on the Future of Energy Consumption

Predicting energy consumption is a key input that influences the macro policies in various sectors of energy and economics. By calculating the geometric mean of elasticity together with the GDP growth in coming years based on predictions by organizations for economic forecasting and statistics center of the country ("Norway - Gross domestic product (GDP) growth rate 2020 | Statistic," n.d.), it is possible to calculate the approximate value of energy consumption (Subhes, 2011). According to Figure 4, GDP and energy consumption will have a direct relationship with each other in the future. This is calculated through the elasticity formula (6), and Table 3 shows estimated energy consumption by Eq. (6) from 2016 to 2020:

Table 3. Estimation of energy consumption by Eq. (6) from 2016 to 2020.

Year	GDP Growth Rate	GDP (us$)	Energy Consumption (ktoe)	Energy Consumption Growth Rate
2015	1.57	5.08E+11	20415	
2016	1	5.13E+11	20450.11	0.17
2017	1.52	5.21E+11	20503.57	0.261
2018	1.8	5.30E+11	20567.05	0.30
2019	2.09	5.41E+11	20640.99	0.359
2020	2.11	5.53E+11	20715.90	0.362

Figure 4. Energy consumption estimation from 2016 to 2020

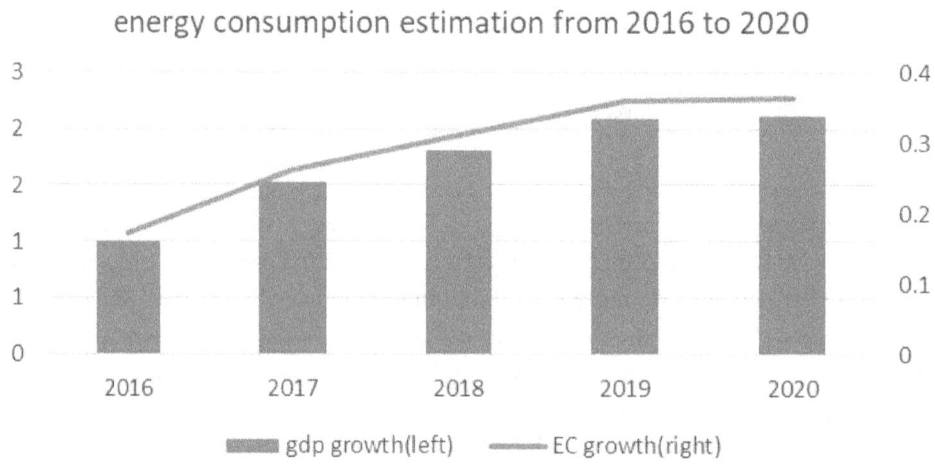

$$e_t = \frac{\frac{EC_{t+1} - EC_t}{EC_t}}{\frac{GDP_{t+1} - GDP_t}{GDP_t}} \quad (6)$$

Power Markets in Norway

As a case of deregulation in the electricity industry the Nordic power market can be pointed, which is the largest market in the industrial world. Likewise, in all Scandinavian countries ownership of this market is provided by the government, and the private sector is involved as well. It can be recognized from the Nord Pool that there exist bilateral contracts outside so that there is no compulsion to participate (Nilsson & Chair of NordREG, 2014). Seeking the new Energy Act of 1990, Norway was one of the first countries to reform its electricity sector following deregulation (Bye & Hope, 2005). But the deregulation of the electricity market had consequences such as a decline in investment, lower tariffs, and an increase in the rate of return on investment. After enactment, the new Energy Act, and electricity market reforms, Norway's electricity market was changed to a market-based system, which was developed based on a basic competitive framework. In a market-based system, regional and local transmission/distribution was carried out by regional and local companies (often municipalities). This structure can be clearly seen in Figure 5. This means producers and consumers do not consider the underlying physics of the network. Consequently, their trading partners are totally free (Rud, 2009).

The Retail Market

In electricity supply industry, Retailer who offers standardized retail products by the Norwegian water resources and energy directorate (NVE). In order to increase consumer choice, switching retail market competition as a national market in Norway

Figure 5. Competitive power markets (Rud, 2009)

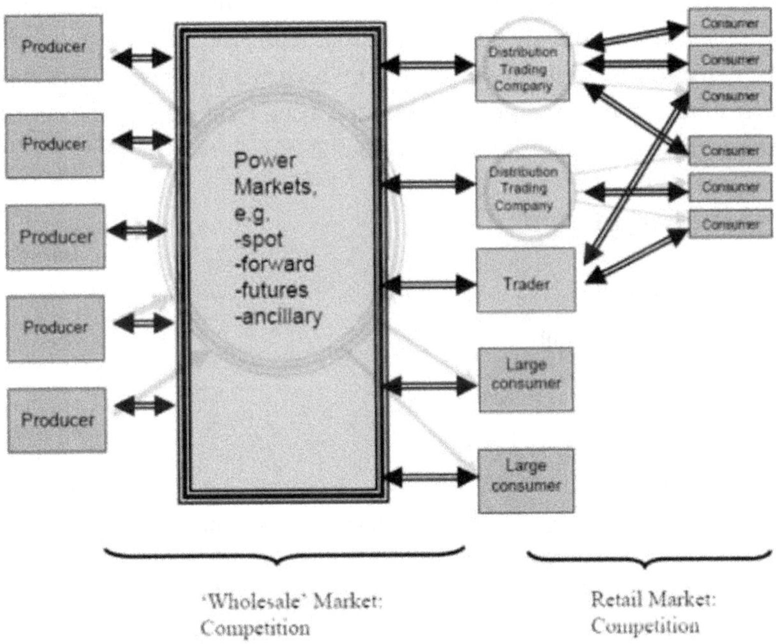

has been encouraged (Bye & Hope, 2005). Before the reform, the retail market was managed by distribution companies that were originally owned by the public. So that was classified to the small company with 150-200 consumers and the large company with 280,000 consumers (Rud, 2009).

Wholesale Market

Wholesale market indicates the electricity consumption thoroughly, so a significant share of electricity trade process is conducted there (King, 2014). In Norway, a competitive electricity market can be sensitive on the demand side because there are large power intensive customers, and of traders that represented retail end-users in the Norwegian wholesale market. At the time of the reform, distribution companies operated at local levels, as a kind of retail power trading.

Capacity Market

In Norway, due to the high density of population, transmission capacity is one of the essential problems. Therefore, the Norwegian system operator, Statnett, has solved this issue by the determination of price areas based on the price mechanism in the spot market. In this way, first, the country is divided into two or more geographical bidding areas to designate the maximum transmission capacity between areas. Then, according to data taken from Statnett, all market participants of the bidding areas are apprized by the Nord Pool to apply for the following week. The number of the areas changes depending on grid conditions, and the link between supply and demand. As for the area relating to the price, there are four price areas at present. As the investment in transmission capacity has been reduced, capacity constraints have become more pronounced (Bye & Hope, 2005).

Norwegian Market Structure

The three major functions of the power supply in Norway are generation, transmission and distribution, and trading.

Generation

An important characteristic of the structure of Norway's electricity supply, which makes it competitive, is its large number of participants. Organization of the power supply sector in Norway consists of large-scale public ownership and Public participants. Though, competition and surveillance policies on the structure of Norway's electricity supply are an acceptable policy. The number of producers before the reform was quite large, with nearly 70 production/wholesale companies that generally were owned by the public (Rud, 2009). Statkraft, as a state-owned company provides the major share of electricity from hydropower in Norway. In the pre-reform period, Statkraft and private firms provided around 30 percent and 10 percent of Norway's production capacity, respectively. Municipalities and counties produced the remainder (Bye & Hope, 2005).

Transmission and Distribution

In Norway, the transmission system operator (TSO), Statnett, has the responsibility for both the security of supply and the transmission grid. The transmission grid has a voltage level between 300kV to 420kV. The distribution grid is a local power grid with a voltage between 22kV to 230V("Nord Pool," n.d; Bartes & Wasenden, 2014). One of the main problems in the transmission and distribution is emissions, caused by the transmission process including production, installation, maintenance, and over transmission (Jorge & Hertwich, 2013)._

Trading of Power

The trading of power consists of companies which buy to sell again. The Power traders, who buy or sell power and own the electricity in the market and typically distribution companies provide the remaining demand. Thus, the trader is in a different role of demand and supplier in both wholesale and retail market (Rud, 2009; Claes, 2010). However, before the reform, traders in the wholesale market consisted of generators, distribution companies, and local distribution companies implemented large end-users and retail.

CONCLUSION

Energy plays a fundamental role in a country's economy, and economic growth relies on energy consumption. Hence, energy security has become a major concern for most societies, especially countries dependent on energy exports. Norway is country rich in energy resources such as oil and hydropower, and as a result of proper macroeconomic management these energy resources has contributed to it attaining a high standard of living. Regarding the energy issue, oil price fluctuations have been one of the significant challenges in the Norwegian energy sector. Forasmuch as Norway is one of the largest exporters of oil, the oil price shocks have affected the Norwegian economy including macroeconomic parameters. So that, falling oil prices have had negative impacts on macroeconomics parameters and economic growth as well.

Considering the economic analysis of the energy sector, the relation between GDP and oil price shows that oil price does not have an extensive effect on GDP and the elasticity of the energy consumption to GDP is not significant. Also, the effect of energy consumption elasticity in the industrial sector on GDP is slight. But the change in the industrial energy consumption causes a change in GDP and its large elasticity. It is concluded that a reduction of energy consumption in the industrial sector, will result in undesirable effects in the economic sector, such as unemployment.

From the mean value of elasticity in the past ten years, the effects of GDP on future energy consumption can be estimated. The mentioned forecasts show that GDP and energy consumption are positively correlated. As the quantity of energy produced is more than its consumption along with the proper use of renewable energy sources, Norway has fewer challenges in energy sector than other countries. Relating to the challenges including the environmental pollution of the energy sector, such as carbon capture and storage, and the reduction of greenhouse gases, some organizations with high budgets have been established and numerous regulations have been approved. Another challenge in Norway is the high dependency on hydropower (96% of the country's electricity is produced from hydropower) which has potentially negative effects on the economy and industry. This problem is most significant in times of drought when there is a concern of power outages. Future research can focus on the identification of policy development in Norway. Indeed, the approach used in the current work can be implemented in other research.

REFERENCES

Aarrestad, K., & Hatlen, L. M. (2014). *Fact 2015 energy and water resources in Norway*. Ministry of Petroleum and Energy.

Aslani, A., Antila, E., & Wong, K.-F. V. (2012). Comparative analysis of energy security in the Nordic countries: The role of renewable energy resources in diversification. *Journal of Renewable and Sustainable Energy*, *4*(6), 62701. doi:10.1063/1.4765695

Bartes, I. E., & Wasenden, O.-H. B. (2014). *Electricity regulation in Norway: Overview*. Academic Press.

Baumeister, C., & Peersman, G. (2013). Time-varying effects of oil supply shocks on the US economy. *American Economic Journal Macroeconomics*, *5*(4), 1–28. doi:10.1257/mac.5.4.1

Beckmann, J., & Czudaj, R. (2012). *Oil price and US dollar exchange rates dynamics*. University of Duisburg-Essen.

Bergholt, D., & Larsen, V. H. (2016). *Business cycles in an oil economy: Lessons from Norway*. Academic Press.

Bodenstein, M., & Guerrieri, L. (2011). *Oil efficiency, demand, and prices: a tale of ups and downs*. Federal Reserve Board.

Bukarica, V., & Tomsic, Z. (2017). Design and Evaluation of Policy Instruments for Energy Efficiency Market. *IEEE Transactions on Sustainable Energy*, *8*(1), 354–362. doi:10.1109/TSTE.2016.2599584

Bye, T., & Hope, E. (2005). Deregulation of Electricity Markets: The Norwegian Experience. *Economic and Political Weekly*, *40*(50), 5269–5278.

Cashin, P., Mohaddes, K., Raissi, M., & Raissi, M. (2014). The differential effects of oil demand and supply shocks on the global economy. *Energy Economics*, *44*, 113–134. doi:10.1016/j.eneco.2014.03.014

Chen, G. S. (2010). New energy development system in a low-carbon economy paradigm: Policies, technologies and funds. *2010 International Conference on Mechanic Automation and Control Engineering, MACE2010*, 4436–4439. 10.1109/MACE.2010.5536639

Chontanawat, J., Hunt, L. C., & Pierse, R. (2008). Does energy consumption cause economic growth?: Evidence from a systematic study of over 100 countries. *Journal of Policy Modeling*, *30*(2), 209–220. doi:10.1016/j.jpolmod.2006.10.003

Claes, D. H. (2010). Global Energy Security: Resource Availability, Economic Conditions and Political Constrains. *Contextualizing Energy Security and Transition. Concepts, Framing and Empirical Evidence, SGIR 7th P*, 1–26.

Cunado, J., & Perez de Gracia, F. (2005). Oil prices, economic activity and inflation: Evidence for some Asian countries. *The Quarterly Review of Economics and Finance*, *45*(1), 65–83. doi:10.1016/j.qref.2004.02.003

Du, L., Yanan, H., & Wei, C. (2010). The relationship between oil price shocks and China's macro-economy: An empirical analysis. *Energy Policy*, *38*(8), 4142–4151. doi:10.1016/j.enpol.2010.03.042

Esfahani, H. S., Mohaddes, K., & Pesaran, M. H. (2014). An empirical growth model for major oil exporters. *Journal of Applied Econometrics*, *29*(1), 1–21. doi:10.1002/jae.2294

Fløttum, K., Dahl, T., & Rivenes, V. (2016). Young Norwegians and their views on climate change and the future: Findings from a climate concerned and oil-rich nation. *Journal of Youth Studies*, *6261*(November), 1–16. doi:10.1080/13676261.2016.1145633

Gass, O. O. G. (2016). *OG21 STRATEGY DOCUMENT Issued for comments*. Author.

Gavenas, E., Rosendahl, K. E., & Skjerpen, T. (2015). CO2-emissions from Norwegian oil and gas extraction. *Energy*, *90*(806), 1956–1966. doi:10.1016/j.energy.2015.07.025

Gullberg, A. T., Ohlhorst, D., & Schreurs, M. (2014, August). Towards a low carbon energy future - Renewable energy cooperation between Germany and Norway. *Renewable Energy*, *68*, 216–222. doi:10.1016/j.renene.2014.02.001

Hamilton, J. D. (2009). *Causes and Consequences of the Oil Shock of 2007-08*. National Bureau of Economic Research. doi:10.3386/w15002

Hamlehdar, M., & Aslani, A. (2017). Analysis of Energy System in Norway with Focus on Energy Consumption Prediction. *Management of Sustainable Development*, *9*(1), 5–14. doi:10.1515/msd-2017-0008

Hanabusa, K. (2009). Causality relationship between the price of oil and economic growth in Japan. *Energy Policy*, *37*(5), 1953–1957. doi:10.1016/j.enpol.2009.02.007

Hemmings, P. (2016). *Vassiliki Koutsogeorgopoulou; Piritta Sorsa*. OECD Economic Surveys.

IEA. (2011). *NorwayEnergy Policies of IEA Countries: Norway2011 Review*. IEA.

International Energy Agency. (2014a). Energy Supply Security 2014 Part 3. *Energy Supply Security: The Emergency Response of IEA Countries - 2014 Edition*, 1–105.

International Energy Agency. (2014b). *European Union 2014 Review, 312*. IEA.

Jayaraman, T. K., & Choong, C. K. (2009). Growth and oil price: A study of causal relationships in small Pacific Island countries. *Energy Policy*, *37*(6), 2182–2189. doi:10.1016/j.enpol.2009.01.025

Jorge, R. S., & Hertwich, E. G. (2013). Environmental evaluation of power transmission in Norway. *Applied Energy*, *101*, 513–520. doi:10.1016/j.apenergy.2012.06.004

Kilian, L. (2009). Not all oil price shocks are alike: Disentangling demand and supply shocks in the crude oil market. *The American Economic Review*, *99*(3), 1053–1069. doi:10.1257/aer.99.3.1053

Kilian, L., & Murphy, D. P. (2012). Why agnostic sign restrictions are not enough: Understanding the dynamics of oil market VAR models. *Journal of the European Economic Association*, *10*(5), 1166–1188. doi:10.1111/j.1542-4774.2012.01080.x

Kim, S., & Willett, T. D. (2000). Is the negative correlation between inflation and growth real? An analysis of the effects of the oil supply shocks. *Applied Economics Letters*, *7*(3), 141–147. doi:10.1080/135048500351681

King, C. S. (2014). *Mapping Demand Response in Europe Today 2015. SEDC.* Smart Energy Demand Coalition.

Kopytin, I. A. (2014). Influence of oil prices on stock market indexes in Russia and Norway. *Studies on Russian Economic Development*, *25*(1), 99–110. doi:10.1134/S1075700714010109

Larsen, B. M., & Nesbakken, R. (1997). Norwegian Emissions of CO2 1987-1994. *Environmental and Resource Economics*, *9*(3), 275–290. doi:10.1007/BF02441400

Lippi, F., & Nobili, A. (2012). Oil and the macroeconomy: A quantitative structural analysis. *Journal of the European Economic Association*, *10*(5), 1059–1083. doi:10.1111/j.1542-4774.2012.01079.x

Mehrara, M. (2007). Energy consumption and economic growth: The case of oil exporting countries. *Energy Policy*, *35*(5), 2939–2945. doi:10.1016/j.enpol.2006.10.018

Mehrpooya, M., Mohammadi, M., & Ahmadi, E. (2018). Techno-economic-environmental study of hybrid power supply system: A case study in Iran. *Sustainable Energy Technologies and Assessments*, *25*, 1–10. doi:10.1016/j.seta.2017.10.007

Mohaddes, K., & Pesaran, M. H. (2016). Country-specific oil supply shocks and the global economy: A counterfactual analysis. *Energy Economics*, *59*, 382–399. doi:10.1016/j.eneco.2016.08.007

Mohaddes, K., & Raissi, M. (2016). *The US oil supply revolution and the global economy*. Academic Press.

Mohammadi, M., Ghasempour, R., Astaraei, F. R., Ahmadi, E., Aligholian, A., & Toopshekan, A. (2018). Optimal planning of renewable energy resource for a residential house considering economic and reliability criteria. *International Journal of Electrical Power & Energy Systems*, *96*, 261–273. doi:10.1016/j.ijepes.2017.10.017

Nakov, A., & Pescatori, A. (2010). Oil and the great moderation. *Economic Journal (London)*, *120*(543), 131–156. doi:10.1111/j.1468-0297.2009.02302.x

Narayan, P. K., Narayan, S., & Prasad, A. (2008). Understanding the oil price-exchange rate nexus for the Fiji islands. *Energy Economics*, *30*(5), 2686–2696. doi:10.1016/j.eneco.2008.03.003

Nilsen, Ø. A. (2015). *The relationship between oil Supervisor. Evidences from Norway*. Norwegian School of Economics.

Nilsson, A. V. (2014). *Nordic Market Report 2014*. NordREG.

Noorollahi, Y., Itoi, R., Yousefi, H., Mohammadi, M., & Farhadi, A. (2017). Modeling for diversifying electricity supply by maximizing renewable energy use in Ebino city southern Japan. *Sustainable Cities and Society*, *34*, 371–384. doi:10.1016/j.scs.2017.06.022

Noorollahi, Y., Saeidi, R., Mohammadi, M., Hosseinzadeh, M., & Amiri, A. (2017). The effects of Ground Heat Exchanger Parameters Changes on Geothermal Heat Pump Performance - A Review. *Applied Thermal Engineering*. doi:10.1016/j.applthermaleng.2017.10.111

Norway, R. C. (2013). Work programme for the DEMO 2000 Programme. Oslo: Author.

OECD. (2016). *OECD Economic Surveys: Norway, 2016*. OECD. doi:10.1787/eco_surveys-chn-2013-en

OED. (2013). *National Renewable Energy Action Plan under Directive 2009 / 28 / EC Norway*. OED.

Ozturk, I., Feridun, M., & Kalyoncu, H. (2008). Do Oil Prices Affect the USD/YTL Exchange Rate: Evidence from Turkey. *Privredna Kretanja I Ekonomska Politika, 18*(115), 49–61.

Peersman, G., & Stevens, A. (2013). *Analyzing oil demand and supply shocks in an estimated dsge model.* Ghent University Unpublished Manuscript.

Programme, L. (2013). *Work Programme for the PETROMAKS programme (2010 – 2013)*. About the Programme.

Rafiq, S., Salim, R., & Bloch, H. (2009). Impact of crude oil price volatility on economic activities: An empirical investigation in the Thai economy. *Resources Policy*, *34*(3), 121–132. doi:10.1016/j.resourpol.2008.09.001

Rautava, J. (2004). The role of oil prices and the real exchange rate in Russia's economy - A cointegration approach. *Journal of Comparative Economics*, *32*(2), 315–327. doi:10.1016/j.jce.2004.02.006

Secretariat, E. C. (2015). *International Energy Security: Common Concept for Energy Producing, Consuming and Transit Countries*. Energy Charter Secretariat.

Shahbaz, M., Zeshan, M., & Afza, T. (2012). Is energy consumption effective to spur economic growth in Pakistan? New evidence from bounds test to level relationships and Granger causality tests. *Economic Modelling*, *29*(6), 2310–2319. doi:10.1016/j.econmod.2012.06.027

Shehu, U. R. A. (2009). Impact of oil price shock and exchange rate volatility on economic growth in Nigeria: An empirical investigation. *Munich Personal RePEc Archive*, *16319*(16319). doi:10.5897/JAERD12.088

Stern, D. I. (2004). Article. *Economic Growth and Energy*, *2*, 35–51.

Subhes, C. B. (2011). *Energy Economics-concepts, Issues, Markets and Governance*. Springer Science & Business Media.

Tønne, T., & Tonne, T. (1983). Energy policy: A Norwegian perspective. *Nw. J. Int'l L. & Bus.*, *5*(4), 722.

US Chamber of Commerce. (2015). Assessing Risk in A Global Energy Market international index of energy security risk. Assessing risk in a global energy market. *Institute for 21 St Century Energy*, 1–115.

World Bank. (2015). Retrieved from www.data.worldbank.org/country/norway

Yildirim, E., Sukruoglu, D., & Aslan, A. (2014). Energy consumption and economic growth in the next 11 countries: The bootstrapped autoregressive metric causality approach. *Energy Economics*, *44*, 14–21. doi:10.1016/j.eneco.2014.03.010

Yousefi, H., Hamlehdar, M., Tabasi, S., & Noorollahi, Y. (2017). Economic and Thermodynamic Evaluations of Using Geothermal Heat Pumps in Different Climate Zone (Case Study: Iran). *Proceedings Geothermal Reservoir Engineering*. Retrieved from https://pangea.stanford.edu/ERE/db/GeoConf/papers/SGW/2017/Yousefi2.pdf

Yousefi, H., Roumi, S., Tabasi, S., & Hamlehdar, M. (2017). Economic and Air Pollution Effects of City Council Legislations on Renewable Energy Utilization in Tehran. *International Journal of Ambient Energy*. (accepted for publication)

Yousefi, H., Tavakkoli-Moghaddam, R., Oliaei, M., Mohammadi, M., & Mozaffari, A. (2017). Solving a bi-objective vehicle routing problem under uncertainty by a revised multi-choice goal programming approach. *International Journal of Industrial Engineering Computations*, *8*(3), 283–302. doi:10.5267/j.ijiec.2017.1.003

Zhang, Y., Rahbari-Asr, N., Duan, J., & Chow, M. Y. (2016). Day-ahead smart grid cooperative distributed energy scheduling with renewable and storage integration. *IEEE Transactions on Sustainable Energy*, *7*(4), 1739–1748. doi:10.1109/TSTE.2016.2581167

Zilberfarb, B. Z., & Adams, F. G. (1981). The energy-GDP relationship in developing countries. Empirical evidence and stability tests. *Energy Economics*, *3*(4), 244–248. doi:10.1016/0140-9883(81)90025-6

ADDITIONAL READING

Chen, G. S. (2010). New energy development system in a low-carbon economy paradigm: Policies, technologies and funds. *2010 International Conference on Mechanic Automation and Control Engineering, MACE2010*, 4436–4439. 10.1109/MACE.2010.5536639

Hamlehdar, M., & Aslani, A. (2017). Analysis of Energy System in Norway with Focus on Energy Consumption Prediction. *Management of Sustainable Development*, *9*(1), 5–14. doi:10.1515/msd-2017-0008

IEA. (2011). *Norway Energy Policies of IEA Countries:Norway2011 Review*. IEA.

Zilberfarb, B. Z., & Adams, F. G. (1981). The energy-GDP relationship in developing countries. Empirical evidence and stability tests. *Energy Economics*, *3*(4), 244–248. doi:10.1016/0140-9883(81)90025-6

KEY TERMS AND DEFINITIONS

Energi21: Energi21 is the Norwegian national strategy for research, development, demonstration, and commercialization of new energy technology. It is the Ministry of Petroleum and Energy's permanent strategic body for research, development and demonstration in the energy sector.

Enova: Enova SF is owned by the Norwegian Ministry of Petroleum and Energy and contributes to reduced greenhouse gas emissions, development of energy and climate technology and a strengthened security of supply.

Gassnova: Gassnova SF is the Norwegian state enterprise for carbon capture and storage. Gassnova stimulates technology research, development and demonstration and contributes to the realization of technology in industrial, full-scale pioneer plants.

GDP: GDP is the total value of everything produced by all the people and companies in the country. It doesn't matter if they are citizens or foreign-owned companies. If they are located within the country's boundaries, the government counts their production as GDP.

OECD: The Organization for Economic Co-operation and Development (OECD). The OECD provides a forum in which governments can work together to share experiences and seek solutions to common problems.

OG21: Norway's technology strategy for the petroleum sector (OG21) was established in 2001 to identify technology priorities for efficient and environmentally responsible petroleum activities on the Norwegian continental shelf.

RENERGIX: The RENERGIX program is designed to provide support for the long-term, sustainable restructuring of the energy system in order to accommodate a greater supply of new renewable energy, improve efficiency and flexibility, and facilitate closer energy integration with Europe, with due consideration given to environmental perspectives.

Chapter 12
Maintaining a Collaborative Environment Between Turkey and Israel on the Issue of Energy Resources:
The Crisis Management Role of the United States

Devrim Şahin
Eastern Mediterranean University, Cyprus

Ahmet Sözen
Eastern Mediterranean University, Cyprus

ABSTRACT

The discovery of energy sources in the Eastern Mediterranean region, while providing opportunities, further complicates Turkey-Israeli relationship. If Israel and Turkey can cooperate on energy, they can revitalize their relationship to the extent when the military elites were strong in Turkey and the relationship between two countries was established with the hands of generals. The 2016 Israel-Turkey agreement, which ended years of tension, provides Israel and Turkey with the opportunity to cooperate in energy areas. This collaboration, in turn, could generate the eventual emergence of the new ruling elites that would fill the vacuum created by the decline of the military's role in Turkey. It was the crisis management experience of the US that made the agreement between two countries possible in June 2016. Yet, any normalization process between Israel and Turkey will not be easy. US policy in the Mideast influences Turkey-Israel relations, and Turkey-Israel relations, in turn, affect the future of the Middle East. This obliges the US to bear a tremendous responsibility.

DOI: 10.4018/978-1-5225-4203-2.ch012

INTRODUCTION

The problems between Turkey and Israel in recent decades have frequently been marked by a lack of mutual trust. The rift in the relations of the two countries dates back to Israel's Operation Cast Lead in Gaza in July 2009. Turkey's mistrust stems from the days when Erdoğan was seen as a prominent person that hold a mediatory position in the Middle Eastern affairs. While Erdoğan was trying to arbitrate in the Syrian issue, Israel invaded Gaza Strip in 2009. Israeli-Turkey diplomatic relations had broken down in May 2010 when Israeli naval commandos intercepted the Mavi-Marmara flotilla, a Turkish boat that was attempting to carry aid to Gaza, resulting in the deaths of nine Turkish activists. The escalation of the Mavi Marmara crisis by Turkey was due to politics to a great degree. Otherwise, the risks were easy to predict when the Mavi Marmara flotilla was sent to Gaza. Gaza has always been one of the so-called red lines of Israel regarding the international sea law and its public law. To send the Mavi Marmara on a risky journey was nevertheless a response by Turkey to Israel's prolonged aggressive attitude towards Gaza/Palestine issue. The Mavi Marmara crisis revealed the lack of trust in the Israel-Turkey relationship.

An agreement to start the process of normalization of ties between Israel and Turkey could only be achieved after long negotiations. Turkey stipulated that Israeli officially apologize for the deaths of Turkish nationals, offer compensation to their families and a lifting of the blockade on Gaza. It was only on June 27, 2016 that an agreement between Israeli and Turkey ended years of tension. The agreement included an official Israeli apology to Turkey for the deaths of Turkish nationals, compensation for their families and a softening of the Israeli blockade on Gaza that would allow Turkey to send aid to the Gaza Strip only through Israel. Israeli Energy Minister Yuval Steinitz has evaluated this 2016 Israeli -Turkey agreement as "opening relations with one of the strongest countries in the Middle East" (Baker, 2017). Yet, this agreement would not have been achieved without the United States (US) experience of crisis management and its capacity to prevent future conflicts. As Steinitz remarked in 2016: "We were in a very negative course with the Turks," and "two or three years ago, some in Israel even thought we might have armed conflict" (Baker, 2017).

Thanks to the June 2010 discovery of hydrocarbon sources in the Eastern Mediterranean's Leviathan region in Israeli territorial waters, Israel moved from being an energy dependent country to being an energy supplier country. The ability to export this natural gas through different pipelines to the Middle East region or to Europe became critical. This need also offered potential to Israel, including the improvement of its relations with neighboring countries – even with those that were not good friends of Israel, namely, Jordan and Turkey. Israel has already signed an energy deal to export to Jordan and talks continue for exporting the natural gas to Turkey via a pipeline. Michael Leight, a senior fellow at the US-based German Marshall Funding, which focuses on natural gas resources in the Eastern Mediterranean, has said that the Turkish option is available because relations between the two countries have entered into a normalization phase. He added that both Israel and Turkey want to materialize the normalization process and move to a more concrete level of economic cooperation. He pointed to the field of energy as a potential area for political cooperation as there is need for energy partners to develop and economically use these natural gas fields. Israel needs to supply natural gas to the international markets as soon as possible, and needs to choose the safest route to do so. In this regard, Turkey is the most inexpensive and effective way and has a very attractive natural gas market.

Turkey needs to diversify its energy sources as it currently meets a large part of its natural gas needs from Russia. This would also benefit the European market. If Turkey becomes an energy center based on its energy links to suppliers, it will also profit from this role. Turkey is seeking to refashion itself as an energy center between the demand side of Europe and the various resource suppliers such as Israel. These roles that Israel and Turkey envisage for themselves, the former as energy supplier and the latter as an energy center, are complementary to each other. Nevertheless, the entanglement of Israeli-Turkey relationship with the energy issues under the regional circumstances of the Middle East requires the US involvement as a mediating force between the two countries.

Since the end of the Cold War, multiple perils have provoked instability in the Mideast. Rogue states in the region have become transnational threats that emerged initially within their borders and then became a threat to other nations. The risky and protracted US wars, including the 2001 Afghanistan War that began in order to address rogue states and nation-building and the 2003 Iraq war to expand democracy, were disastrous and caused a huge budget deficit in the US economy. The crisis of the US mortgage financing industry in 2007 generated a fast-moving chain reaction that provoked global political and socio-economic disorder that when coupled with the disastrous Iraq war and the rise of Asia, "chastened America, ending its period of dominance in world affairs" –according to Kaplan (2008), an American author who has published controversial essays about the nature of US power. The diminishing US hegemony in the Middle East triggered the global turmoil and heightened power struggle and rivalries for regional leadership in the Middle East. The 2011 Arab Spring further complicated the rivalries in the Middle East: Iraq, Syria and Libya are still fighting civil wars. While Iran's hegemony expands throughout the Middle East, Russia has already been involved militarily in Syria. Turkey's military presence in Iraq, Syria and Qatar further complicates the network of relations in the Middle East. Former US National Security Advisor Brzezinski (1997) defines Turkey as "not only an important geopolitical pivot," but her "own internal condition is of critical importance to the fate of the region." Brzezinski (1997) also argues that "recognition of global turmoil as the basic challenge of our time requires confronting complexity." The simultaneous existence of these significant political and economic challenges in the Mideast causes global turmoil, which is in a state of disorder, adds to the complexity of the challenge.

The survival of the global political economy depends on the energy sources exported from the Middle East. It is timely, then, to reconsider the US role in the Middle East as a hegemonic country that applies a high level of capacity and experience to manage conflicts. Also, the US should use its experience to manage its rifts with Turkey like the 2017 visa crisis which is a part of broader deterioration in relationship between two NATO allies and has potential for devastating results in the Middle East. This article discusses the US role in Israeli-Turkish relations in order to explain the profound complexity of international relations as practiced today. This analysis explains how the US involvement in Israeli-Turkey relations is determinative of their recent actions as well as of their future relationship. The US should not miss the opportunities to prevent a miscalculated escalation of a crisis between Israel and Turkey into a military conflict. The US should immediately mobilize diplomatic efforts for effective crisis management, but not in the way that supports Israel automatically. On the contrary, it should apply the same pressure on Israel to avoid violence as it applies to other countries in the region. US involvement in Israeli-Turkish relations also affects the relationships of the US with other Middle Eastern countries.

BACKGROUND OF THE US-ISRAEL-TURKEY RELATIONSHIP

The current status quo in the Middle East, by which prolonged instability and wars dominates the borders of unstable countries that are ruled by dictatorships, was shaped according to the interests of Israel during the Cold War. Since her establishment as a security state, Israel has formulated a national strategy to maintain her security by having a capacity to defeat any possible Arab coalition organized against its existence. With this goal, it established alliances with the important Middle East countries such as Egypt and Saudi Arabia. During the Cold War years, Turkey was the only democratic and secular Muslim country in the Middle East. Israeli officials came to realize the western yearnings evident in Turkey's foreign policy. They thought that cooperation with Turkey would improve their relations with the West. Also, there were Israeli officials who thought that with the assistance of Turkey, the problems between Israel and the Arabs could be solved through diplomatic channels (Nachmani, 1987).

Turkey has always been important for Israel. It was the first Muslim country that recognized Israel and has long had a geostrategic importance in the Middle East region. The statehood of Israel was declared on May 14, 1948, which triggered a war between Israel and the regional Arab states, including Egypt, Jordan, Syria and Iraq. Diplomatic relations between Israel and Turkey were established in March, 1949, when Turkey recognized the independence of Israel. However, because of the increasing tension between the Arab world and Israel, Israeli-Turkish diplomatic communications increased very gradually. Turkey's first diplomatic mission in Israel was a "legation," which is a diplomatic representative office lower than an embassy.

Since the mid-1950s, there have always been ups and downs in the relations between Israel and Turkey. One of the first declines in the relationship was the establishment of the Baghdad Pact between Turkey, Iraq, Iran, Pakistan and Britain in 1955, which was inspired by then-US Secretary of State John Foster Dulles. The Baghdad Pact, which emerged as part of the US global initiative to contain the USSR, was intended to prevent the expansion of the Soviet Union's influence in the Middle East. Nevertheless, Israel responded to Baghdad Pact by claiming that the pact promotes Arab sentiments and aggression against Israel, and perceived Turkey's position as a change in attitude away from Israel toward the Arabs. This tension, coupled with the Israeli-British-French invasion of Egypt in 1956, after the nationalization of the Suez Canal by Egyptian leader General Gamal Abdel Nasser in July 1956, brought Israeli-Turkish relations to a deadlock. In response, Turkey reduced its diplomatic representation in Israel to the level of second secretary, one of the lowest diplomatic ranks. Some years later, in July 1963, the representation of Turkey was upgraded in Israel to the ambassador level and on January 1, 1980 was upgraded to embassy level. Another dip in the relations of Israel and Turkey was in 1980 when Israel annexed East Jerusalem and when Jerusalem was declared as the capital of Israel. At that time, Turkey again reduced its diplomatic representation in Israel to the level of the second secretary. Diplomatic relations were reintroduced at the embassy level in 1990.

For the most part, Turkey's relationships with other countries operated in line with its security concerns during the Cold War period. Olson (2015) reminds that Turkey was also crucial to Israel's security concerns. Despite the ups and downs of the diplomatic relationship, two countries cooperated on security and intelligence issues. The regional dynamics in the Middle East were shaped according to the Cold-war polarization. During the Cold War, the superpower competition intruded into all regions in the world through the bi-polar international system. This provided the US with an opportunity to play a determinative role in the Israel-Turkey relationship by facilitating Turkey's respect for US foreign policy decisions, including the US policy towards Israel. Acting in line with Israel's priorities allowed Turkey to gain political support from both Israel and the US for its political priorities, including its struggle against the Kurdish Workers Party (PKK), its efforts to protect its national interests in Cyprus and to prevent recognition of the so-claimed Armenian Genocide. Influential circles in US foreign policy, including at the Pentagon, had always supported Turkey's position when there was an issue regarding these priorities of Turkey. And, Israel's strong lobbying power and influence within the US policy making circles allowed Turkey to gain political support in her political priorities, which was conditioned on Turkey's acting in line with Israel's preferences.

In the early 1990s, the relationship between two countries reached to the level of strategic partnership in order to maintain the status-quo in the polarized Middle East. During this time, Israel-Turkey bilateral relations gained momentum in military, political and economic affairs. Israel implemented Turkey's modernization program of F-4 jets (long-range supersonic jet interceptors and fighter-bombers used extensively during the Vietnam War) and F-16 jets (single-engine supersonic multirole fighter aircraft). These modernization programs were implemented within the context of the Israel-Turkey Defense Industry Cooperation Agreement in 1996. In 1998, a search and rescue exercise titled Reliant Mermaid was operated with the participation of the US, Israel and Turkish vessels. These bilateral relations between Israel and Turkey continued at the highest levels in economic and military fields until the beginning of 2000s. When Ariel Sharon came to power in Israel in March, 2001 amid increasing violence in the West Bank and in Jerusalem on the Palestinian issue, the Israeli-Turkish relationship was damaged. Nevertheless, modernization of Turkey's 170 M-60 tanks (primary battle tanks which had become the US Army's primary tank during the Cold War.) was granted to an Israel firm on March 30, 2002, and bilateral economic relations continued despite the political tensions.

In the aftermath of Cold War, the international community by and large expected the appearance of a globally cooperative system by which all the power struggles would be solved and Western-type democracy would be extended in many parts of the world under the leadership of US. During the Clinton administration in the 1990s, the US was active in joint efforts with Turkey to conclude successfully the Baku-Tbilisi-Ceyhan (BTC) pipeline project, which was intended to become an energy corridor for oil and gas mines in the Caucasus and the Central Asia (Said, 2005). There was also a close working relationship between Turkey, the US and Israel on each country's security concerns. For the first time, Clinton began officially referring Turkey as a strategic partner of the US and an important member state of the North Atlantic Treaty Organization (NATO), an intergovernmental political and military alliance committed to providing mutual security to its North American and European member states. In his speech at Turkey's Grand National Assembly on November 15, 1999, Clinton addressed the strategic partnership between the US and Turkey. In his words, the "partnership has become even more important. Together we are adapting NATO to the demands of a new century. We are partners for peace in the Balkans and the Middle East. We are developing new sources of energy to help the entire region" (Clinton, 1999). Similarly, Mark Parris (2010), the US Ambassador to Turkey from 1997 to 2000, also highlighted Turkey's importance when he maintained that Turkey's relationship to the US is as "important as during the Cold War" due to "Ankara's capabilities and readiness to shoulder responsibility as a security producing nation."

Unlike the initial post-Cold War expectations for a more peaceful international environment, the IR system experienced a different reality, one that Brzezinski (2012) defines as: "a protracted stage of rather inconclusive realignments of both global and regional power." Since the end of the Cold War, Brzezinski (2012) contends, the IR system has experienced various organized rivalries "in a setting of international uncertainty and even of potentially fatal risks to global well-being". Following the collapse of the Soviet Union in 1989, US interests in Middle East began to gradually decline due to the lack of superpower competition that had previously determined the dynamics of the Cold War era. The decline of the US interests in the Middle East turned Israel adrift as it was the US that has always sustained and guaranteed the latter's existence. Brzezinski (2012) claims that the world will be dangerously unstable in an age of decline of US influence in the Middle East, in his words: "the security of a number of weaker states located geographically next to major regional powers also depends on the international status quo reinforced by America's global preeminence and would be made significantly more vulnerable in proportion to America's decline". Brzezisnki (2012) includes Israel among those states whose exposed geopolitical positions make them "today's geopolitical equivalents of nature's most endangered species". Thus, the deterioration of US hegemony threatens the securitization of Gaza and how both Israel and Turkey address energy issues. The struggle of these two countries to extend their own influence in the Middle East is a zero–sum game, which means the strategies formulated by the two countries towards each other by which each party attempts to increase its influence at the expense of the other.

GLOBAL FACTORS

Another iteration of the IR system is the four terrorist attacks against the US in September 11, 2001 (9/11) by the radical Islamic terrorist organization al-Qaeda, which triggered a security quagmire. After 9/11, the IR system experienced hyper-securitization by Western democracies, which took extreme security measures in the struggle against transnational terrorism, at a cost of their society's individual liberties. The widely held view among the European and American public was that these attacks were linked with radical Islam, which resulted in a mutual hyper-securitization of identity between Western and Muslim countries. As the securitization theory explains it, the US started to perceive Islam as threat against the existence of the West. This perception of threat was clear in the rhetoric of the former US President George W. Bush, who was in office between January 20, 2001 and January 20, 2009, when he referred to crusaders as he declared war against Iraq. The Muslim countries also began to perceive the US as a threat against their existence. The 2003 Iraq war facilitated securitization throughout the Middle East and created a power vacuum that not only Turkey but also Saudi Arabia, Qatar, Iran, Russia and Israel are now struggling to fill.

In the post-9/11 era, US foreign policy drifted towards unilateral behavior, that is, unilateral actions without the collaboration and support of its former allies. This US attitude was a key factor in declaring war against Iraq in 2003, which damaged the image of the US throughout the world. McAllister (2003) maintains that the Iraq war led to widely held misunderstanding about US motives throughout the world, which then led to the simplification of US intentions as an interest in oil. "76% of Russians, 75% of French, 54% of Germans and 44% of British believe the desire to control Iraq's oil lies behind Bush's bellicosity". Although Iraq's oil reserves are the second largest in the world and the cheapest to extract, the economic burdens of the war were too high, including the investment required to increase oil production, the high level of debts of the Iraqi government, the need for investment in infrastructure and building democracy and, obviously, the cost of US military action. The primary motivation behind the 2003 Iraq war was not simply oil revenue but to secure the energy supplies: as Vice President Cheney warned, Saddam might "seek domination of the entire Middle East" and "take control of a great portion of the world's energy supplies" Sanger (2002). The 2003 Iraq war obviously demonstrates how the struggles for control over energy fuels global conflicts as the US attempt to secure the energy supplies in the Middle East generated a hyper-securitization that weakened the US image among her European allies as well as Turkey.

As a result of the 2003 Iraq war, the Kurdish administration in northern Iraq became stronger and many in Turkey came to believe that the US no longer supported Turkey's struggle with the PKK, supporting an independent Kurdistan instead. Also, many NATO members were reluctant to support Turkey's military operations against the PKK. The PKK was allowed to open offices in some NATO member countries, such as Denmark. As a result, the US-Turkey relationship experienced a kind of breakdown which included Turkey's loss of respect for US foreign policy preferences, including its policy toward Israel. Simultaneously, Israel formulated and implemented a policy to widen and deepen its influence in the region by extending its ties with Masoud Barzani, who has been President of the Iraqi Kurdistan Region since 2005. This policy included using Israel's own population of 50,000 Jewish Kurds, who had emigrated from northern Iraq, to influence the behavior of Kurds in Iraq or Syria. Articles were published in the Israeli press, including one titled "Let the Kurds be," and supported the establishment of a Kurdish quasi–state in northern Iraq (Jerusalem Post, 2004). There were also credible claims that Israel's Mistaravim Commandos, which are counter-terrorism units of the Israeli Special Forces, had trained the Peshmerges, which number about 50,000 and were the biggest armed group of Iraq in the post-Saddam period. The information that Israel was involved in training of groups that are adversarial to Turkey emerged when the training coordination of the Israeli special security company was leaked to the Israeli press. So it was that Turkey became suspicious of the reliability and credibility of the US, Israel and NATO to guarantee its security against threats coming from the Middle East, including those fighting for a Kurdish quasi–state.

The 2003 Iraq War is one of the key global factors that helped generate Turkey's deepening mistrust in its former allies, and increased its engagement in the Middle East. Since it now lacked a reliable multi-lateral security framework, Turkey began to implement a unilateral foreign policy in the Middle East that was formulated independently from the US and NATO. Meanwhile, the Justice and Development Party (AKP) government coming into power in Turkey struggled against the influence of Turkey's military on its foreign policy. During the initial years of the AKP administration, Turkey moved from an authoritarian and totalitarian regime, in which the military had a dominant place in its bureaucracy, into a democratic state where public opinion plays a crucial role in the country's decisions and choices. The AKP implemented foreign policies that differed from previous policies regarding the balance between security and freedom. According to the *Economist* (2011), "in the past, keeping Israel close has allowed the [Turkish] army to cozy up to America". As the role of the military declined, the old approach of keeping Israel close faded as well, which led to the deterioration of relations with Israel, which in turn clung to the status-quo against the dynamic developments in the Middle East. This is similar to what happened in Egypt in the beginning of 2010s when the notion of *security-uber-alles* (security above everything else) declined and more authentic political sentiments took over, which meant that Egypt's partnership with Israel, which had been based almost exclusively on security needs, not a general friendship of like-minded allies, was no longer justified. However, as *Economist* (2011) puts it, the rise to power of "Mr. Erdogan's mildly Islamist Justice and Development (AK) party and the exposure of mischief by soldiers—from coup plotting and corruption to incompetence in the field—has changed that".

In the aftermath of the 2003 Iraq War, as a result of the increased securitization in the Middle East, Turkey began to solidify its leadership position in the region through unilateral engagements with Muslim/Arab countries. Nevertheless, the ability of Turkey to unilaterally expand these called for espousing an anti-Israeli policy. The Muslim/Arab world welcomed the new image of Turkey as a counterbalance to Israel that resisted US hegemony and its support for Israel. The AKP wanted to establish warm relations with the Arab/Muslim world by cooling relations with Israel and implementing a proactive Palestinian policy. Israel's aggressive policy in Gaza by its excessive use of military power angered Turkey's citizens. During his prime ministry, Turkey's President Tayyip Erdoğan, walked out of the 2007 Davos, Switzerland, meeting of world leaders after a controversy erupted between him and Shimon Peres, the then-President of Israel, about Israel's treatment of the Palestinians. Following Israel's 2010 Marmara flotilla attack, in which nine Turks were killed, the Palmer Report and apology crisis, and ultimately the cessation of security cooperation between the two countries, diplomatic relations between the two countries had weakened considerably. Erdoğan's tough stance against Israel has impressed Arabs. In September 2011, during his visit to Egypt aimed to advancing Turkey's status as a regional leader, Erdogan received a hero's welcome from the Egyptians. In effect, Erdoğan has offered Turkey as a model for an Arab world in turmoil, including his advising Middle Eastern countries to balance secularism and Islam in order to challenge Israel.

Öniş (2014) points out that the worsening relationship between Israel and Turkey then turned into confrontation, by which Mavi Marmara crisis was generated, as Turkey sought to forge a closer relationship with Egypt and with other actors to realign the balance of power in the Middle East. Turkey's new course of diplomacy in the Middle East certainly clashed with Israel's desire to continue their previous relationship shaped by her national interests. Israel also used her lobbying power in the US and her technological military power as "political bribery." However, by the 2007 the US Mortgage Crisis, the international uncertainty caused by the decline of US hegemony in the Middle East has changes the circumstances in the region. The US now has more limited control of and leverage on Turkey; Turkey is now a more independent- minded and self-confident country where public opinion plays a crucial role in Turkey's decisions and choices.

DOMESTIC FACTORS

As a very dramatic consequence of the 2003 Iraq War, the international system has shifted from a community of power led by the US until the Bush administration to a balance of power marked by regionalized international security. In addition, the 2007 US mortgage crisis accompanied the global financial crisis, whose negative effects were felt throughout the world. These events not only weakened US domination in the Middle East, they resulted in regional powers solidifying their regional leadership positions. Very slowly and very gradually, the mantle of leadership of international security has slipped from shoulders of the US, and the regional level of the provision of security has become more obvious. Evidence of the rise of regional security leadership is more revealing especially throughout the Middle East where the regional powers have more room to maneuver. The decline of US hegemony in the region has generated a power vacuum – a condition that exists when one country loses control of power and no other country replaces them. While the US domination in the Middle East has faced increased challenges, Turkey's Middle East policy has more explicitly begun to challenge the continuation of the previous order – which was shaped according to Israel's interests. Turkey has begun to formulate and implement provocative policies against Israel, including the aforementioned Mavi Marmara crisis, which is evidence of Turkey's increasingly ambitious behavior in the region.

In 2006, Israel established a naval blockade of the Gaza Strip to prevent the entry of weapons into Gaza. In June 2010, a number of ships set sail from Turkey, some manned by Turkish activists, seeking to break the Israeli blockade. Turkey's naval force, called the Mavi Marmara Flotilla, was confronted by Israeli commandos outside Israel's territorial waters, resulting in an armed confrontation that killed 9 Turkish citizens. As a result of this flotilla attack, the flotilla was captured by Israeli soldiers and failed short of its mission's goal. As a product of the escalation by both sides, the Mavi Marmara crisis led to a serious military confrontation between Turkey and Israel that also threatened regional stability. As a deterrence policy against Turkey's provocative policies, Israel then created alliances with countries that had troubled relations with Turkey in the past, including the Greek Cypriot Administration of Southern Cyprus and Greece. Israel took advantage of the economic crisis of Greece, which followed the US mortgage crisis in 2007, according to its own interests. Thus, the tensions between Israel and Turkey have become more complicated. Greece in its current weakened state has had less influence on Cyprus, as did the EU, which, in turn, has caused a power vacuum that Israel is now filling.

Following the Israeli Prime Minister Benjamin Netanyahu's visit to Athens in 2010, the Israeli and Greek governments agreed to cooperate on security and intelligence issues. They decided to establish a joint committee to coordinate strategic collaboration and to also establish their collaboration in the arenas of politics and economy. In the meantime, in 2011, pro-Palestinian activists were organizing a second flotilla, to break the blockade of Gaza, in which nine or ten ships from various countries came together and set to sail from Greece. For Netanyahu, this was a nightmare that had to be prevented at all costs. Greece prevented the ship from leaving Greek territorial waters and thus prevented Israeli forces from clashing with the political activists – which saved Israel from facing the potential impact of this event. This Greek support was part and parcel of the Netanyahu government's message to convince major US Jewish lobbies and community groups and Israeli corporations to invest in Greece. So it was that Israel's strong lobbying power in the US and its technological military power became an important inspiration for Greece in her Cyprus issue with Turkey. The year 2011 was very challenging for President Dimitris Christofias's government in Cyprus. In that year, an explosion at the Evangelos Florakis Naval Base knocked out the island's electricity plant and sparked protests against former Christofias's government by the opposition. When the legitimacy problem of governance was coupled with economic problems, Christofias's government began exploratory drilling for oil and gas in its territorial waters and also sought to establish a security alliance with Israel. This further increased Israel's influence and counterbalanced Turkey's influence throughout the Eastern Mediterranean.

Following Israel's Marmara flotilla attack and ultimately the cessation of security cooperation between Israel and Turkey, diplomatic relations between the two countries became even weaker. A reconciliation agreement to normalize diplomatic relations between İsrael and Turkey could only be achieved after long negotiations. Turkey stipulated that Israel officially apologize for the deaths of Turkish nationals, compensation for their families and a lifting of the blockade on Gaza. The official Israeli apology to Turkey came from Netanyahu in 2013, at the time when the former US President Barack Obama made his first visit to Israel at the beginning of his second presidential term. It was only then that an agreement between Israel and Turkey could be achieved on compensation. Since then, negotiations on the Gaza blockade issue stretched till the 2016 agreement between Israel and Turkey. According to the agreement, Israel's Gaza blockade is not lifted but softened specifically for Turkey. In fact, at the beginning of the crisis, Turkey stipulated only an apology and compensation conditions. It later stipulated one more condition that required Israel to lift the blockade of Gaza. This caused the widely held perception, especially among the Israeli and in US circles that Turkey was complicating the reconciliation with Israel. And, the fact that Israel's apology coming during Obama's visit compelled many to think that Obama coerced Netanyahu to apologize in order to pave the way for reconciliation.

As a reaction against Israel's alliances with countries that have troubled relations with Turkey, the latter has pursued a vision of a realigned Mideast and an alliance with Egypt to encourage a new order in a region roiled by the Arab Spring in December 2010, whose primary purpose was to highlight the economic and social hardships challenging people in the lowest layers of society and the irresponsibility of governing elites. Turkey's policy of increased engagement with the Middle East was welcomed warmly in the Arab world. In the wake of the Arab Spring, Turkey's exceptionality as an indigenous Muslim Democracy offered a model of reform for the entire Middle East. Its economic and political capital was a tangible economic incentive for the development of democracy and peace in the region. Turkey became an inspiration for the Muslim/Arab world as it sought the solution to problems not with the use of hard power through military tools, but with the use of soft power through diplomacy and establishing dialogue with all actors. This has resulted in a greater regional engagement for Turkey, as the Arab world welcomed its increased role on several accounts. These include the symbolic meaning of Turkey's increasing interest in the region, its neutrality and readiness to undertake an active and stable dialogue, its role as a counterbalance in the polarized Mideast, its new image as resisting US hegemony, its negotiations with the EU and explicit criticisms of Israel, its rapid economic growth and its increasing trade bonds with Arab countries, and Turkey's readiness to use her hard power when necessary.

When Israel began moving closer to Greece and the Greek Cypriots, Turkey increased its engagement with the Middle East. In other words, while Israel increased its engagement with the countries that Turkey has problematical diplomatic relations, Turkey changed its attitude towards Israel in favor of the Arabs. Consequently, the bidding between Israel and Turkey has shifted into a more confrontational mode. Rivalry between the two countries took on the characteristic of a zero-sum game, as each attempted to increase its influence at the expense of the other, both in order to strengthen its overall global position and to protect specific security interests in the Middle East. Turkey's Middle East policy gradually and more explicitly has begun to challenge the continuation of the previous order which was shaped according to Israel's interests.

THE RESULT

Since the end of the Cold War, the Middle East has experienced a period of fundamental change. This change was facilitated first, when the US unilaterally declared the 2003 Iraq War as a reaction against the 9/11 attacks and second, when the Arab Spring emerged in an era of global socio-economic and political crises. This regional change includes the appearance of rivalries for regional hegemony, as well as for control of energy pipelines, and securitization thanks to the emergence of transnational terrorism as a threat against long-defined territorial borders and the energy ways in the Middle East. Under changing international and regional circumstances, Israel tried to protect her status quo and has never hesitated to use methods even if it they hurt former allies such as Turkey. (Israel's attack on the Mavi Marmara flotilla attack certainly proved Israel's commitment to deterrence.)

Nevertheless, the roots of the crisis in Israeli-Turkish relations goes far beyond Israel's commitment to deterrence and the domestic and regional pressures on Turkey to play a leadership role in the Middle East. Turkey's and Israel's choices were made not only under the circumstances rationally chosen by them. Turkey's and Israel's choices are shaped under the regional as well as global circumstances.

There were clouds on the horizon in the post-Cold war era. These clouds became darker because of the reactions produced by the 9/11 attacks and the global economic crisis. As discussed above, both Turkey and Israel have attempted to take advantage of the recent power vacuum in the Middle East, and have expected to increase their influence at the expense of each other in a zero-sum game. Both have sought to unilaterally strengthen their own overall global position and protect their own specific security interests in the Middle East.

The issue of US hegemony is intertwined with the shifting circumstances that have been experienced in the IR system in recent years. The end of the Cold war allowed former US allies, such as Turkey, to act more independently from the US. Then, the 9/11 attacks triggered the 2003 Iraq war and further damaged US hegemony. The global economic crisis exhibited increased the challenges against the US domination in the Middle East. The main reason for the rift in Israel-Turkey relations in the aftermath of Mavi Marmara flotilla crisis was not only their recent actions, but also the international uncertainty caused by the decline of US hegemony. However, the US still can be the primary actor in the management of regional conflicts if it chooses to apply its capacity for and experience with effective conflict management.

Thanks to the Obama administration's effective crisis management, Turkey and Israel escaped from Mavi Marmara crisis with little or no harm. It was the crisis management experience of the US and its use of coercion that made the agreement between two countries possible in June 2016. Despite this US intervention into the Israeli-Turkish crisis, any normalization process between Israel and Turkey would not be easy. Israeli-Turkish relations had experienced the disappearance of mutual trust that prevents further cooperation between them. Turkey was now suspicious that Israel does not support its struggle against the PKK and PYD. That mistrust of Israel was increased as a result of Israel's 2009 Gaza attack during Erdoğan's efforts to mediate in the Syrian case. Meanwhile, Israel was suspicious of the 2011 Arab Spring and Erdoğan's ambitions to take the lead of the Muslim/Arab world. Israeli public opinion is also deeply concerned about Turkey's relations with Hamas, the Sunni-Islamic fundamentalist organization that is an important political influence in Palestine.

The relationship between Israel and Turkey continues to be vulnerable to new crises. The warning signs of these political crises are abundant, including, above all, the Palestinian issue. The relationship between the two countries could be adversely affected by, for example, US President Trump's decision in December 2017, to recognize Jerusalem as the capital of Israel. US Senator Sheldon Whitehouse has suggested that Trump's decision could harm the US influence in the Middle East region and could corrupt relations with Turkey, the location of strategically important US military installations. Whitehouse said that "after what we've seen from, in particular the Saudis, and the Turkish prime minister, we have some real diplomatic issues. And if they really decide that this is a big deal and that they need to push back against us, then they can hurt us in important ways" (Persons, 2017). In words of Whitehouse, "what's really dangerous" is "what it means for our relationship with Turkey, which we still depend on for a lot of basing for our military efforts in the Middle East, and with the Saudis" (Persons, 2017). Turkey's immediate reactions to Trump's decision on Jerusalem proved Whitehouse's claims. Upon Trump's Jerusalem decision, Erdoğan accused Israel of operating a state of terror: "Jerusalem is the red line of Muslims. This could go as far as [our] breaking diplomatic relations with Israel" (Reuters, 2017).

Another challenge for the relationship between Israel and Turkey is the former's behavior toward groups with which Turkey has troubled relations. The only state leader in the world who explicitly supported the referendum for the establishment of an independent Kurdish state in northern Iraq was Israeli Prime Minister Netanyahu, who said in 2017: "We have very great sympathy for their desires and the world needs to concern itself with their safety and with their future" (The Times of Israel, 2017). Another Israeli contribution to already frayed Israeli-Turkish relations came from Former Israeli Defense Forces Deputy Chief Major General Yair Golan. In a conference speech in Washington in 2017, he expressed his support for an independent Kurdish state. Golan said, "the PKK is not a terrorist organization" and "when you look at Iran in the east, when you look at the instability in the region, a stable and unified Kurdish entity in the middle of this swamp is not a bad idea" (Haaretz, 2017). Needless to say, these statements of high-ranking Israeli officials were not welcomed in Turkey. Subsequently, Turkey would increasingly move away from Israel and cooperate with Israel's adversaries, including Iran.

The relations between Israel and Turkey at this time are more than extensions of their internal politics. Turkey has been experiencing a political transformation, which includes her vying for a major role in the regional hegemony, and Israel is resisting this change. It is doubtful that the Israeli-Turkey relations will ever return to its previous level of the 1990s. The recent decline ın the role of the military in Turkey's foreign policy has changed the old approach of keeping Israel close. In addition, public opinion has come to play a very significant role in shaping Turkey's foreign policy: Turkey's public follows the developments in Palestine issue with deep concerns. At the least, public opinion affects the political rhetoric used by Erdoğan, whose rhetoric was very influential in Turkey's response to the Mavi Marmara crisis. Still, Erdoğan's recent rhetoric when US President Trump announced his decision to move the capital of Israel to Jerusalem accused Israel of being a state of terror, language which surely was targeted to rouse Turkish nationalist opinion. On the other hand, Israel fears that greater anti-Israeli movements in the Middle East could be triggered by increasingly extreme public opinion. The raising of such public voices both within other Muslim/Arab countries in the region as well as within Turkey has increased the pressure on Erdoğan to voice anti-Israeli sentiments.

Israel's increased diplomatic relations with countries that Turkey has problematical diplomatic relations with also challenges further cooperation between Israel and Turkey. Israel signed energy deals with Cyprus and Greece while her diplomatic relations with Turkey were worsening. Any conflict between Turkey and Cyprus or Greece now threatens to draw Israel into the conflict. The exploration for hydrocarbons in Cyprus has already provoked confrontations by which Erdoğan has threatened that any exploration of hydrocarbon sources in Cyprus without the consent of Turkey would be retaliated by Turkey and trigger further devastating results. Clearly, the stability of the Middle East is in greater flux today than it has been for some time, and any resolution of current issues depends on effective US involvement.

While the energy issue potentially endangers the security of the region, it is a two-sided issue. The discovery of gas in Israel's Leviathan parcel in 2010 offers a window of opportunity for regional cooperation based on mutual benefits. Cooperation between Israel and Turkey on energy would have important consequences for the energy security of the Middle East. The low prices of oil and gas, the US becoming an oil exporter for the first time in her history and Iran entering into the market require that Israeli gas be sold in regional markets. When the high gas prices in Turkey are taken ınto account, Turkey is a very profitable market as well as a safe passage for Israeli gas. Cooperation between Israel and Turkey in the energy sector might enhance the size of their bilateral trade relationship.

Moreover, as Figure 1 shows above, the diplomatic crises experienced since the AKP came into power in 2002 have not influenced economic relations between the two countries. The economic volume of Israeli-Turkish trade has not deteriorated, in fact, it has rather grown. After the ratification of the Free Trade Agreement between Turkey and Israel on May 1, 1997, the volume of trade between two countries increased significantly: the trade volume was 1 billion US dollars in 2000; it reached its maximum value of about 5.8 billion US dollars in 2014 and decreased to 4.342 billion dollars in 2016. The stakes involved in the energy issue are not the only motivation for further cooperation between Israel and Turkey. Both countries need each other to tackle the complexity of the new regional circumstances that are coupled with the decline of US hegemony in the Middle East. US domination in the Middle East is increasingly being challenged: both Iran and Russia are extending their influence in the region. Turkey and Israel have many reasons for becoming allies. They could establish a new alliance as a part of the regional security and stability apparatus. Neither country wants the situation in Syria to continue to create instability. And, they do not want Iran to occupy a dominant position in Syria.

Figure 1. Israel-Turkey foreign trade values (thousands of American dollars) Source: Turkish Statistical Institute

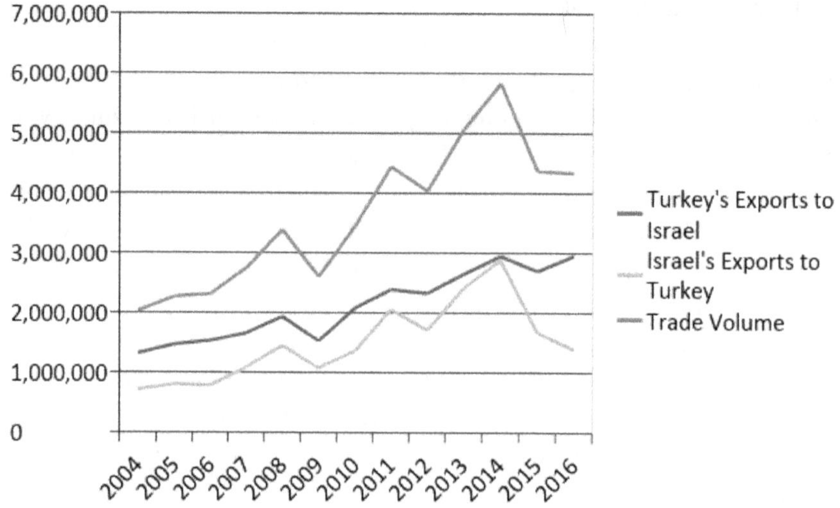

CONCLUSION

The recent Mavi Marmara crisis experienced between Israel and Turkey from 2011 to 2016 did not occur only because the AKP's rise to power in Turkey included its seeking a leadership role in the Middle East. Nor was the crisis caused by Israel's commitment to deterrence during the crisis. Overlooking the complexity of international system might lead to zero-sum strategies. Instead, Israel and Turkey need positive-sum strategies. A collaborative environment on the issue of the natural resources of Israel, for instance, might have important consequences in terms of the security and diversification of energy sources of Turkey. It is possible to revitalize relations between Israel and Turkey to the extent when the military elites were strong and the relationship was established with the hands of generals. Collaboration on energy could generate the eventually emergence of new ruling elites that would fill the vacuum created by the decline of the military's role in Turkey. The 2016 Israel-Turkey agreement provides Israel and Turkey with the opportunity to restore cooperation in security and intelligence areas as well as to strengthen bilateral trade by cooperation in energy areas. Turkey and Israel signed this agreement with a heavy heart: they couldn't find any more rational way to advance their interests. The agreement was also in line with Turkey's Prime Minister Binali Yıldırım's policy to strengthen friendships while decreasing adversaries.

Nevertheless, the 2016 Israeli-Turkey agreement seems to have been achieved, at least to some extent, by US coercion. This means that there is potential for new crises. Israel and Turkey could not focus on their mutual interests as a result of disappearance of trust between each other. Despite the enormous stakes of cooperation in the areas of energy and trade, the Israeli-Turkish relationship has not exhibited a positive (absolute) sum approach by which all parties win. Rather, the two countries have opted for zero-sum strategies by focusing on their own unilateral interests. This zero-sum approach has generated retaliation between the two countries which, in turn, turned into confrontation. On the one side, Turkey changed its attitude towards the Palestinian problem in favor of the Arabs. On the other side, Israel moved closer to the groups or countries having troubled relations with Turkey – such as the PKK, the PYD, Kurds, Greece and the Greek Cypriots. One of the most important consequences of the Mavi Marmara crisis was the disappearance of mutual trust between Israel and Turkey. The lack of trust in the Israel-Turkey relationship plays a key role in their problems with each other and could negatively affect regional challenges, including the Kurdish, Palestinian, Cyprus, and hydrocarbon issues, which still provoke uncertainties in the relationship of two countries. The return of a reasonable amount of trust in their relationship will take a long time. During this transition process, the US should do its best to help manage this relationship.

The nature of the challenges in the Middle East has changed since the twentieth century. To meet these challenges, the US should not hesitate to help prevent potential crisis emerging in the Middle East. As the only democracies in the Middle East, Turkey and Israel are crucial to the region's security. It is the US that operates with a systemic moral conscience that can assist both countries to manage their relations. Confronting complexity requires enhancing dialogue and cooperation, employing multilateralism and soft power and building a number of enhanced strategic partnerships. This complexity requires that the US use its effective crisis management and coercive abilities to address challenges. The behavior of the US towards Israeli-Turkish relations will be determinative of the future of the Middle East.

US involvement with an unbiased attitude is vital and could contribute to the re-building of trust in Israel-Turkey relations. And effective conflict management by the US could help prevent the escalation of any crisis to military conflict that might have devastating effects for the entire Middle East. While it would be complicated for the US to manage any potential crisis in the region, the US would do well to note that Turkish citizens perceive that Israel is state of the US. Hence, the US should mobilize diplomatic efforts for effective crisis management by applying the same level of pressure on both Turkey and Israel. Otherwise, Turkey will not respect the US role in managing any unexpected crisis between Turkey and Israel. Whether we like it or not, the behavior of the US towards Israeli-Turkish relations will be determinative for the future of the Middle East.

REFERENCES

Baker, P. (2017, January 14). For Israel, Energy Boom Could Make Friends Out of Enemies. *The New York Times*.

Brzezinski, Z. (1997). *The Grand Chessboard: American Primacy and its Geostrategic Imperatives*. New York: Basic Books.

Brzezinski, Z. (2012, January). After America. *Foreign Policy*, 26–29.

Buzan, B., Wæver, O., & de Wilde, J. (1998). *Security: A New Framework for Analysis*. Boulder, CO: Lynne Rienner.

Fuat, E. (2010). Turkish Foreign Policy 2009 and Beyond. *Perceptions*, *15*(3), 1–20.

Haaretz. (2017, September 13). *Israeli General Yair Golan, Former IDF Deputy Chief, Says PKK Not a Terror Organization*. Retrieved from https://www.haaretz.com/israel-news/ex-idf-deputy-chief-says-pkk-not-terror-group-1.5450379

Kaplan, R. (2008, December 17). The US May Be in Decline, but Only to a Degree. *The Washington Post*.

Knudsen, E. L. (2003). The Quagmire of Northern Iraq: The Clash of US, Turkish and Iraqi Kurdish Interests. *Journal of South Asian and Mideastern Studies*, *26*(4), 11–36.

Mcallister, J. F. O. (2003, January 20). Mad at America. *Time Europe, 161*(3). Retrieved February 14, 2003 from http://www.time.com/time/europe/magazine/printout/ 0,13155,901030120-407287,00.html

Nachmani, A. (1987). *Israel, Turkey and Greece, Uneasy Relations in the East Mediterranean*. London: Frank Cass.

Olson, R. (2015, January 25). US, Turkey, Israel and new challenges in the Middle East. Zaman Newspaper.

Öniş, Z. (2014). Turkey and Arab Revolutions: Boundaries of Regional Power Influence in a Turbulent Middle East. *Mediterranean Politics*, *19*(2), 203–219.

Parris, M. (2003). *Starting Over: US-Turkish Relations in the Post-Iraq War Era*. The Washington Institute for Near East Policy.

Persons, S. (2017, December 7). Sen. Sheldon Whitehouse calls Jerusalem decision a political tactic by Trump. *The Washington Times*.

Reuters. (2017, December 5). *Turkey could break off ties with Israel over Jerusalem.* Retrieved from https://www.reuters.com/article/us-israel-usa-turkey/turkey-could-break-off-ties-with-israel-over-jerusalem-erdogan-idUSKBN1DZ13R

Said, Y. (2005). Oil and Activism. In Global Civil Society. Sage.

Sanger, E. D. (2002, September 1). The World: First Among Evils? The Debate Over Attacking Iraq Heats Up. *The New York Times.*

The Economist. (2011, September 10). *Israel and Turkey: Can it get worse?* Retrieved from https://www.economist.com/node/21528687/comments#sort-comments

The Jerusalem Post. (2004, May 28). *Let the Kurds be.* Author.

The Times of Israel. (2017, October 24). *Netanyahu: World must take care of Kurds' future.* Retrieved from https://www.timesofisrael.com/netanyahu-world-must-take-care-of-kurds-future/

William, C. J. (2001). *Public Papers of the Presidents of the United States.* Washington, DC: U.S. Government Publication.

Compilation of References

Aarrestad, K., & Hatlen, L. M. (2014). *Fact 2015 energy and water resources in Norway*. Ministry of Petroleum and Energy.

ADB. (2004). *Asian Development Outlook 2004*. Retrieved from http://www.adb.org/Documents/Books/ADO/2004/default.asp

AfDB, Organization for Economic Cooperation and Development (OECD), United Nations Development Programme (UNDP), & Economic Commission for Africa (ECA). (2012). Youth Employment: Five Challenges for North Africa. Paper for the regional conference: Promoting Youth Employment in North Africa, Tunis, Tunisia.

African Development Bank (AfDB). (2012). Inequality and Arab Spring Revolutions in North Africa and the Middle East. *Africa Economic Brief*, *3*(7).

African Development Bank (AfDB). (2012). *Jobs, Justice and the Arab Spring: Inclusive Growth in North Africa*. Author.

African Development Bank (AfDB). (2012). *Political Transitions and New Socioeconomic Bargains in North Africa*. Economic Brief.

Agiomirgianakis, G., Asteriou, D., & Papathoma, K. (2003). *The Determinants of Foreign Direct Investment: A Panel Data Study for the OECD Countries*. Retrieved from http: //www.city. ac.uk/economics/ dps/discussion_papers/0306.pdf

Alfaro, L., Chanda, A., Kalemli-Ozcan, S., & Sayek, S. (2004). FDI and Economic Growth: The Role of Local Financial Markets. *Journal of International Economics*, *64*(1), 89–112. doi:10.1016/S0022-1996(03)00081-3

Alimov, M. (2014). *Oil despotism: Black Gold" Kazakhstan: blessing or curse?* Central Asia Monitor. Retrieved from http: //camonitor.com/14029-neftyanaya-despotiya.html

Altun, İ. H. (1996). Turkey's hydropower potential and development status. In *TMMOB 1.Energy Symposium Proceeding Book* (pp. 9-14). Ankara: EMO Publications.

Amineh, M. P., & Crijns-Graus, W. H. J. (2014). Rethinking EU energy security considering past trends and future prospects. *Perspectives on Global Development and Technology*, *13*(5-6), 757–825. doi:10.1163/15691497-12341326

Amineh, M. P., & Guang, Y. (2012). *The Geopolitics of Energy Paths of China and the European Union. Secure Oil and Alternative Energy*. Leiden, Boston: Brill.

Ancharaz, V. D. (2003). Determinants of Trade Policy Reform in Sub-Saharan Africa. *Journal of African Economies*, *12*(3), 417–443. doi:10.1093/jae/12.3.417

Applied Energy Partner. (2018). Retrieved from http://www.appenergy.com/2017/01/23/10-factors-that-affect-the-cost-of-energy/

APS Review Oil Market Trends. (2014, October 20). *Uzbekistan - the Gas Production System*. Author.

Asiedu, E. (2013). *Foreign Direct Investment, Natural Resources and Institutions*. Working Paper, International Growth Center, 1-47.

Asiedu, E. (2002). On the Determinants of Foreign Direct Investment to Developing Countries: Is Africa Different? *World Development*, *30*(1), 107–119. doi:10.1016/S0305-750X(01)00100-0

Aslani, A., Antila, E., & Wong, K.-F. V. (2012). Comparative analysis of energy security in the Nordic countries: The role of renewable energy resources in diversification. *Journal of Renewable and Sustainable Energy*, *4*(6), 62701. doi:10.1063/1.4765695

Asseburg, M. (2012). Protest, Revolt and Regime Change in the Arab World: Actors, Challenges, Implications and Policy Options. German Institute for International and Security Affairs.

Atilgan, İ. (2000). Overview of Turkey's Energy Potential. *Gazi University Journal of Engineering and Architecture*, *15*, 1.

Aydın, M. Z. (2010). *Yatırımcıların İhtiyaçları ve Liberizasyon Sürecinde Arz Güvenliği. In Enerji Platformu 2010: Avrupa'nın Enerji Güvenliğinde Türkiye'nin Rolü* (pp. 79–84). Ankara: Konrad-Adenauer-Stiftung.

Azernews. (2016, January 6). *Ashgabat can face challenges in gas transportation*. Author.

Baker, P. (2017, January 14). For Israel, Energy Boom Could Make Friends Out of Enemies. *The New York Times*.

Balasubramanyam, V. N., Salisu, M., & Sapsford, D. (1996). FDI and growth in EP and IS countries. *Economic Journal (London)*, *106*, 92–105. doi:10.2307/2234933

Bareiss, T. (2010). *Enerji Güvenliği-İşbirliği ve Çeşitlendirme. In Enerji Platformu 2010: Avrupa'nın Enerji Güvenliğinde Türkiye'nin Rolü* (pp. 19–25). Ankara: Konrad-Adenauer-Stiftung.

Barrell, R., & Pain, N. (1999). Domestic institutions, agglomerations and foreign direct investment in Europe. *European Economic Review*, *43*(4-6), 925–934. doi:10.1016/S0014-2921(98)00105-6

Bartes, I. E., & Wasenden, O.-H. B. (2014). *Electricity regulation in Norway: Overview*. Academic Press.

Baumeister, C., & Peersman, G. (2013). Time-varying effects of oil supply shocks on the US economy. *American Economic Journal Macroeconomics*, *5*(4), 1–28. doi:10.1257/mac.5.4.1

Baumeister, Ch., & Lutz, K. (2016). Forty Years of Oil Price Fluctuations: Why the Price of Oil May Still Surprise Us. *The Journal of Economic Perspectives*, *30*(1), 139–160. doi:10.1257/jep.30.1.139

Beckmann, J., & Czudaj, R. (2012). *Oil price and US dollar exchange rates dynamics*. University of Duisburg-Essen.

Belkin, P. (2008). The Europen Union's energy security challenges. *Connections: Quarterly Journal*, *7*(1), 76–103. doi:10.11610/Connections.07.1.05

Bemberg, N. (2010). Bir Enerji Konjonktürü Analizi: Türkiye'de Yabancı Enerji Yatırımlarının Geleceği. In Enerji Platformu 2010: Avrupa'nın Enerji Güvenliğinde Türkiye'nin Rolü (pp. 73-78). Academic Press.

Bergholt, D., & Larsen, V. H. (2016). *Business cycles in an oil economy: Lessons from Norway*. Academic Press.

Bevan, A., Estrin, S., & Meyer, K. (2004). Foreign investment location and institutional development in transition economies. *International Business Review*, *13*(1), 43–64. doi:10.1016/j.ibusrev.2003.05.005

Bilgin, M. (2010). Energy and Turkey's Foreign Policy: State Strategy, Regional Cooperation and Private Sector Involvement. *Turkish Policy Quarterly*, *9*(2), 81–92.

Bilgin, M. (2015). Turkey's energy strategy: Synchronizing geopolitics and foreign policy with energy security. *Insight Turkey*, *17*(2), 67–81.

Birchfield, V. L., & Duffield, J. S. (2011). *Toward a Common European Union Energy Policy*. Springer. doi:10.1057/9780230119819

Blomstrom, M., Lipsey, R. E., & Zejan, M. (1994). What explains developing country growth? In W. J. Baumol (Ed.), *Convergence of Productivity: Cross-National Studies and Historical Evidence* (9th ed.). New York: Oxford University Press, Incorporated.

Bloomberg, H. T. (2016). *2015 tüm zamanların en çok elektrik tüketilen yılı oldu, 2015–2016*. Retrieved from http://www.bloomberght.com/haberler/haber/1846242-2015-tum-zamanlarin-en-cok-elektrik-tuketilen-yili-oldu

Bloomberg. (2016). *Cash-Strapped Iraqi Kurds to Start Gas Exports to Turkey in 2016*. Retrieved from http://www.bloomberg.com/news/articles/2016-01-13/cash-strapped-iraqi-kurds-to-start-gas-exports-to-turkey-in-2016

Bodenstein, M., & Guerrieri, L. (2011). *Oil efficiency, demand, and prices: a tale of ups and downs*. Federal Reserve Board.

Bordoff, J. (2016). The New Geopolitics of Declining Oil Demand. *Wall Street Journal*. Retrieved from https://blogs.WSJ.com

Borensztein, E., De Gregorio, J., & Lee, J.-W. (1998). How Does Foreign Direct Investment Affect Economic Growth? *Journal of International Economics*, *45*(1), 115–135. doi:10.1016/S0022-1996(97)00033-0

Botric, V., & Skuflic, L. (2006). Main Determinants of Foreign Direct Investment in the Southeast European Countries. *Transition Studies Review*, *13*(2), 359–377. doi:10.100711300-006-0110-3

Bouoiyour, J. (2003). *The Determining Factors of Foreign Direct Investment in Morocco*. Retrieved from http: //www.erf.org.eg/tenthconf/Trade_Background/Bouoiyour.pdf

Bourguignon, F., & Morrison, Ch. (2002). Inequality among World Citizens: 1820–1992. *The American Economic Review*, *92*(4), 727–744. doi:10.1257/00028280260344443

Brzezinski, Z. (1998). *The Grand Chessboard: American Primacy and its Geostrategic Imperatives*. New York: Basic Books.

Brzezinski, Z. (2012, January). After America. *Foreign Policy*, 26–29.

Brzezinski, Z. (2016, April 17). Toward a Global Realignment. *The American Interest*.

Bukarica, V., & Tomsic, Z. (2017). Design and Evaluation of Policy Instruments for Energy Efficiency Market. *IEEE Transactions on Sustainable Energy*, *8*(1), 354–362. doi:10.1109/TSTE.2016.2599584

Buşe, M. (2017). Energy Security-Priority of the European Union. *International Scientific Conference "Strategies XXI: Bucharest,"* 170-175.

Busse, M., & Hefeker, C. (2007). Political risk, institutions and foreign direct investment. *European Journal of Political Economy*, *23*(2), 397–415. doi:10.1016/j.ejpoleco.2006.02.003

Buzan, B. (1991). *People, States and Fear: an Agenda for International Security Studies in the Post-Cold War*. London: Hasvester Wheatsheaf.

Buzan, B. (1998). *Security – A New Framework for Analysis*. Boulder, CO: Lynne Rinner Publishers, Inc.

Buzan, B., Wæver, O., & de Wilde, J. (1998). *Security: A New Framework for Analysis*. Boulder, CO: Lynne Rienner.

Bye, T., & Hope, E. (2005). Deregulation of Electricity Markets: The Norwegian Experience. *Economic and Political Weekly*, *40*(50), 5269–5278.

C.E. (2017). *11 Leadership Quotes From JFK That Still Resonate Today*. Retrieved from https://www.cashthechecks.com/11-leadership-quotes-jfk-still-resonate-today/

Cambridge Dictionary. (2017). Retrieved from http://dictionary.cambridge.org/dictionary/english-chinese-traditional/power-vacuum

Çapanoğlu, İ. K. (2010). Avrasya Enerji Arenasının Merkezi Türkiye. In *Enerji Platformu 2010: Avrupa'nın Enerji Güvenliğinde Türkiye'nin Rolü* (pp. 61-67). Ankara: Konrad-Adenauer-Stiftung.

Cashin, P., Mohaddes, K., Raissi, M., & Raissi, M. (2014). The differential effects of oil demand and supply shocks on the global economy. *Energy Economics*, *44*, 113–134. doi:10.1016/j.eneco.2014.03.014

Cassou, S. P. (1997). The Link between Tax Rates and Foreign Direct Investment. *Applied Economics*, *29*(10), 1295–1301. doi:10.1080/00036849700000019

CEIC. (2018). *Azerbaijan*. Retrieved from https://www.ceicdata.com/indicator/azerbaijan/data/real-gdp-yoy

Çelikpala, M. (2013). *Enerji Politikası. In Türkiye-Avrupa Birliği İlişkileri* (pp. 385–406). Ankara: Turhan.

Chazan, G. (2013, June 12). Changing Sources of Energy. *Financial Times*.

Chen, G. S. (2010). New energy development system in a low-carbon economy paradigm: Policies, technologies and funds. *2010 International Conference on Mechanic Automation and Control Engineering, MACE2010*, 4436–4439. 10.1109/MACE.2010.5536639

Chenoy, A. (2007, September). India's Foreign Policy Shifts and the Calculus of Power. *Economic and Political Weekly*.

Chester, L. (2010). Conceptualizing energy security and making explicit its polysemic nature. *Energy Policy*, *38*(2), 887–896. doi:10.1016/j.enpol.2009.10.039

Chevron. (2016). Retrieved from http://www.chevron.com/

China Daily. (2016, January 27). *Slump in oil prices fuels increase in China's crude imports*. Author.

Chontanawat, J., Hunt, L. C., & Pierse, R. (2008). Does energy consumption cause economic growth?: Evidence from a systematic study of over 100 countries. *Journal of Policy Modeling*, *30*(2), 209–220. doi:10.1016/j.jpolmod.2006.10.003

CIA - The World FactBook. (2016a). *Kazakhstan, Last Updated on May 5*. Author.

CIA - The World FactBook. (2016b). *Kyrgyzstan, Last Updated on May 5*. Author.

CIA - The World FactBook. (2016c). *Tajikistan, Last Updated on May 5*. Author.

CIA-Kazakhstan, Crude Oil and Natural Gas Information for Kazakhstan. (n.d.). Retrieved from https://www.cia.gov/library/publications/the-world-factbook/geos/kz.html

CIA-Tajikistan, Crude Oil and Natural Gas Information for Tajikistan. (n.d.). Retrieved from https://www.cia.gov/library/publications/the-world-factbook/geos/ti.html

CIA-Turkmenistan, Crude Oil and Natural Gas Information for Turkmenistan. (n.d.). Retrieved from https://www.cia.gov/library/publications/the-world-factbook/geos/tx.html

CIA-Uzbekistan, Crude Oil and Natural Gas Information for Uzbekistan. (n.d.). Retrieved from https://www.cia.gov/library/publications/the-world-factbook/geos/uz.html

Çınar, B. (2008). Tarihte Üçüncü Güç ve Orta Asya Enerji Savaşları. *Güvenlik Stratejileri Dergisi, 8*.

Civil Code of the Republic of Azerbaijan. (n.d.). Retrieved from http://isb.az/wp-content/uploads/2015/03/mulki_mecelle.pdf

Claes, D. H. (2010). Global Energy Security: Resource Availability, Economic Conditions and Political Constrains. *Contextualizing Energy Security and Transition. Concepts, Framing and Empirical Evidence, SGIR 7th P*, 1–26.

Commission Staff Working Paper: Energy Roadmap 2050: Impact Assesment Part 1, European Commission. (n.d.). Retrieved from https://ec.europa.eu/energy/sites/ener/files/documents/sec_2011_1565_part1.pdf

Contessi, N. (2016). Central Asia in Asia: Charting growing trans-regional linkages. *Journal of Eurasian Studies*, *7*(1), 3–13. doi:10.1016/j.euras.2015.11.001

Çukurçayır, M., & Sağır, H. (2007). Energy Problem, Environment and Alternative Energy Sources. *Selcuk University Journal of the Institute of Social Sciences*, *20*, 257.

Culem, C. G. (1988). The Locational Determinants of Direct Investment among Industrialized Countries. *European Economic Review*, *32*(4), 885–904. doi:10.1016/0014-2921(88)90051-7

Cunado, J., & Perez de Gracia, F. (2005). Oil prices, economic activity and inflation: Evidence for some Asian countries. *The Quarterly Review of Economics and Finance*, *45*(1), 65–83. doi:10.1016/j.qref.2004.02.003

Daly, J. (2015, July 10). Kashagan Oil Field Comes Back to Life. *Silk Road Reporters*.

Danesh, A. (2010). In the oil and gas industry to strengthen the role of the state, and one of the levers - the strengthening of KMG and its mining "daughter" continues. *Panorama*. Retrieved from http: //www.nomad .su /? a = 4-201009060034

Davis, J. (2006). The Changing World of Oil: An Analysis of Corporate Change and Adaptation. *Environmental Politics*, *18*, 159–160.

De Mello, L. R. (1997). FDI in Developing Countries and Growth: A Selective Survey. *The Journal of Development Studies*, *34*(1), 1–34. doi:10.1080/00220389708422501

De Mello, L. R. (1999). Foreign Direct Investment-Led Growth: Evidence from time series and panel data. *Oxford Economic Papers*, *51*(1), 133–151. doi:10.1093/oep/51.1.133

Demirbilek, S. (2010). *Avrupa'nın Enerji Güvenliği Bağlamında Türkiye'nin Konumu. In Enerji Platformu 2010: Avrupa'nın Enerji Güvenliğinde Türkiye'nin Rolü* (pp. 27–35). Ankara: Konrad-Adenauer-Stiftung.

Demirhan, E., & Masca, M. (2008). Determinants of Foreign Direct Investment Flows to Developing Countries: A Cross-Sectional Analysis. *Prague Economic Papers*, *4*(4), 356–369. doi:10.18267/j.pep.337

Dikmen, A. Ç. (2009). *Locating in Turkey's Future Renewable Energy Sources for Sustainable Development Framework* (Doctoral Thesis). University of Ankara, Ankara, Turkey.

Dirioz, A. O. (2013). Energy security, politics, markets, peace. *Journal of Foreign Policy and Peace*, *1*(1), 90–97.

DTM. (2010). *Energy Production and Consumption in Turkey*. Undersecretariat of Foreign Trade. Retrieved from http://www.dtm.gov.tr/dtmadmin/upload/EAD/KonjonkturIzlemeDb/teut.do

Duarte, P. (2016). Ahuyentando sus Espantapájaros: Una evaluación de la (in) seguridad alimentaria y hídrica en China. *Revista Geográfica Venezolana*, *57*(2), 298–317.

Duarte, P. (2017). *Pax Sinica*. Lisbon: Chiado Editora.

Du, L., Yanan, H., & Wei, C. (2010). The relationship between oil price shocks and China's macro-economy: An empirical analysis. *Energy Policy*, *38*(8), 4142–4151. doi:10.1016/j.enpol.2010.03.042

Dunning, J. H. (1988). *Explaining International Production*. London: Harper Collins Academic.

Durkheim, E., & Swain, J. W. (2015). *The Elementary Forms of the Religious Life, a Study in Religious Sociology – Scholars Choice Edition*. Scholars Choice.

EBRD. (2000). *Transition Report 2000: Employment, Skills, and Transition*. London: European Bank for Reconstruction and Development.

EC. (2009). *The impact of renewable energy policy on economic growth and employment in the European Union.* European Commission Directorate General, Energy and Transport. Retrieved from http://ec.europa.eu/energy/renewables/studies/doc/renewables/2009_employ_res_summary.pdf

EIA. (n.d.). *U.S Energy Information Administration, What is Energy?* Retrieved from https://www.eia.gov/energyexplained/index.cfm/index.cfm?page=about_home

EMRA. (2011). *Electricity Market Regulatory Agency*. Retrieved from www.epdk.gov.tr

EMRA. (2013). *Electricity Market Regulatory Agency, Electricity Market Licences.* Retrieved from http://www.emra.org.tr/tr/web/electricity-market/licenses

Energy 2020, A Strategy for Competetive, Sustainable and Secure Energy, European Commission, COM (2010), 639 Final, Brussels.

Energy 2050. (n.d.). Retrieved from http://ec.europa.eu/energy/en/topics/energy-strategy-and-energy-union/2050-energy-strategy

Energy Roadmap 2050, European Commission, COM (2011). 855 Final, Brussels. Retrieved from http://eur-lex.europa.eu/legal content/EN/TXT/PDF/?uri=CELEX:52011DC0885&from=EN

Energy Strategy-2030, European Commission. (n.d.). Retrieved from http://ec.europa.eu/energy/en/topics/energy-strategy-and-energy-union/2030-energy-strategy

Engdahl, W. (2016, May 15). Anglo-American Oil Market Is Broken. *New Eastern Outlook*. Retrieved from https://journal-neo.org/2016/05/15/anglo-american-oil-market-is-broken/

ENI. (2015). *World Oil and Gas Review*. Retrieved from https://www.eni.com/docs/en_IT/enicom/media/press-release/2016/11/Eni_presents_15_edition_World_Oil_and_Gas_Review.pdf

Eriş, Ö. Ü. (2016). *Enerji Politikası. In Avrupa Birliği* (pp. 617–637). Ankara: Seçkin.

Erkişi, K. (2017). Strategic Use of Economic Power: Geo-economics. *The Journal of International Social Research Volume*, *10*(52), 1025–1035. doi:10.17719/jisr.2017.19

Esfahani, H. S., Mohaddes, K., & Pesaran, M. H. (2014). An empirical growth model for major oil exporters. *Journal of Applied Econometrics*, *29*(1), 1–21. doi:10.1002/jae.2294

Esteban, M., & Otero-Iglesias, M. (2015). What are the prospects for the new Chinese-led Silk Road and Asian Infrastructure Investment Bank? *Elcano Royal Institute.*, *23/2015*, 1–10.

European Commission Energy DG. (n.d.). *2020 Energy Strategy*. Retrieved from https://ec.europa.eu/energy/en/topics/energy-strategy-and-energy-union/2020-energy-strategy

European Commission Progress Report Turkey. (2016). Retrieved from http://www.ab.gov.tr/files/5%20Ekim/20161109_report_turkey.pdf

European Commission. (2011, November). *Report From The Commission To The European Parliament, The Council And The European Economic And Social Committee, Implementation of the Communication on Security of Energy Supply and International Cooperation and of the Energy Council Conclusions of November 2011*. Author.

European Commission. (2016). EU Reference Senario 2016. Retrieved from https://ec.europa.eu/energy/sites/ener/files/documents/ref2016_report_final-web.pdf

European Union, Energy Supplier Countries. (2017). Retrieved from http://ec.europa.eu/energy/en/topics/imports-and-secure-supplies/supplier-countries

European Union, Gas and Oil Supply Routes. (2017). Retrieved from http://ec.europa.eu/energy/en/topics/imports-and-secure-supplies/gas-and-oil-supply-routes

European Union, International Cooperation on Energy. (2017). Retrieved from http://ec.europa.eu/energy/en/topics/international-cooperation

European Union, Overview of the secondary EU legislation that falls under the legislative competence of DG ENER and that is currently in force. (2016). Retrieved from https://ec.europa.eu/energy/sites/ener/files/documents/Overview%20of%20ENER-related%20legislation%20%28by%20policy%20areas%29%20update%2014.03.2016.pdf

Eurostat. (n.d.). *Extra-EU Imports*. Retrieved from http://ec.europa.eu/eurostat/statistics-explained/images/7/71/Percentage_of_extra-EU_imports_%28entries%29_of_natural_gas_by_country_of_origin.png

Eurostat. (n.d.). *Gross Inland Consumption*. Retrieved from http://ec.europa.eu/eurostat/statisticsexplained/images/e/e3/National_shares_of_fuels_in_gross_inland_energy_consumption%2C_2015%2C_percentage_F6_update.png

Eurostat. (n.d.). *Primary Energy Imports*. Retrieved from http://ec.europa.eu/eurostat/statisticsexplained/images/9/92/Main_origin_of_primary_energy_imports%2C_EU-28%2C_2005-2015_%28%25_of_extra_EU28_imports%29_YB17.png

Eurostat-Energy Dependency. (n.d.). Retrieved from http://ec.europa.eu/eurostat/statisticsexplained/index.php/Energy_trends#Energy_dependency

Evans, R. L. (2008). *Fueling Our Future: An Introduction to Sustainable Energy*. Cambridge University Press.

Facts & Details. (2015). *China's City Dwellers Now Outnumber the Rural Population*. Retrieved from http://factsanddetails.com/china/cat11/sub72/item151.html

Farchy, J. (2015, October 14). China's Great Game: In Russia's backyard. *Financial Times*.

Fingar, T. (2016). *The New Great Game: China and South and Central Asia in the Era of Reform*. Stanford, CA: Stanford University Press.

Fløttum, K., Dahl, T., & Rivenes, V. (2016). Young Norwegians and their views on climate change and the future: Findings from a climate concerned and oil-rich nation. *Journal of Youth Studies*, *6261*(November), 1–16. doi:10.1080/13676261.2016.1145633

FocusEconomics. (2016). Turkey. *Economic Outlook*, *15*(June). Retrieved from http://www.focus-economics.com/countries/turkey

Fuat, E. (2010). Turkish Foreign Policy 2009 and Beyond. *Perceptions*, *15*(3), 1–20.

Gass, O. O. G. (2016). *OG21 STRATEGY DOCUMENT Issued for comments*. Author.

Gault, J. (2002). *The European Union: Energy Security and the Periphery*. Occasional Paper Series. No.40.

Gavenas, E., Rosendahl, K. E., & Skjerpen, T. (2015). CO2-emissions from Norwegian oil and gas extraction. *Energy, 90*(806), 1956–1966. doi:10.1016/j.energy.2015.07.025

Gazetesi, K. (2016). *Türkiye'nin enerji faturası 10.2 milyar dolar azaldı*. Retrieved from http://www.karar.com/ekonomi-haberleri/turkiyenin-enerji-faturasi-102-milyar-dolar-azaldi-295487#

Gazetesi. (2016). Türk Akımı'nın ertelenmesi ile enerjide kaynak çeşitliliği hedefleniyor. *Focus*, (8).

Geopolitics of Oil Hearings before the Committee of Energy and Political Resources (1980). *US Senate, 96 Congress, the 2nd Session. Pt. 1–2. Wash.: US gov., print. off.*

Georgetown Security Studies Review (GSSR). (2017). *Opportunities and Challenges for Europe's Energy Union*. Feb.26, 2017. Retrieved from http://georgetownsecuritystudiesreview.org/2017/02/26/opportunities-and-challenges-for-europes-energy-union/

Giles, C. (2014). Winners and losers of oil price plunge. *The Financial Times*. Retrieved from http://inosmi.ru/world/20141217/224991616.html#ixzz3by68oQVq

Global Business Reports (2012). *Special Report: Caspian Region*. Author.

Global Security. (2013). Retrieved from http://www.globalsecurity.org/military/world/centralasia/tajik- energy.htm

Globes, Israel's Business Area. (2015). *New report sees 20% less gas reserves in Leviathan*. Retrieved from http://www.globes.co.il/en/article-new-report-sees-20-less-gas-reserves-in-leviathan-1001041847

Godoy, E. (2011). *World Bank Calibrating its Measurement of Sustainability*. Global Geopolitics and Political Economy. WordPress.

Goldsbrough, D. G. (1979). The Role of Foreign Direct Investment in the External Adjustment Process. *Staff Papers, 26*(4), 725–754. doi:10.2307/3866968

Goodreads. (2018). *Theodore Roosevelt*. Retrieved from https://www.goodreads.com/quotes/63681-character-in-the-long-run-is-the-decisive-factor-in

Gorst, I. (2006). Struggle for Gas Intensifies. Petroleum Economist, 73(5).

Graaf, T. V. (2013). *The Politics and Institutions of Global Energy Governance*. New York: Palgrave. doi:10.1057/9781137320735

Green Growth Platform (GGP). (2014). *Advisory Council I: Report to Ministers Energy Security & the 2030 Climate & Energy Package, The Big Picture: A Low Carbon Energy Union Targets Security, Climate, Jobs & Growth*. University of Cambridge.

Guardian. (2014). *Shale and non-Russian gas imports at heart of new EU energy strategy*. Retrieved from https://www.theguardian.com/environment/2014/may/28/shale-gas-russia-eu-renewables-ukraine-crisis

Gullberg, A. T., Ohlhorst, D., & Schreurs, M. (2014, August). Towards a low carbon energy future - Renewable energy cooperation between Germany and Norway. *Renewable Energy*, *68*, 216–222. doi:10.1016/j.renene.2014.02.001

Gutiérrez, D. G. (2007). The Politics of the Interstices: Reflections on Citizenship and Non-Citizenship at the Turn of the Twentieth Century. *Race/Ethnicity. Multidisciplinary Global Contexts*, *1*(1), 89–120.

Haaretz. (2017, September 13). *Israeli General Yair Golan, Former IDF Deputy Chief, Says PKK Not a Terror Organization*. Retrieved from https://www.haaretz.com/israel-news/ex-idf-deputy-chief-says-pkk-not-terror-group-1.5450379

Haghighi, S. S. (2007). *Energy Security: The External Legal Relations of the EU with Major Oil and Gas Supplying Countries*. Bloomsbury Publishing.

Hamilton, J. D. (2009). *Causes and Consequences of the Oil Shock of 2007-08, NBER Working Paper no. 15002*. Washington, DC: Brookings

Hamilton, J. D. (2009). *Causes and Consequences of the Oil Shock of 2007-08*. National Bureau of Economic Research. doi:10.3386/w15002

Hamlehdar, M., & Aslani, A. (2017). Analysis of Energy System in Norway with Focus on Energy Consumption Prediction. *Management of Sustainable Development*, *9*(1), 5–14. doi:10.1515/msd-2017-0008

Hanabusa, K. (2009). Causality relationship between the price of oil and economic growth in Japan. *Energy Policy*, *37*(5), 1953–1957. doi:10.1016/j.enpol.2009.02.007

Hankel, W., & Isaak, R. (2011, May). Megatrends of the Global Economy of Tomorrow. *The World Financial Review*.

Hart, M. (1992). *The 100: A Ranking of The Most Influential Persons in History*. New York, NY: Kensington Press.

Hartman, D. G. (1994). Tax Policy and Foreign Direct Investment in the United States. *National Tax Journal*, *37*(4), 475–488.

Heather, P. (2012). *Continental European gas hubs: are they fit for purpose?* Oxford Institute for Energy Studies. Working paper. Retrieved from http://www.oxfordenergy.org/wpcms/wp-content/uploads/2012/06/NG-63.pdf

Heinz, A. (2011, November 27). Henry Kissinger: "If You Can't Hear the Drums of War You Must Be Deaf". *The Daily Squib*.

Hemmings, P. (2016). *Vassiliki Koutsogeorgopoulou; Piritta Sorsa*. OECD Economic Surveys.

Herberg, M. (2011). *China's Energy Rise and the Future of U.S.- China Energy Relations*. New America Foundation.

Herdem Attorneys At Law. (2014). *2013 Turkey Energy Report*. Retrieved from http://herdem.av.tr/energy-sector-turkey-central-importance-epias

Hobson, E. (2011). *After the Arab spring Revolutions: 7 reasons why LED offers a major solution to North Africa's economic problems*. Local Economic Development Network of Africa (LEDNA). Retrieved from www.ledna.org

Huang, C. (2016, February 2). China's Silk Road ventures carry financial risks as well as political rewards. *South China Morning Post*.

Hurriyet Daily News. (2015). *Turkey to raise gas storage capacity by 4 bcm*. Retrieved from http://www.hurriyetdailynews.com/turkey-to-raise-gas-storage-capacity-by-4-bcm-minister.aspx?pageID=238&nID=89963&NewsCatID=348

Hürriyet Daily News. (2016). *Turkey, Russia strike strategic Turkish Stream gas pipeline deal.* Retrieved from http://www.hurriyetdailynews.com/turkey-russia-strike-strategic-turkish-stream-gas-pipeline-deal.aspx?pageID=238&nID=104822&NewsCatID=348

Ibrahim, D., Basir, S.A., & Rahman, A. A. (2011). Sustainable Economic Development: Concept, Principles and Management from Islamic Perspective. *European Journal of Soil Science, 24*(3).

İdiz, S. (2012, October 30). Erdoğan's 'Ankara Criteria'. *Hürriyet Daily News.*

IEA. (2006). *Renewable Energy: RD&D Priorities.* Paris: IEA Publications.

IEA. (2011). *NorwayEnergy Policies of IEA Countries: Norway2011 Review.* IEA.

Ikanberry, J. (2015). *America China and the Struggle for World Order: Ideas, Traditions, Historical Legacies and Global Visions.* New York: Palgrave. doi:10.1057/9781137508317

International Energy Agency. (2009). *How The Energy Sector Can Deliver On A Climate Agreement.* Copenhagen: IEA Publications.

International Energy Agency. (2014). World Energy Outlook 2014. *International Energy Agency.* Retrieved from https://www.iea.org/publications/freepublications/publication/WEO2014.pdf

International Energy Agency. (2014a). Energy Supply Security 2014 Part 3. *Energy Supply Security: The Emergency Response of IEA Countries - 2014 Edition,* 1–105.

International Energy Agency. (2014b). *European Union 2014 Review, 312.* IEA.

International Energy Agency. (2015). *China. International energy data and analysis.* IEA.

International Energy Agency. (2016). *World Energy Outlook.* Paris, France: IEA.

İşcan, İ.H. (2004). Uluslararası İlişkilerde Klasik Jeopolitik Teoriler ve Çağdaş Yansımaları. *Uluslararası İlişkiler, 1*(2).

Jackson, S., & Markowski, S. (1995). The Attractiveness of Countries to Foreign Direct Investment. *Journal of World Trade, 29,* 159–180.

Jayaraman, T. K., & Choong, C. K. (2009). Growth and oil price: A study of causal relationships in small Pacific Island countries. *Energy Policy*, *37*(6), 2182–2189. doi:10.1016/j.enpol.2009.01.025

Johnson, K. (2015, May). China Tops U.S. as Biggest Oil Importer. *Foreign Policy*.

Jordaan, J. C. (2004). *Foreign Direct Investment and Neighbouring Influences* (Unpublished doctoral thesis). University of Pretoria.

Jorge, R. S., & Hertwich, E. G. (2013). Environmental evaluation of power transmission in Norway. *Applied Energy*, *101*, 513–520. doi:10.1016/j.apenergy.2012.06.004

Kalın, İ. (2011, March 7). Türkiye ve Japonya: Değerli Bir Ortaklık [Turkey and Japan: A Precious Partnership]. *Today's Zaman.*

Kaplan, R. (2008, December 17). The US May Be in Decline, but Only to a Degree. *The Washington Post*.

Kaplan, R. D. (2014). Geopolitics of Energy. *Forbes*. Retrieved in April 2018 https://www.forbes.com/sites/stratfor/2014/04/04/the-geopolitics-of-energy/#61be93a93b39

Karl, T. L. (2007). Oil-Led Development: Social, Political, and Economic Consequences. *CDDRL Working Papers,* No 80. Retrieved from https://cddrl.fsi.stanford.edu/sites/default/files/No_80_Terry_Karl_-_Effects_of_Oil_Development.pdf

KazEnergy. (2013). *Social responsibility of the oil and gas industry of the Republic of Kazakhstan in 2012-2013*. Astana: Association KAZENERGY.

Kemsley, D. (1998). The Effect of Taxes on Production Location. *Journal of Accounting Research*, *36*(2), 321–341. doi:10.2307/2491480

Keohane, R. (2000). *Realism, Neorealism and the Study of World Politics*. Basic Books.

Khwaja, A. (2003). Pakistan and The New Great Game. *IPRI Paper*, *5*(April), 5.

Kilian, L. (2009). Not all Oil Price Shocks Are Alike: Disentangling Demand and Supply Schocks in Crude Oil Market. *The American Economic Review*, *93*(3), 1053–1069. doi:10.1257/aer.99.3.1053

Kilian, L., & Murphy, D. P. (2012). Why agnostic sign restrictions are not enough: Understanding the dynamics of oil market VAR models. *Journal of the European Economic Association*, *10*(5), 1166–1188. doi:10.1111/j.1542-4774.2012.01080.x

Kim, S., & Willett, T. D. (2000). Is the negative correlation between inflation and growth real? An analysis of the effects of the oil supply shocks. *Applied Economics Letters*, *7*(3), 141–147. doi:10.1080/135048500351681

King, C. S. (2014). *Mapping Demand Response in Europe Today 2015. SEDC.* Smart Energy Demand Coalition.

King, M. (2009). Whyte Paradoxes of China's Economic Boom. *Annual Review of Sociology*, *35*(1), 371–392. doi:10.1146/annurev-soc-070308-115905

Kinninmont, J. (2012). *'Bread, Dignity and Social Justice': The Political Economy of Egypt's Transition*. Chatham house briefing paper. Retrieved from www.chathamhouse.org

Knudsen, E. L. (2003). The Quagmire of Northern Iraq: The Clash of US, Turkish and Iraqi Kurdish Interests. *Journal of South Asian and Mideastern Studies*, *26*(4), 11–36.

Kok, R., & Ersoy, B. A. (2009). Analyses of FDI Determinants in Developing Countries. *International Journal of Social Economics*, *36*(1/2), 105–123. doi:10.1108/03068290910921226

Kopytin, I. A. (2014). Influence of oil prices on stock market indexes in Russia and Norway. *Studies on Russian Economic Development*, *25*(1), 99–110. doi:10.1134/S1075700714010109

Kouddous, S. A. (2011). Crackdown in Cairo. *The Pulitzer Center.* Retrieved from http://pulitzercenter.org/articles/cairo-egypt-habib-al-adly-trial-postponed-tahrir-square-revolution?format=print

Krifa-Schneider, H., Matei, I., & Matei, I. (2010, November). Business Climate, Political Risk and FDI in Developing Countries: Evidence from Panel Data. *International Journal of Economics and Finance*, *2*(5), 54–65. doi:10.5539/ijef.v2n5p54

Kruyt, B., Vuuren, D. P., Vries, H. J. M., & Groenenberg, H. (2009). Indicators for energy security. *Energy Policy*, *37*(6), 2166–2181. doi:10.1016/j.enpol.2009.02.006

Kuban & Uyar. (2007). Renewable Energy Regulations and Turkey. *The European Union and Turkey Renewable Energy Regulatory Workshop*, 283-290.

LaBelle, M. C. (2017). In pursuit of Energy Justice. *Energy Policy*, *107*, 615–620. doi:10.1016/j.enpol.2017.03.054

Larsen, B. M., & Nesbakken, R. (1997). Norwegian Emissions of CO2 1987-1994. *Environmental and Resource Economics*, *9*(3), 275–290. doi:10.1007/BF02441400

Lau, L., Qian, Y., & Roland, G. (2000). Reform without losers: An interpretation of China's dual-track approach to transition. *Journal of Political Economy*, *108*(1), 120–143. doi:10.1086/262113

Lemann, N. (2001, July 16). Dreaming about War. *The New Yorker*. Retrieved from: http://www.comw.org/qdr/0107lemann.html

Lenz, N. V., & Grgurev, I. (2017). Assessment of Energy Poverty in New European Union Member States: The Case of Bulgaria, Croatia and Romania. *International Journal of Energy Economics and Policy*, *7*(2), 1–8.

Leung, G., Cherp, A., Jewell, J., & Wei, Y.-M. (2014). Securitization of energy supply chains in China. *Journal of Applied Energy.*, *123*, 316–326. doi:10.1016/j.apenergy.2013.12.016

Licklider, R. (1988). The Power of Oil: The Arab oil Weapon and the Netherlands, the United Kingdom, Canada, Japan, and the United States. *International Studies Quarterly*, *32*(5), 205–226. doi:10.2307/2600627

Lim, D. (1983). Fiscal Incentive and Direct Foreign Investment in Less Developed Countries. *The Journal of Development Studies*, *19*(2), 207–212. doi:10.1080/00220388308421859

Lim, T. (2016). *China's One Belt One Road Initiative*. Imperial College Press. doi:10.1142/p1076

Ling, W. (2013, October 9). A community of common destiny. *China Daily*.

Linke, K., & Vietor, V. (2010). *Beyond Turkey. The EU's Energy Policy and the Southern Corridor*. Bonn: Friedrich Ebert Foundation. Retrieved from http://www.fes.kiev.ua/Dokument/beyond_turkey.pdf

Lippi, F., & Nobili, A. (2012). Oil and the macroeconomy: A quantitative structural analysis. *Journal of the European Economic Association*, *10*(5), 1059–1083. doi:10.1111/j.1542-4774.2012.01079.x

Lunn, J. (1980). Determinants of U.S. Direct Investment in the E. E. C.: Further Evidence. *European Economic Review*, *13*(1), 93–101. doi:10.1016/0014-2921(80)90048-3

Luo, Y. (2009). Political risk and country risk in international business. Concepts and measures. In The Oxford Handbook of International Business (2nd ed.). Oxford University Press.

Lynch, D. (2008). Tension in Egypt shows potency of food crisis. *USA Today*. Retrieved from http://www.usatoday.com/money/world/2008-04-29-egypt-food-shortage_N.htm

MacKenzie, D. (2011). Can complexity theory explain Egypt's crisis? *New Scientist*. Retrieved from http://www.newscientist.com/article/7dn20082-can-complexity-theory-explain-egypts-crisis.html

Mackinder, H. (2004). The geographical pivot of history. *The Geographical Journal*, *170*(4), 298–321. doi:10.1111/j.0016-7398.2004.00132.x

Macmillan Endcyclopedia of Energy. (2000). *Energy Economics*. Retrieved from https://www.encyclopedia.com/environment/encyclopedias-almanacs-transcripts-and-maps/energy-economics

Majeed, M. T., & Ahmad, E. (2009). An Analysis of Host Country Characteristics that Determine FDI in Developing Countries: Recent Panel Data Evidence. *The Lahore Journal of Economics*, *14*(2), 71–96.

Makinen, H. (2010). The future of natural gas as the European Union's energy source – risks and possibilities. *Pan-European Institute*. Retrieved from www.tse.fi/pei

Maltby, T. (2013). European Union energy integration: A case of European Commission policy enterpreneurship and increasing supranationalism. *Energy Policy*, *55*, 435–444. doi:10.1016/j.enpol.2012.12.031 PMID:24926115

Markusen, J.R., & Venables, A. J. (1999). Foreign direct investment as a catalyst for industrial development. *European Economic Review*, *43*(2), 335-356.

Markusen, J. R. (2002). *Multinational firms and the Theory of International Trade*. Cambridge, MA: MIT Press.

Marten, K. (2007). Russia Efforts to Control Kazakhstan's Oil: The Kumkol Case. *Post-Soviet Affairs*, *23*(1), 18–37. doi:10.2747/1060-586X.23.1.18

Mashrab, F. (2012, August 9). Tajikistan holds back on cheering huge oil, gas find. *Asia Times*.

Mason, P. (2011). Twenty reasons why it's kicking off everywhere. *BBC News*. Retrieved from http://www.bbc.co.uk/blogs/newsnight/paulmason/2011/02/twenty_reasons_why_its_kicking.html

Maza, C. (2015, September 24). Central Asia's pipeline politics: a quest for energy independence. *The Christian Science Monitor*.

Mcallister, J. F. O. (2003, January 20). Mad at America. *Time Europe, 161*(3). Retrieved February 14, 2003 from http://www.time.com/time/europe/magazine/printout/ 0,13155,901030120-407287,00.html

McAllister, F. 0. (2003, January 20). Mad at America. *Time*.

Mearshmeimer, J. (2015). *The Rise of China and the Decline of the US Army*. San Diego, CA: Didactic Press.

Mehr News Agency. (2015). *Export of Iran's gas to Turkey soars*. Retrieved from http://en.mehrnews.com/news/112622/Export-of-Iran-s-gas-to-Turkey-soars

Mehrara, M. (2007). Energy consumption and economic growth: The case of oil exporting countries. *Energy Policy*, *35*(5), 2939–2945. doi:10.1016/j.enpol.2006.10.018

Mehrpooya, M., Mohammadi, M., & Ahmadi, E. (2018). Techno-economic-environmental study of hybrid power supply system: A case study in Iran. *Sustainable Energy Technologies and Assessments*, *25*, 1–10. doi:10.1016/j.seta.2017.10.007

Meinhardt, H. I. (2002). Cooperative Decision Making in Common Pool Situation. Lecture Notes in Economics and Mathematical Systems. Springer Berlin Heidelberg.

Merrill, K. R. (2007). *The Oil Crisis of 1973 – 1974: A Brief History with Documents*. Bedford: St. Martin's.

Metais, R. (2013). *Ensuring Energy Security in Europe: The EU between a Market-based and a Geopolitical Approach. EU Diplomacy Paper (03 / 2013)*. College of Europe.

Miller, J. B. (2014, March 18). Japan's Strategic Push with Turkey. *The Diplomat.*

Ministry of EU (MEU). (2017). Retrieved from http://www.ab.gov.tr/_80.html

Ministry of Foreign Affairs of Japan. (2017). Retrieved from http://www.mofa.go.jp/region/middle_e/turkey/data.htm

Mobarak, A. (2005, May). Democracy, Volatility and Economic Development. *The Review of Economics and Statistics*, *87*(2), 348–361. doi:10.1162/0034653053970302

Mohaddes, K., & Raissi, M. (2016). *The US oil supply revolution and the global economy*. Academic Press.

Mohaddes, K., & Pesaran, M. H. (2016). Country-specific oil supply shocks and the global economy: A counterfactual analysis. *Energy Economics*, *59*, 382–399. doi:10.1016/j.eneco.2016.08.007

Mohammadi, M., Ghasempour, R., Astaraei, F. R., Ahmadi, E., Aligholian, A., & Toopshekan, A. (2018). Optimal planning of renewable energy resource for a residential house considering economic and reliability criteria. *International Journal of Electrical Power & Energy Systems*, *96*, 261–273. doi:10.1016/j.ijepes.2017.10.017

Molina-Garcia, A., Hansen, A. D., Muljadi, E., Gevorgian, V., Fortmann, J., & Gomez-Lazaro, E. (2017). *International Requirements for Large Integration of Renewable Energy Sources*. London, UK: The Institution of Engineering and Technology. Retrieved from http://digital-library.theiet.org/content/books/10.1049/pbpo098e_ch2

Montano-De La Cruz, E., Magallona-Flores, M., Angeles-Guevara, E., & Ponce De Leon-Ladera, H. (2002). *Dimensions in Learning English: A Series for Philippine Secondary Schools*. Retrieved from https://books.google.co.ma/books?id=gN74sdoJSkQC&pg=PA282&lpg=PA282&dq=%22The+true+wealth+of+a+nation+lies+not+in+its+gold+or+silver+but+in+its+learning,+wisdom+and+in+the+uprightness+of+its+sons.%22+Gibran+Khalil+Gibran&source=bl&ots=EyQyBucsFI&sig=ja2oj9U0x2ANh7uj978OlL2fQFc&hl=en&sa=X&ved=0ahUKEwji9ZOFzPPaAhWCTd8KHV-5DekQ6AEIJjAB#v=onepage&q=%22The%20true%20wealth%20of%20a%20nation%20lies%20not%20in%20its%20gold%20or%20silver%20but%20in%20its%20learning%2C%20wisdom%20and%20in%20the%20uprightness%20of%20its%20sons.%22%20Gibran%20Khalil%20Gibran&f=false

Morgenthau, H. (1978). *Politics Among Nations: The Struggle for Power and Peace* (5th ed.). New York: Alfred A. Knopf.

Muhammad, M.T.S. (2010). Principals of sustainable development in ibn Khaldoun's economic thought. *Malaysian Journal of Real Estate, 5*(1).

Musayev, E. (2012). Securities Market In Azerbaijan: State And Tendencies Of Development. *Russian Business, 10*, 53-65.

Mütercimler, E. (2010). *Ertuğrul Faciası, Ertuğrul Disaster*. İstanbul: Alfa Printing House.

Nachmani, A. (1987). *Israel, Turkey and Greece, Uneasy Relations in the East Mediterranean*. London: Frank Cass.

Nakov, A., & Pescatori, A. (2010). Oil and the great moderation. *Economic Journal (London), 120*(543), 131–156. doi:10.1111/j.1468-0297.2009.02302.x

Narayan, P. K., Narayan, S., & Prasad, A. (2008). Understanding the oil price-exchange rate nexus for the Fiji islands. *Energy Economics, 30*(5), 2686–2696. doi:10.1016/j.eneco.2008.03.003

National Bureau of Economic Research (NBER). (2005). *Forms of Democracy, Policy and Economic Development*. Working Paper 11171, March 2005.

Nigh, D. (1985). The Effect of Political Events on United States Direct Foreign Investment: A Pooled Time-Series Cross-Sectional Analysis. *Journal of International Business Studies*, *16*(1), 3–17. doi:10.1057/palgrave.jibs.8490439

Nilsen, Ø. A. (2015). *The relationship between oil Supervisor. Evidences from Norway*. Norwegian School of Economics.

Nilsson, A. V. (2014). *Nordic Market Report 2014*. NordREG.

Noorollahi, Y., Itoi, R., Yousefi, H., Mohammadi, M., & Farhadi, A. (2017). Modeling for diversifying electricity supply by maximizing renewable energy use in Ebino city southern Japan. *Sustainable Cities and Society*, *34*, 371–384. doi:10.1016/j.scs.2017.06.022

Noorollahi, Y., Saeidi, R., Mohammadi, M., Hosseinzadeh, M., & Amiri, A. (2017). The effects of Ground Heat Exchanger Parameters Changes on Geothermal Heat Pump Performance - A Review. *Applied Thermal Engineering*. doi:10.1016/j.applthermaleng.2017.10.111

Nopens, P. (2014). The Impact of the Withdrawal from Afghanistan on Russia's Security. Security Policy Brief, n.º 54, March, Ergmont Royal Institute for International Relations, 1-10.

Nordal, K. B. (2001). Country risk, country risk indices and valuation of FDI: A real options approach. *Emerging Markets Review*, *2*(3), 197–217. doi:10.1016/S1566-0141(01)00017-6

Norway, R. C. (2013). Work programme for the DEMO 2000 Programme. Oslo: Author.

ODI. (1997). *Foreign Direct Investment Flows to Low-Income Countries: A Review of the Evidence*. Retrieved from http: //www.odi.org.uk/publications/briefing/3_97.html

OECD. (2016). *OECD Economic Surveys: Norway, 2016*. OECD. doi:10.1787/eco_surveys-chn-2013-en

OECD-FAO. (2011). *OECD-FAO Agricultural Outlook 2011-2020*. Retrieved from http://www.agrioutlook.org/dataoecd/23/56/48178823.pdf

OED. (2013). *National Renewable Energy Action Plan under Directive 2009 / 28 / EC Norway*. OED.

Oil & Gas Monitor. (2015, September 30). *The World's Largest Oil and Gas Projects – What You Should Know*. Author.

Olah, Goeppertve, & Prakash. (2006). *Beyond Oil and Gas: The Methanol Economy*. Wiley-VCH Publications.

Olson, R. (2015, January 25). US, Turkey, Israel and new challenges in the Middle East. Zaman Newspaper.

Öniş, Z. (2014). Turkey and Arab Revolutions: Boundaries of Regional Power Influence in a Turbulent Middle East. *Mediterranean Politics*, *19*(2), 203–219.

Online Business Dictionary. (2007). *Markets*. Retrieved from http://www.businessdictionary.com/definition/market.html#ixzz3zBQnuZCK

OPEC. (n.d.). *World Proven Oil Reserves, 2007-2016*. Retrieved from http://www.opec.org/opec_web/en/data_graphs/331.htm

Ozturk, I., Feridun, M., & Kalyoncu, H. (2008). Do Oil Prices Affect the USD/YTL Exchange Rate: Evidence from Turkey. *Privredna Kretanja I Ekonomska Politika, 18*(115), 49–61.

Parris, M. (2003). *Starting Over: US-Turkish Relations in the Post-Iraq War Era*. The Washington Institute for Near East Policy.

Peak Oil Review. (2015). Retrieved from http://peak-oil.org/peak-oil-review-13-jul-2015/

Peersman, G., & Stevens, A. (2013). *Analyzing oil demand and supply shocks in an estimated dsge model.* Ghent University Unpublished Manuscript.

Persons, S. (2017, December 7). Sen. Sheldon Whitehouse calls Jerusalem decision a political tactic by Trump. *The Washington Times.*

Petsinger, M. (2016). *What is Geo-economics*. Retrieved in April 2018 from https://www.chathamhouse.org/system/files/publications/twt/WiB%20YQA%20 Geoeconomics.pdf

Peyrouse, S. (2015, March 18). China as an Economic Actor in Central Asia: Between Development And Concerns. Testimony before the U.S.-China Economic and Security Review Commission.

Peyrouse, S. (2016). Discussing China: Sinophilia and sinophobia in Central Asia. *Journal of Eurasian Studies*, *7*(1), 14–23. doi:10.1016/j.euras.2015.10.003

Plummer, R. (2013, August 1). Alba alliance ambitions lay bare Latin trade confusion. *BBC News*. Retrieved from http://www.bbc.com/news/business-23515979

Policy Framework for 2020-2030, European Commission, COM (2014) 15 Final, Brussels.

Popkostova, Y. (2016). *Disruptive change: Europe's energy security and the US LNG*. Retrieved from https://guests.blogactiv.eu/2016/09/08/disruptive-change-europes-energy-security-and-the-us-lng/

Porcano, T. M., & Price, C. E. (1996). The Effects of Government Tax and Non-Tax Incentives on Foreign Direct Investment. *Multinational Business Review*, *4*, 9–20.

Portrait of Japan's Main Political Parties. (2012, 17 December). *Financial Times*.

Powell, F. (2008). Civil Society, Social Policy and Participatory Democracy: Past, Present and Future. *Social Policy and Society*.

Programme, L. (2013). *Work Programme for the PETROMAKS programme (2010 – 2013)*. About the Programme.

Proliferation Issues. (1992). JPRS Report. JPRS-TND-92-006. *Foreign Broadcast Information Service*. Retrieved from http://www.dtic.mil/dtic/tr/fulltext/u2/a333831.pdf

Prontera, A. (2017). *The New Politics of Energy Security in the European Union and Beyond: States, Markets, Institutions*. Routledge.

Qiang, S., & Xiaokun, L. (2010, November 24). Russia quit dollar. *China Daily*. Retrieved from http://www.chinadaily.com.cn/china/2010-11/24/content_11599087.htm

Questions that Abe's Article 9 Pitch Will Raise. (2017, May 28). *Japan Times.*

QuotesA. Z. (n.d.). Retrieved from http://www.azquotes.com/quote/434274

Rafiq, S., Salim, R., & Bloch, H. (2009). Impact of crude oil price volatility on economic activities: An empirical investigation in the Thai economy. *Resources Policy*, *34*(3), 121–132. doi:10.1016/j.resourpol.2008.09.001

Ramachandran, S. (2015). *China-Pakistan Economic Corridor: Road to Riches?* China Brief.

Rautava, J. (2004). The role of oil prices and the real exchange rate in Russia's economy - A cointegration approach. *Journal of Comparative Economics*, *32*(2), 315–327. doi:10.1016/j.jce.2004.02.006

Reeves, J. (2014). *SCO: A Tenable Provider of Security in Post-2014 Central Asia.* Asia-Pacific Center for Security Studies.

Republic of Turkey: Ministry of Foreign Affairs. (2017). Retrieved from http://www.mfa.gov.tr/turkey_s-commercial-and-economic-relations-with-japan.en.mfa

Reuters. (2015, May 11). *China becomes world's top crude buyer despite economy stuttering*. Author.

Reuters. (2017, December 5). *Turkey could break off ties with Israel over Jerusalem.* Retrieved from https://www.reuters.com/article/us-israel-usa-turkey/turkey-could-break-off-ties-with-israel-over-jerusalem-erdogan-idUSKBN1DZ13R

Riley, A. (2014). *The Western Balkans and EU energy security: protecting Europe's Flank.* The Institute for Statecraft. Retrieved from http://www.statecraft.org.uk/research/western-balkans-and-eu-energy-security

Robert, L., & King, S. (2000). *Discovering the Leader in You- A Guide to Realizing Your Personal Leadership Potential.* Center for Creative Leadership.

Rolland, N. (2015). *China's New Silk Road.* The National Bureau of Asian Research.

Root, E., & Ahmed, A. (1979). Empirical Determinants of Manufacturing: Direct Foreign Investment Developing Countries. *Economic Development and Cultural Change*, *27*(4), 758–767. doi:10.1086/451139

Rugman, A. M. (2009). *The Oxford Handbook of International Business* (2nd ed.). Oxford University Press. doi:10.1093/oxfordhb/9780199234257.001.0001

Rumley, D., & Chaturvedi, S. (2015). *Geopolitical Orientations, Regionalism and Security in the Indian Ocean*. Taylor & Francis. Retrieved from http://www.tandfebooks.com/doi/view/10.4324/9781315689487

Sabah.com.tr. (2016). *Gazprom'dan flaş Türk Akımı açıklaması*. Retrieved from http://www.sabah.com.tr/ekonomi/2016/12/30/gazpromdan-flas-turk-akimi-aciklamasi

Said, Y. (2005). Oil and Activism. In Global Civil Society. Sage.

Said, Y. (2005). *Oil and Activism, Global Civil Society*. Sage. doi:10.4135/9781446211908.n5

Sanger, D. E. (2002, November 1). The debate over attacking Iraq heats up. *The New York Times*.

Sanger, E. D. (2002, September 1). The World: First Among Evils? The Debate Over Attacking Iraq Heats Up. *The New York Times*.

Schaffer, T. C. (2002). Building a New Partnership with India. *The Washington Quarterly*, *25*(2), 31–44. doi:10.1162/016366002528201l7

Schneider, F., & Frey, B. (1985). Economic and Political Determinants of Foreign Direct Investment. *World Development*, *13*(2), 161–175. doi:10.1016/0305-750X(85)90002-6

Scholvin, S. (2016). *Geopolitics. An Overview Of Concepts And Empirical Examples From International Relations*. FIIA Working Paper.

Schuss, H. (2010). *Türkiye – Avrupa Ortak Enerji Politikası: Hedeflerde Uyum, Çatışma veya Siyasi Blokaj. In Enerji Platformu 2010: Avrupa'nın Enerji Güvenliğinde Türkiye'nin Rolü* (pp. 41–47). Ankara: Konrad-Adenauer-Stiftung.

Schwartz, H. M. (1994). *States versus Markets*. New York: Palgrave. doi:10.1007/978-1-349-13536-3

Secretariat, E. C. (2015). *International Energy Security: Common Concept for Energy Producing, Consuming and Transit Countries*. Energy Charter Secretariat.

Sekkat, K., & Veganzones-Varoudakis, M. A. (2007). Openness, Investment Climate, and FDI in Developing Countries. *Review of Development Economics, 11*(4), 607–620. doi:10.1111/j.1467-9361.2007.00426.x

Sencar, M., Pozeb, V., & Krope, T. (2014). Development of EU energy market agenda and security of supply. *Energy, 77*, 117–124. doi:10.1016/j.energy.2014.05.031

Senkyr, J. (2010). *Sonuç. In Enerji Platformu 2010: Avrupa'nın Enerji Güvenliğinde Türkiye'nin Rolü* (pp. 85–87). Ankara: Konrad-Adenauer-Stiftung.

Sevim, C. (2012). Küresel Enerji Politiği ve Enerji Güvenliği. *Journal of Yasar University, 26*(7).

Shah, A. (2011). Middle East and North African unrest. *Global Issues*. Retrieved from http://www.globalissues.org/article/792/mideast-north-africa-unrest

Shahbaz, M., Zeshan, M., & Afza, T. (2012). Is energy consumption effective to spur economic growth in Pakistan? New evidence from bounds test to level relationships and Granger causality tests. *Economic Modelling, 29*(6), 2310–2319. doi:10.1016/j.econmod.2012.06.027

Shehu, U. R. A. (2009). Impact of oil price shock and exchange rate volatility on economic growth in Nigeria: An empirical investigation. *Munich Personal RePEc Archive, 16319*(16319). doi:10.5897/JAERD12.088

Shiells. (2003). *FDI and the Investment Climate in the CIS Countries*. IMF Policy Discussion Paper, PDP/05/03, 1-35.

Sigot, M. (2013). *Energy Security and the EU: between independency priorities, strategic vulnerabilities and sustainability needs* (Master Thesis). Stockholm University.

Singh, H., & Jung, K. W. (1995). *Some new evidence on Determinants on Foreign Direct Investment in Developing Countries*. Policy Research Working Paper 1531, World Bank.

Smith, J., & High, M. (2017). Exploring the Anthropology of Energy: Ethnography, Energy and Ethics. *Energy Research & Social Science*, *30*, 1–6. doi:10.1016/j.erss.2017.06.027

Solmaz & Solmaz. (2001). Concept of Geothermal Energy Potential and Environmental Effects in Bursa. *Symposium on Renewable Energy Resources*, 283-290.

South China Morning Post. (2015, December 21). *China's growing upper middle class to drive consumption by 2020, research finds*. Author.

Stegen, K., & Kusznir, J. (2015). Outcomes and strategies in the 'New Great Game': China and the Caspian states emerge as winners. *Journal of Eurasian Studies*, *6*(2), 91–106. doi:10.1016/j.euras.2015.03.002

Stern, D. I. (2004). Article. *Economic Growth and Energy*, *2*, 35–51.

Subhes, C. B. (2011). *Energy Economics-concepts, Issues, Markets and Governance*. Springer Science & Business Media.

Suphi, D. (2010). *Avrupa'nın Enerji Güvenliğinde Hazar'daki Doğalgaz Projelerinin Rolü. In Enerji Platformu 2010: Avrupa'nın Enerji Güvenliğinde Türkiye'nin Rolü* (pp. 37–40). Ankara: Konrad-Adenauer-Stiftung.

Swenson, D. L. (1994). The Impact of U. S. Tax Reform on Foreign Direct Investment in the United States. *Journal of Public Economics*, *54*(2), 243–266. doi:10.1016/0047-2727(94)90062-0

T.C. Enerji ve Tabii Kaynaklar Bakanlığı (ETKB). (2016). Dünya ve Ülkemiz Enerji ve Tabii Kaynaklar Görünümü. Strateji Geliştirme Başkanlığı, 12.

T.C. Enerji ve Tabii Kaynaklar Bakanlığı, Petrol İşleri Genel Müdürlüğü. (2016). *Turkish Petroleum Law*. T.C. Enerji ve Tabii Kaynaklar Bakanlığı. Retrieved from http://www.pigm.gov.tr/index.php/english/turkish-petroleum-law

TANAP. (2016). *Major Shareholder in the Project of The Century*. Retrieved from http://www.tanap.com/media/press-releases/major-shareholder-in-the-project-of-the-century/

Taşyürek & Acaroğlu. (2012). Emission Reduction in Biofuels (Biomotor) and Global Warming Effect. *Productivity Review*, 162.

TDK. (n.d.). *Enerji*. Retrieved from http://www.tdk.gov.tr/index.php?option=com_gts&kelime=ENERJ%C4%B0

TEIAS. (2011a). *Turkish Electricity Production and Transmission Statistics 2011*. Retrieved from www.teias.gov.tr

TEIAS. (2011b). *Electricity Power Production Projection for Turkish Electrcity Energy 2011-2012*. Retrieved from www.teias.gov.tr

Tekin, A., & Williams, P. A. (2011). *Geo-Politics of the Euro-Asia Energy Nexus: The European Union*. Palgrave Macmillan. doi:10.1057/9780230294943

Tethys Petroleum. (2016). *Bokhtar Production Sharing Contract Area*. Retrieved from http://www.tethyspetroleum.com/operations/bokhtar-psc-area

The Economist. (2011, September 10). *Israel and Turkey: Can it get worse?* Retrieved from https://www.economist.com/node/21528687/comments#sort-comments

The Guardian. (2014, December 12). *China's water diversion project starts to flow to Beijing*. Author.

The Jerusalem Post. (2004, May 28). *Let the Kurds be*. Author.

The Petroleum Pipeline Corporation (BOTAŞ). (2015). *2015 Sektör raporu*. Retrieved from http://www.botas.gov.tr/docs/raporlar/tur/sektorap_2015.pdf

The Petroleum Pipeline Corporation (BOTAŞ). (2016). *2015 Annual Report*. Retrieved from http://botas.gov.tr/docs/raporlar/faalrap_2015.pdf

The Times of Israel. (2017, October 24). *Netanyahu: World must take care of Kurds' future*. Retrieved from https://www.timesofisrael.com/netanyahu-world-must-take-care-of-kurds-future/

The World Factbook, Central Intelligence Agency (CIA). (n.d.). *Country Comparison by Natural Gas Proved Reserves*. Retrieved from https://www.cia.gov/library/publications/the-world-factbook/rankorder/2253rank.html

Thieffry, P. (2017). *Environmental Protection and European Union Energy Policy: Energy Transition After the Paris Agreement*. Springer.

Tønne, T., & Tonne, T. (1983). Energy policy: A Norwegian perspective. *Nw. J. Int'l L. & Bus.*, *5*(4), 722.

Trading Economics. (2018a). *Azerbaijan Foreign Direct Investment*. Retrieved from https://tradingeconomics.com/azerbaijan/foreign-direct-investment

Trading Economics. (2018b). *Азербайджан - ВВП на душу населения ППС*. Retrieved from. https://ru.tradingeconomics.com/azerbaijan/gdp-per-capita-ppp

Trading Economics. (2018c). *Azerbaijan Consumer Confidence*. Retrieved from https://tradingeconomics.com/azerbaijan/consumer-confidence

Trading Economics. (2018d). *Azerjaiban Disposable Personal Income*. Retrieved from https://tradingeconomics.com/azerbaijan/disposable-personal-income

Trading Economics. (2018e). *Azerbaijan Consumer Spending*. Retrieved from https://tradingeconomics.com/azerbaijan/consumer-spending

Trading Economics. (2018f). *Azerbaijan Consumer Price Index CPI*. Retrieved from https://tradingeconomics.com/azerbaijan/consumer-price-index-cpi

Traywick, C. & Tobben, S. (2017, April 4). China Surpasses Canada as Top Buyer of U.S. Crude. *Bloomberg*.

Tribune, F. (2016). *KRG Gas Export to Turkey in 2016*. Retrieved from http://financialtribune.com/articles/energy/34341/krg-gas-export-turkey-2016

Tsai, P. (1994). Determinants of Foreign Direct Investment and its Impact on Economic Growth. *Journal of Economic Development*, *19*(1), 137–163.

Turkey, Iran set to increase energy ties (2007, February 22). *Hurriyet*.

Turkey, Iran vow to boost bilateral, regional cooperation. (2006, December 4). *Hurriyet*.

Turkish Petroleum Corporation (TPAO). (2013). *2012 Annual Report*. Retrieved from http://www.tpao.gov.tr/eng/docs/report/tp-2012-annual-report.pdf

Turkish Stream. (n.d.). *Turk Stream*. Retrieved from http://turkstream.info/project/

Turksen, U., & Vihma, A. (2015). The geoeconomics of the south stream pipeline project. *Journal of International Affairs*, *69*(1), 34.

Tvarozkova, E. (2010). *Türkiye ve Avrupa Enerji Güvenliği – İmkanlar ve Güçlükler. In Enerji Platformu 2010: Avrupa'nın Enerji Güvenliğinde Türkiye'nin Rolü* (pp. 49–52). Ankara: Konrad-Adenauer-Stiftung.

U.S. Energy Information Administration (2015c). *Kazakhstan. International energy data and analysis*. Author.

U.S. Energy Information Administration. (2014). *Annual Energy Outlook*. Retrieved on August 22, 2014 from https://www.eia.gov/outlooks/aeo/#pet_oli

U.S. Energy Information Administration. (2015a). *International energy data and analysis. Uzbekistan*. Author.

U.S. Energy Information Agency. (2017). *Country Analysis Brief: Turkey*. Retrieved from https://www.eia.gov/beta/international/analysis.cfm?iso=TUR

Ubiznes.ru. (2015). *Top 5 countries for oil production in the world*. Retrieved from http: //ubiznes.ru/top/reyting-stran-po-dobyche-nefti-v-mire.html

Umbach, F. (2010). Global energy security and the implications for the EU. *Energy Policy*, *38*(3), 1229–1240. doi:10.1016/j.enpol.2009.01.010

United Nations Economic and Social Commission for Western Asia (UNESCWA). (2011). Youth Exclusion in the ESCWA Region: Demographic, Economic, Educational and Cultural Factors. Population Report, issue no. 5. United Nations.

United Nations Economic and Social Commission for Western Asia (UNESCWA). (2012). Promoting Participatory Democracy in the ESCWA Region. *Social Development Bulletin*, *4*(2).

US Chamber of Commerce. (2015). Assessing Risk in A Global Energy Market international index of energy security risk. Assessing risk in a global energy market. *Institute for 21 St Century Energy*, 1–115.

Van der Linde, C. (2007). *External energy policy: Oil fears and new dilemmas in a larger Union*. INIS.

Veli, S. V., & Sarıkaya, E. (2004, February 5). Yüzyılın kurtarma operasyonu [The rescue operation of the century]. *Sabah Newspaper.*

Vidakis, I., & Baltos, G. (2015). Security Aspects of "Geoenergeia" and the Significance of Energy Resources Management in International Politics. *Geopolitics of Energy*, *37*(3), 1-16.

Wang, H. (2016, January 15). China's Triple Wins: The New Silk Roads. *Forbes*.

Wang, Z. (2015). *Securing Energy Flows from Central Asia to China and the Relevance of the Energy Charter Treaty to China*. Energy Charter Secretariat.

Weitz, R. (2013). Beijing Builds its Eurasian Transportation Network. *China Brief*, *13*(8), 6–9.

William, C. J. (2001). *Public Papers of the Presidents of the United States*. Washington, DC: U.S. Government Publication.

Winzer, C. (2012). Conceptualizing energy security. *Energy Policy*, *46*, 36–48. doi:10.1016/j.enpol.2012.02.067

WIR. (2017), *World Investment Reports.* Division on Investment and Enterprise United Nations Conference on Trade and Development.

World Bank Archives. (n.d.). *Global Geopolitics & Political Economy*. Author.

World Bank. (2015). Retrieved from www.data.worldbank.org/country/norway

World Economic Forum. (2014). *The Global Competitiveness Report 2014- 2015*. Author.

World Economic Forum. (2015). *The Africa Competitiveness Report 2015*. Author.

World Economic Forum. (2015). *The Global Competitiveness Report 2015- 2016*. Author.

World Economic Forum. (2016). *The Global Competitiveness Report 2016- 2017*. Author.

Xiaochuan, Z. (2009). Reform the International Monetary System. *BIS Review*. Retrieved from https://www.bis.org/review/r090402c.pdf

Xu, Q., & Chung, W. (2016). *China's Energy Policy from National and International Perspectives: The Energy Revolution and One Belt One Road*. City University of HK Press.

Yang, M. (2016). Understanding the One Belt One Road Initiative: China's Perspective. In China's One Belt One Road: Initiative, Challenges and Prospects. Vij Books India Pvt Ltd. Delhi.

Yazar, Y. (2010). *Açılış ve Selamlama. In Enerji Platformu 2010: Avrupa'nın Enerji Güvenliğinde Türkiye'nin Rolü* (pp. 13–18). Ankara: Konrad-Adenauer-Stiftung.

Yergin, D. (1991). *The Prize: The Epic Quest for Oil, Money and Power*. New York, NY: Simon and Schuster.

Yildirim, E., Sukruoglu, D., & Aslan, A. (2014). Energy consumption and economic growth in the next 11 countries: The bootstrapped autoregressive metric causality approach. *Energy Economics*, *44*, 14–21. doi:10.1016/j.eneco.2014.03.010

Yolcular, S. (2009). Hydrogen Production for Energy Use in European Union Countries. In Recovery, Utilization and Environmental Effects (pp. 1329-1337). Taylor & Francis.

Yousefi, H., Hamlehdar, M., Tabasi, S., & Noorollahi, Y. (2017). Economic and Thermodynamic Evaluations of Using Geothermal Heat Pumps in Different Climate Zone (Case Study: Iran). *Proceedings Geothermal Reservoir Engineering*. Retrieved from https://pangea.stanford.edu/ERE/db/GeoConf/papers/SGW/2017/Yousefi2.pdf

Yousefi, H., Roumi, S., Tabasi, S., & Hamlehdar, M. (2017). Economic and Air Pollution Effects of City Council Legislations on Renewable Energy Utilization in Tehran. *International Journal of Ambient Energy*. (accepted for publication)

Yousefi, H., Tavakkoli-Moghaddam, R., Oliaei, M., Mohammadi, M., & Mozaffari, A. (2017). Solving a bi-objective vehicle routing problem under uncertainty by a revised multi-choice goal programming approach. *International Journal of Industrial Engineering Computations*, *8*(3), 283–302. doi:10.5267/j.ijiec.2017.1.003

Zakaria, F. (2011). How democracy can work in the Middle East. *TIME Magazine*. Retrieved from http://www.time.com/time/magazine/article/0,9171,2046038,00.html

Zhang, H. (2014). Confronting China's Water Insecurity. *RSIS Commentaries, S. Rajaratnam School of International Studies*, (57), 1-2.

Zhang, Y., Rahbari-Asr, N., Duan, J., & Chow, M. Y. (2016). Day-ahead smart grid cooperative distributed energy scheduling with renewable and storage integration. *IEEE Transactions on Sustainable Energy*, *7*(4), 1739–1748. doi:10.1109/TSTE.2016.2581167

Zhiqin, S., & Yang, L. (2016, December 21). *The Benefits and Risks of the China-Pakistan Economic Corridor*. Carnegie-Tsinghua. Center for Global Policy.

Zilberfarb, B. Z., & Adams, F. G. (1981). The energy-GDP relationship in developing countries. Empirical evidence and stability tests. *Energy Economics*, *3*(4), 244–248. doi:10.1016/0140-9883(81)90025-6

About the Contributors

Aisha Ansari is currently completing doctoral studies in the School of Public Administration Leadership, DPA, Capella University, Minneapolis, MN, USA. MPH, also completed at Capella , and the author has diversity in public administration which includes business and finance, healthcare, education, and multiculturalism, for example Jordan, Pakistan and Sudan spanning more than thirty years.

Alireza Aslani is an assistant professor of department of renewable energy and environment at the university of Tehran. His research interests include energy modelling and security, and energy modelling. He is also adjunct lecturer of Maastricht School of Management, Netherland.

Selma Aytüre completed her undergraduate education at Gazi University in Ankara. She received a Master's Degree and PhD in Ankara University, Institute of Social Sciences on European Union. She worked at the Ministry of Food, Agriculture and Livestock for many years with the position of Turkey-EU Relations and Legislation Branch Manager, from time to time acted as the Chairman of the Department. During this mission she attended many national/ international meetings related to EU, represented the Ministry and Turkey. At the beginning of 2012, she started her academic career as Assistant Professor at Aksaray University. During the years 2012-2014, she was the Director of Şereflikoçhisar School of Applied Technology and Management and head of departments. She continues her academic life as Associate Professor at Yıldırım Beyazıt University, Faculty of Applied Sciences, Department of International Trade and Logistics Management. Her published books are: "Turkey's Sugar Production and Trade Policies in terms of EU Compliance", "FEOGA and Applicability in Turkey (with S.Tan and B. Şener)" , "Future of Agriculture and Agricultural Policy in the World and Turkey (with M. Acar)" and "Agriculture in Turkey and Agricultural Transformation Directors in 1980's (with M. Acar)". She has many articles and reports in Turkey and abroad. She is married and has two children.

Tuğba Aydın Azalov is Assistant Professor at Artvin Coruh University, Department of Political Science and Public Administration. She obtained her BA and MA degrees in Public Administration at Akdeniz University and PhD in Political Theory at Luiss Guido Carli University. Her academic interests are in the fields of nationalism, migration, citizenship, European Union and Central Asian relations.

Paulo Duarte holds a PhD and a Masters in Political and Social Sciences, awarded by the Catholic University of Louvain, Belgium. He is an expert in China's One Belt One Road, on China, World Politics, and specifically Central Asia, where he carried out an extensive on-the-ground doctoral research (Kazakhstan, Kyrgyzstan and Tajikistan) with a scholarship awarded from the Calouste Gulbenkian Foundation. He is the author of more than 40 articles published in peer-review journals. Columnist at Revista Sociedade Militar. Author of Metamorfoses no Poder: rumo à hegemonia do dragão? with preface and presentation by Marcelo Rebelo de Sousa, current President of the Portuguese Republic, and a long Postfacium by Professor Doctor Armando Marques Guedes. Recent publications include Duarte, Paulo (2017). "Changes in the World Power: The United States within the New Balance of Power", pp. 2-23, in Eurasian Economies in Transition, Cambridge Scholars Publishing, edited by Gencer, A.; Sözen, I. & Sari, S. Author of Duarte, Paulo (2017). Pax Sinica. Lisboa: Chiado Editora [presented on June 20, 2017 by Former Minister of Education and of Finance of Portugal, Excellency Prof. Dr. Guilherme d'Oliveira Martins]. P. Duarte is also the author of a Report for the Ministry of Foreign Affairs of the Republic of China (Taiwan): Duarte, Paulo (2016). South and East China Seas: prospects for a new sino-centric era? Taiwan Fellowship. December, Taipei, 50p. He was Guest researcher (January-December 2016) at the National Chengchi University (Graduate Institute of East Asian Studies), in Taiwan, within the framework of the Program Taiwan Fellowship which brings together researchers of various parts of the world. Also, a Member of China's Observatory in Portugal and is Editor-in-Chief of ELA Journal (IAPSS publication). Member of the Foundation Friends of the New Silk Road, Portugal. Member of the International Association for Political Science Students (IAPSS). Member of the International Political Science Association (IPSA) and of the Portuguese Political Science Association. He is a member of Centre for African, Asian and Latin American Studies (CESA), at the Lisbon School of Economics and Management, University of Lisbon. He has been a Guest lecturer at Coimbra University and at Lisbon University. Paulo Duarte has been awarded a scholarship by the Ministry of Commerce of the People's Republic of China (MOFCOM), at the Normal Beijing University, in Beijing. Mentor of an academic partnership between the National Taiwan University (R.O.C.) in association with the Chinese Academy of Social Sciences (headquartered in Beijing) and the Observatory of China (in Portugal), in order to set up an Oral History project about Chinese studies in Portugal.

Amal Nagah Elbeshbishi is an International Professional Staff Member holding a Continuing Contract at the United Nations Secretariat, and working as an Economist for the United Nations Economic Commission for Africa (UNECA) - office for North Africa, Rabat- Morocco. She is also a tenured Professor of Economics at Faculty of Commerce- Mansoura University, Egypt [on sabbatical leave (national assignment) to work for UNECA]. Prior to joining UNECA in 2005, Prof. Dr. Elbeshbishi lectured at universities in Bahrain, Egypt and the United States of America. She has also worked with the United Nations Economic and Social Commission for Western Asia (UNESCWA) and the Egyptian Cabinet's Information and Decision Support Center (IDSC). Prof. Dr. Elbeshbishi is the author/co-author of many research papers and books on several economic and social development issues. She holds a Master's and a Ph.D. degree in Economics from Fordham University- New York, United States of America; and a Master's degree in Economics from Mansoura University, Egypt.

Suudan Gökçe GÖK completed her bachelor's degree in Department of Economics in Dokuz Eylül University, master degree in Human Resources Department of Business Management in Ufuk University and doctorate degree in Business Department of Selçuk University. She has extensive studies related with management organization, organizational behavior, renewable energy and ethic matters and she has participated into various congress and symposiums. The author who has different publications in business management engaged into university-industry cooperation and social responsibility projects voluntarily. She is still working as Dr. lecturer in Gazi University and continuing her academic studies.

Hasmet Gokirmak is an assistant professor at Istanbul Sabahattin Zaim University. He received his MA degree in international economics from Georgetown University (Washington, D.C.) in 1991 and his Ph.D. degree in economics from American University (Washington, D.C.) in 1995. Dr. Gokirmak has worked as a professional manager for transportation and media companies for more than twenty-five years. His current areas of research include logistics and transportation, international trade, international finance, and energy economics.

Aida Guliyeva, Associate Professor, PhD of economic science, dean of the faculty "SABAH QROUPS" Azerbaijan State University of Economics, UNEC She was born on August, 28 1959. She graduated from Baku State University, mechanic-mathematical faculty and aspirantura ASUE (UNEC). She had postdoctoral research in Oxford University, UK and MBA courses certificates in GSU, Atlanta, USA. She published more than 75 scientific researches (both in Azerbaijan and abroad) including 4 monographs and 1 manual The spheres of her scientific activity are: economic theory, macroeconomics, economic development, new economy and technologies. She was awarded by the medal of honor of Azerbaijan "TARAGGI."

Maryam Hamlehdar is a student of Energy Engineering in the department of renewable energy and environment at the university of Tehran. Maryam's research interests include energy modelling and economics.

Nurlan Aitkaliuly Nurseiit worked in the Company "Kazakh oil products" in 1987-1988, in the National Bank of Kazakhstan in 1993-2002, in the JSC "KazMunayGas" in 2003-2004. He received PhD in Economics and Finance in 1996, and the degree of Doctor in 2006. He worked as professor in the Kazakh-British Technical University, the Kazakhstan Institute of Oil and Gas, the Kazakh National Research Technical University in 2005-2016, and is currently the Head of Department of the Kazakh-American University. He is alsoa visiting professor in the University of International Business and the Eurasian Innovation Academy of Economics and Management. His research interest is the resource economy, in particular oil and gas producing and financial sectors.

Ulviyye Rzayeva is doctor of philosophy of mathematics, lecturer, Department of "Information economy and technologies" of The Azerbaijan State University of Economics She was born on November, 13 1973. She graduated from Baku State University majoring in the mathematician. She published more than 30 scientific researches (both in Azerbaijan and abroad) including 2 monographs and 1 manual. The spheres of her scientific activity are: intellectual systems in economy, system of pattern recognition, fuzzy logic.

Reza Saeediare is a student of Energy Engineering in the department of renewable energy and environment at the university of Tehran. Reza's research interests include energy modelling and economics.

Devrim Şahin studied B.Sc. in Applied Maths and Computer Science at Eastern Mediterranean University (EMU) and received B.Sc. in Management Information Systems from Girne American University. He finished his MA and is a PhD. Candidate in the International Relations Department at EMU. Since 2007, he has been a Research Fellow in the Cyprus Policy Center – which is a policy oriented think tank. Şahin's areas of research and publications include Complex and Chaotic Systems.

Ahmet Sözen is professor of International Relations and the Provost at Eastern Mediterranean University. Sözen has published extensively on the Cyprus conflict and Turkish foreign policy. Sözen has given dozens of invited speeches on conflict resolution – including two TED talks (TEDxNicosia 2012, TEDxEMUniversity 2014) and the Cyprus conflict in various countries and he frequently appears on both local and Turkish media, as well as international media where he comments on world events.

Index

F

G

H

I

J

M

N

O

P

R

S

T

U

W

Ensure Quality Research is Introduced to the Academic Community

Become an IGI Global Reviewer for Authored Book Projects

The overall success of an authored book project is dependent on quality and timely reviews.

In this competitive age of scholarly publishing, constructive and timely feedback significantly expedites the turnaround time of manuscripts from submission to acceptance, allowing the publication and discovery of forward-thinking research at a much more expeditious rate. Several IGI Global authored book projects are currently seeking highly qualified experts in the field to fill vacancies on their respective editorial review boards:

Applications may be sent to:
development@igi-global.com

Applicants must have a doctorate (or an equivalent degree) as well as publishing and reviewing experience. Reviewers are asked to write reviews in a timely, collegial, and constructive manner. All reviewers will begin their role on an ad-hoc basis for a period of one year, and upon successful completion of this term can be considered for full editorial review board status, with the potential for a subsequent promotion to Associate Editor.

If you have a colleague that may be interested in this opportunity, we encourage you to share this information with them.

www.ingramcontent.com/pod-product-compliance
Lightning Source LLC
LaVergne TN
LVHW081337300926
849161LV00008B/2000